CAMBRIDGE MONOGRAPHS ON APPLIED AND COMPUTATIONAL MATHEMATICS

29 Partial Differential Equation Methods for Image Inpainting

Cambridge Monographs on Applied and Computational Mathematics

The *Cambridge Monographs on Applied and Computational Mathematics* series reflects the crucial role of mathematical and computational techniques in contemporary science. The series publishes expositions on all aspects of applicable and numerical mathematics, with an emphasis on new developments in this fast-moving area of research.

State-of-the-art methods and algorithms as well as modern mathematical descriptions of physical and mechanical ideas are presented in a manner suited to graduate research students and professionals alike. Sound pedagogical presentation is a prerequisite. It is intended that books in the series will serve to inform a new generation of researchers.

A complete list of books in the series can be found at www.cambridge.org/mathematics.

Recent titles include the following:

11. *Generalized Riemann Problems in Computational Fluid Dynamics*, Matania Ben-Artzi & Joseph Falcovitz
12. *Radial Basis Functions*, Martin D. Buhmann
13. *Iterative Krylov Methods for Large Linear Systems*, Henk van der Vorst
14. *Simulating Hamiltonian Dynamics*, Benedict Leimkuhler & Sebastian Reich
15. *Collocation Methods for Volterra Integral and Related Functional differential equations*, Hermann Brunner
16. *Topology for Computing*, Afra J. Zomorodian
17. *Scattered Data Approximation*, Holger Wendland
18. *Modern Computer Arithmetic*, Richard Brent & Paul Zimmermann
19. *Matrix Preconditioning Techniques and Applications*, Ke Chen
20. *Greedy Approximation*, Vladimir Temlyakov
21. *Spectral Methods for Time-Dependent Problems*, Jan Hesthaven, Sigal Gottlieb & David Gottlieb
22. *The Mathematical Foundations of Mixing*, Rob Sturman, Julio M. Ottino & Stephen Wiggins
23. *Curve and Surface Reconstruction*, Tamal K. Dey
24. *Learning Theory*, Felipe Cucker & Ding Xuan Zhou
25. *Algebraic Geometry and Statistical Learning Theory*, Sumio Watanabe
26. *A Practical Guide to the Invariant Calculus*, Elizabeth Louise Mansfield
27. *Difference Equations by Differential Equation Methods*, Peter E. Hydon
28. *Multiscale Methods for Fredholm Integral Equations*, Zhongying Chen, Charles A. Micchelli & Yuesheng Xu

Partial Differential Equation Methods for Image Inpainting

CAROLA-BIBIANE SCHÖNLIEB
University of Cambridge

CAMBRIDGE
UNIVERSITY PRESS

CAMBRIDGE
UNIVERSITY PRESS

Shaftesbury Road, Cambridge CB2 8EA, United Kingdom

One Liberty Plaza, 20th Floor, New York, NY 10006, USA

477 Williamstown Road, Port Melbourne, VIC 3207, Australia

314–321, 3rd Floor, Plot 3, Splendor Forum, Jasola District Centre, New Delhi – 110025, India

103 Penang Road, #05–06/07, Visioncrest Commercial, Singapore 238467

Cambridge University Press is part of Cambridge University Press & Assessment, a department of the University of Cambridge.

We share the University's mission to contribute to society through the pursuit of education, learning and research at the highest international levels of excellence.

www.cambridge.org
Information on this title: www.cambridge.org/9781107001008

First published 2015

A catalogue record for this publication is available from the British Library

Library of Congress Cataloging-in-Publication data
Schönlieb, Carola-Bibiane, 1979–
Partial differential equation methods for image inpainting / Carola-Bibiane Schönlieb, University of Cambridge.
pages cm. – (Cambridge monographs on applied and computational mathematics)
Includes bibliographical references and index.
ISBN 978-1-107-00100-8 (alk. paper)
1. Image reconstruction–Mathematical models. 2. Image processing–Digital techniques. 3. Differential equations, Partial. I. Title.
TA1637.5.S36 2015
006.601′515353–dc23 2015016114

ISBN 978-1-107-00100-8 Hardback

To my love, Bertram

Contents

Preface

The purpose of this book is to provide an introduction to the use of partial differential equations (PDEs) for digital image restoration. It is a way of sharing what I have learned while studying these methods for about ten years, defending a Ph.D. thesis on PDE inpainting, teaching courses in Göttingen and Cambridge on the topic and writing a couple of research papers on the use of PDEs in image inpainting along the way. Let me say what this book is and what it is not. It is:

1. An introduction to inpainting methods that use PDEs and *local* variational approaches to restore lost image contents;
2. An account from an enthusiast on some state-of-the-art inpainting methods from their abilities to their limitations and reference for informed researchers in the field of digital image processing; and
3. A work targeted at readers with basic knowledge in functional analysis, PDEs, measure theory and convex optimisation.
   Therefore, I recommend this textbook only for students from the graduate level onwards.

This book is not:

1. A book that gives credit to the whole wealth of inpainting methods (in particular, this book will focus only on local inpainting methods and will scratch only non-local inpainting methods such as exemplar-based inpainting); or
2. A book for undergraduate students.

One final comment before we go *in medias res*: the more I learn and understand about image inpainting, the more the complexity of the matter and the variety of different problems, each requiring specialised methods, become apparent. Therefore, I feel a bit like Goethe's Faust, who exclaims:

> Habe nun, ach! Philosophie,
> Juristerei und Medizin,
> Und leider auch Theologie
> Durchaus studiert, mit heißem Bemühn.
> Da steh ich nun, ich armer Tor!
> Und bin so klug als wie zuvor;
> Heiße Magister, heiße Doktor gar
> Und ziehe schon an die zehen Jahr
> Herauf, herab und quer und krumm
> Meine Schüler an der Nase herum –
> Und sehe, daß wir nichts wissen können!
> Das will mir schier das Herz verbrennen.

Let me finally mention some important people who have influenced me and this work. This book could not have been written without the support of many of my colleagues and friends.

First of all, I thank Bertram Düring, who accompanied and supported me thoughout the work on this book. I also thank Peter Markowich – without him, I would never have started to write this book, nor finish it without his regular enquiries, 'Und, wie gehts dem Buch?' And I thank Arieh Iserles for inspiring me with his renowned book on the numerical solution of partial differential equations.

Many sections of this book are based on joint work with various co-authors: Wolfgang Baatz, Martin Benning, Andrea Bertozzi, Christoph Brune, Martin Burger, Luca Calatroni, Bertram Düring, Massimo Fornasier, Carsten Gottschlich, Lin He, Rob Hocking, Rien Lagerwerf, Andreas Langer, Jan Lellmann, Peter Markowich, Jean-Michel Morel, Stanley Osher, Kostas Papafitsoros and Bati Sengul. These people have made working on the topic of image inpainting interesting, exciting and fun.

I also owe many thanks to my image-analysis group for proofreading and invaluable comments on the presentation of this book. I especially thank Luca Calatroni, Rob Hocking, Jan Lellmann and Kostas Papafitsoros.

1
Introduction

This book is concerned with digital image processing techniques that use partial differential equations (PDEs) for the task of image inpainting. *Image inpainting* is an artistic term for virtual image restoration or image interpolation whereby missing or occluded parts of images are filled in based on information provided by the intact parts of the image. Computer graphic designers, artists and photographers have long used manual inpainting to digitally restore damaged paintings or manipulate photographs. Today, mathematicians apply powerful methods based on PDEs to automate this task. They operate in much the same way that trained restorers do: they propagate information from the structure around a hole into the hole to fill it in.

Virtual image restoration is an important challenge in our modern computerised society. From the reconstruction of crucial information in satellite images of the Earth to the renovation of digital photographs and ancient artwork, virtual image restoration is ubiquitous. The example in Figure 1.1 is entitled *Mathematical Analysis Can Make You Fly*, and it should give you a first impression of the idea of image inpainting with PDEs. The PDE model used for this example is called TV-H^{-1} *inpainting* and will be discussed in great detail in Section 5.3.

1.1. Digital Image Restoration in Modern Society

Digital images are one of the main sources of information today. The vast number of images and videos that exist in digital form nowadays makes their unaided processing and interpretation by humans impossible. Automatic storage management, processing and analysis algorithms are needed to be able to retrieve only the essence of what the visual world has up its sleeve.

Figure 1.1. 2013 EPSRC science photo competition winner: *Mathematical Analysis Can Make You Fly.* How is this possible? How can Joana – the woman in the picture and a master's student in mathematics – fly? Does she have supernatural powers? The clue to the solution of this mystery can be seen on the blackboard. It is a PDE that can be used for digital image inpainting. Inpainting is the process by which specified parts of an image are filled in based on the remaining part of the image. In this example, we solve this equation numerically and are able to remove the stool on which Joana was sitting originally. She appears to fly! While this may seem like gadgetry, image inpainting has wide-ranging practical applications: from the restoration of satellite images, the enhancement of medical images and the renovation of digital photographs and artwork to special effects in images and videos. As in this photograph, image inpainting is ubiquitious. (Image courtesy of Joana Grah, Kostas Papafitsoros and the author. Winner of the 2013 EPSRC Science Photo Competition in the 'People' category.)

In today's society, we encounter digital images on a daily basis: from everyday life, where analogue cameras have long been replaced by digital ones, to their professional use in medicine, earth sciences, the arts and security applications. In these contexts, we often have to deal with the processing of images, for example, the restoration of images corrupted by noise, blur or intentional scratching. The idea behind *image processing* is to provide methods that improve the quality of these images by postprocessing them. Medical imaging tools such as magnetic resonance imaging (MRI) usually produce noisy or incomplete image data. Satellite images of our Earth are often poorly resolved and blurred. Furthermore, digital image restoration is used in art preservation and restoration, where digital photographs are taken of historical artwork and are digitally restored and stored. As such, they can serve as templates for restorers and can be kept in a database for preservation. In sum, digital image restoration provides an effective tool to recover or complete lost image information. Keywords in this context are *image de-noising*, *image de-blurring*, *image decomposition*, *image inpainting* and *image synthesising*.

Considering the vast number of image restoration applications and problems in this area that have not been completely and satisfactorily resolved yet, it is not surprising that this is a very active and broad field of research. From engineers to computer scientists and mathematicians, a large group of people has been and is still working in this area.

1.2. What is a Digital Image?

To appreciate the following theory and the image inpainting applications, we first need to understand what a digital image is. Roughly speaking, a *digital image* is obtained from an analogue image (representing the continuous world) by sampling and quantisation. Similar to our eye, a digital camera superimposes a regular grid on an analogue image and assigns a value to each grid element. For example, in the case of a digital photograph, each grid element stores the mean brightness in the recorded field encoded in the photon counts. Thus, a digital image can be considered to be a sample of the continuous world.

In the terminology of digital images, the sampling points are grid elements called *pixels* (from *pict*ure *el*ement*s*). The image content is then described by the grey values or colour values prescribed in each pixel. The grey values are usually scalar values ranging between 0 (black) and 255 (white). Colour values are represented by vectors, most commonly (r,g,b), where each channel r,g and b represents the red, green and blue component of the colour, respectively, ranging again from 0 to 255. The mathematical representation of a digital image is a so-called image function u defined (for now) on a two-dimensional (in general, rectangular) grid. Indeed, in some applications, images are three-dimensional (e.g., videos and medical imaging) or even four-dimensional objects involving three spatial dimensions and time. However, in this book we concentrate on the two-dimensional case. Figure 1.2 illustrates the connection between a digital image and its image function for the case of a grey value image.

Typical sizes of digital images range from $2,000 \times 2,000$ pixels in images taken with a simple digital camera to $10,000 \times 10,000$ pixels in images taken

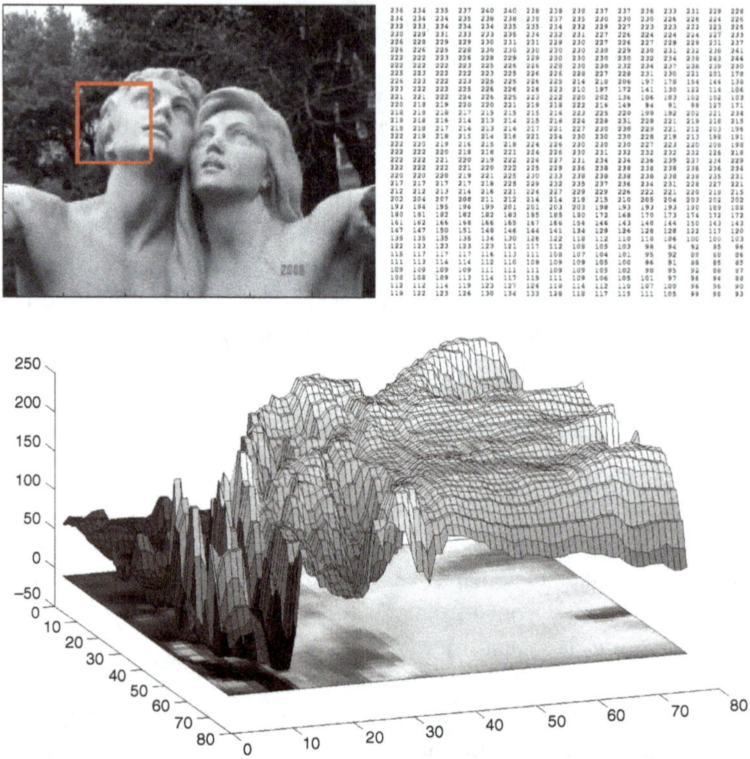

Figure 1.2. Digital image represented as an image function. In the top row on the left, a grey value photograph is shown; on the right, the grey values for a small detail of the digital photograph are displayed in matrix form. In the bottom row, the image function within a small selection of the digital photograph is shown where the grey value $u(x,y)$ is plotted as the height over the (x,y)-plane.

with high-resolution cameras used by professional photographers. The sizes of images in medical imaging applications depend on the task at hand. Positron emission tomography (PET), for example, produces three-dimensional (3D) image data, where a full-length body scan has a typical size of $200 \times 200 \times 500$ pixels.

Since the image function is a mathematical object, we can treat it as such and apply mathematical operations to it. These mathematical operations are summarised by the term *image processing techniques* and range from statistical methods and morphological operations to solving a PDE for the image function (cf. Section 1.1). We are especially interested in the latter, that is, PDE methods for image inpainting.

Remark 1.2.1 Digital Images and Continuous Models Note that although digital images are discrete finite-dimensional objects, most image processing methods discussed in this book are modelled and analysed in infinite-dimensional function space rather than $\mathbb{R}^d$. This reflects our aim of finding and reconstructing from a finite-dimensional sample (the digital image) an image from the physical, continuous world (the analogue image). On a microscopic level, a line in a digital image is only a collection of pixels with common characteristics that allow us to identify it as a continuous line on a macroscopic level where pixels are invisible to us. Refer to Chapter 3 for more discussion on the perception of geometrical objects from point clouds. Moreover, performing the modelling and analysis of image reconstruction in function space allows us to specify finer properties of a desired reconstruction. For instance, it is easy to talk about discontinuities for a function that lives in the continuum, whereas the definition of discontinuities for a finite-dimensional function is unclear. This is similar to analogous situations in many other areas of mathematics, for example, in kinetic theory, where particles are replaced by densities, or in statistics, where finite-dimensional analysis is replaced by infinite-dimensional stochastic analysis.

1.3. Image Inpainting

An important task in image processing is the process of filling in missing parts of a damaged image based on the information obtained from the intact part of the image. It is a type of interpolation called *inpainting*.

Let g represent some given image defined on an image domain Ω. The problem is to reconstruct the original image u in the (damaged) domain $D \subset \Omega$, called the *inpainting domain* or *hole/gap* (Figure 1.3).

The term *inpainting* was invented by art restoration workers [EM76, Wal85, WG04] and first appeared in the framework of digital image restoration in the work of Bertalmio et al. [BSCB00]. Therein the authors designed a discrete inpainting model motivated by a PDE which intends to imitate the restoration work of museum artists. Their method will be explained in more detail in Section 6.1.

To give you a first 'gusto' for real-world problems where digital inpainting might be useful, let us evoke one example from the recent history of art restoration. In August 2012, Cecilia Giménez, an eighty-year-old amateur artist from a small village near Zaragoza, Spain, gained fame by an attempt to restore a wall painting in a local church. She produced the now-famous painting dubbed 'Ecce Mono' ('Behold the Monkey') when aiming to restore

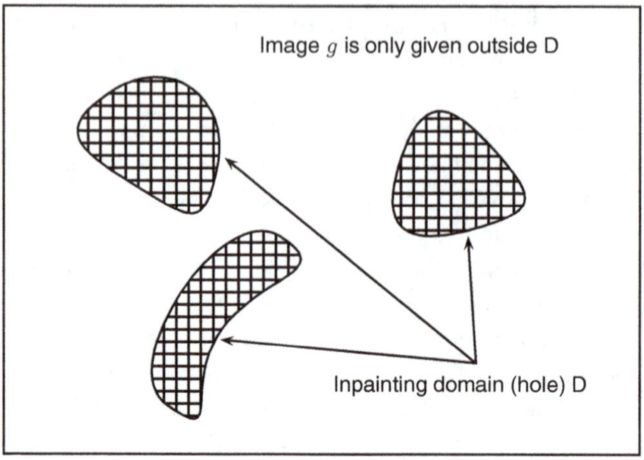

Figure 1.3. The inpainting task.

Figure 1.4. *Ecce Homo* (left) and *Ecce Mono* (right).

the wall painting *Ecce Homo* (*Behold the Man*) by Spanish painter Elías García Martínez (Figure 1.4).

Although Cecilia's restoration clearly follows a conceptual approach, the result seems to us rather suboptimal. This shows the complexity of the image interpolation problem and the challenge of defining and formalising rules which produce a visually appealing restoration result. To see what virtual image restoration methods make of *Ecce Homo*, continue reading (see Figure 2.4) or see [Sch14].

Applications Applications of digital image inpainting are numerous. From automatic scratch removal in old photographs and films [BSCB00, SC02, KMFR95b] to digital restoration of ancient paintings for conservation purposes [BFM$^+$08], text erasing such as the removal of dates, subtitles or publicity from a photograph [BSCB00, BBC$^+$01, SC02, CS01b], special effects such as object removal [BSCB00, CS01b], disocclusion [NMS93, MM98, Mas98, Mas02], spatial/temporal zooming and super-resolution [BBC$^+$01, SC02, Mal00, MG01a, TYJW01], error concealment [WZ98], lossy perceptual image coding [SC02], removal of the laser dazzling effect [CCBT03] and sinogram inpainting in X-ray imaging [GZY$^+$06], to name only a few.

The Beginnings of Digital Image Inpainting The history of digital image inpainting has its beginning in the works of engineers and computer scientists. Their methods were based on statistical and algorithmic approaches in the context of image interpolation [AKR97, KMFR95a, KMFR95b], image replacement [IP97, WL00], error concealment [JCL94, KS93] and image coding [Cas96, LFL96, RF95]. In [KMFR95b], for example, the authors present a method for video restoration. Their algorithm uses intact information from earlier and later frames to restore the current frame and is therefore not applicable to still images. In interpolation approaches for 'perceptually motivated' image coding [Cas96, LFL96, RF95], the underlying image model is based on the concept of 'raw primal sketch' [Mar82]. More precisely, this method assumes that the image consists of mainly homogeneous regions separated by discontinuities, that is, edges. The coded information then just consists of the geometrical structure of the discontinuities and the amplitudes at the edges. Some of these early coding techniques already used PDEs for this task (see, e.g., [Car88, Cas96, CT94]).

Initiated by pioneering works [NMS93, MM98, CMS98a, BSCB00 and SC02], the mathematics community got involved in image restoration using PDEs and variational methods for this task. Their approach and some of their methods shall be honoured in this book.

2

Overview of Mathematical Inpainting Methods

Digital inpainting methods are being designed with the desire for an *automated* and *visually convincing* interpolation of images. In this chapter we give an overview of approaches and trends in digital image inpainting and provide a preview of our discussion in Chapters 4 through 7. Before we start with this, let us raise our consciousness about the challenges and hurdles we might face in the design of inpainting problems.

The first immediate issue of image inpainting is, of course, that we do not know the truth but can only guess. We can make an educated guess, but still it will never be more than a guess. This is so because once something is lost, it is lost, and without additional knowledge (based on the context, e.g., historical facts), the problem of recovering this loss is an ambiguous one. Just look at Figure 2.1, and I ask you: is it a black stripe behind a grey stripe or a grey stripe behind a black stripe? Thus, the challenge of image inpainting is that *the answer to the problem might not be unique*. We will discuss this and strategies to make 'good' guesses based on the way our perception works in Chapter 3.

When inspecting different inpainting methods in the course of this book, you should be aware of the fact that *mathematical inpainting methods are designed for inpainting the image completely automatically*, that is, without intervention (supervision) by the user. Hence, the art of designing efficient and qualitatively high inpainting methods is really the skill of modelling the mechanisms that influence what the human brain can usually do in an instant. At present, we are still far away from a fair competition with the human brain. Digital inpainting methods are currently not (will never be?) as smart as our brain. In particular, *no all-round inpainting model exists* that can solve a variety of inpainting problems with sufficient quality. One of the main shortcomings of inpainting methods is their inability to realistically reconstruct both structure and texture simultaneously (see Section 2.2).

Figure 2.1. Non-uniqueness of image inpainting. The image got lost inside the hatched domain. Should the reconstructed image show a black stripe behind a grey stripe or a grey stripe behind a black stripe?

Finally, let us also emphasise that the *difficulty of an image inpainting problem increases radically with the size of the damage* (the inpainting domain) in the image. The inpainting of large, connected gaps is more challenging than the inpainting of a collection of small holes – even if both their total areas cover the same amount of pixels. Intuitively, this is so because the larger the inpainting domain, the larger the distance across which image contents have to be interpolated and the less the known image information suggested by the intact image weighs for inpainting the inner parts of the inpainting domain. Lower-order inpainting methods – as we will discuss in Chapter 4 – may be sufficient for the inpainting of small holes because the interpolation model can act more locally there. For large gaps, more sophisticated inpainting models must be consulted, in particular, non-linear partial differential equations (PDEs) of higher order, as discussed in Chapter 5, and transport inpainting, as discussed in Chapter 6, as well as exemplar-based inpainting methods. One could loosely say that the sophistication of an interpolation method must increase with the size of the holes. Figure 2.2 shows two examples of harmonic inpainting, that is, interpolation from the grey values from the boundary of the inpainting domain ∂D by harmonic extension, as discussed in Section 4.2. This is one of the most basic inpainting methods. It constitutes a second-order linear elliptic PDE whose properties are

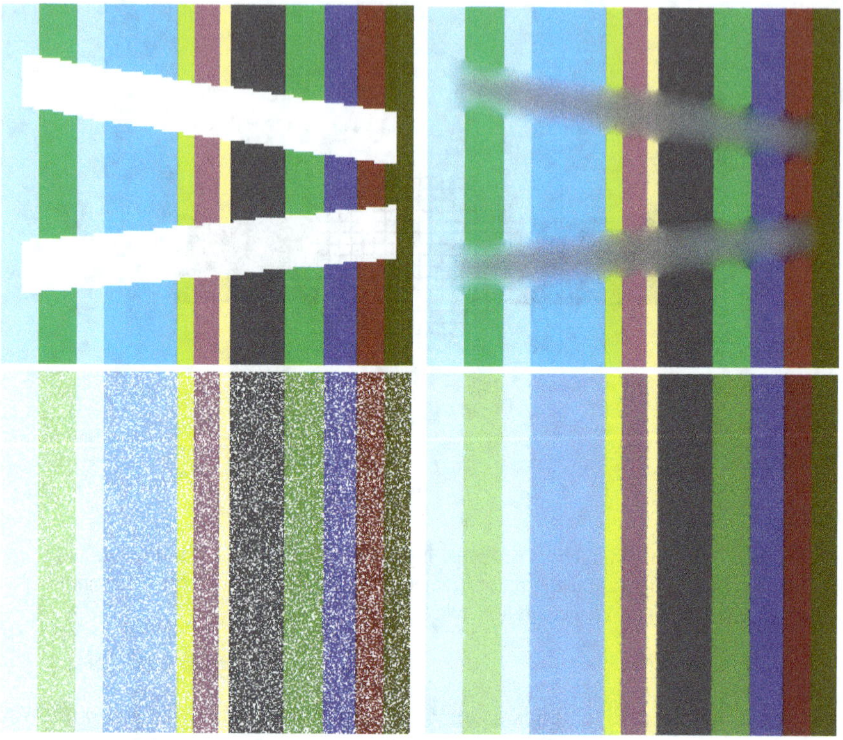

Figure 2.2. Large versus small gap image restoration using harmonic inpainting (see Section 4.2). On the left, the damaged images with inpainting domain marked in white. On the right, the corresponding restored images with harmonic inpainting. The percentage of damaged pixels is 30 per cent in both examples.

very well understood [Eva98] and for whose numerical solution very efficient solvers exist, for instance, iterative methods, spectral solvers and many more (cf. [Ise09]). This method performs very well for the inpainting of many small points of damage. If the inpainting domain, however, is one large occlusion, it fails to return a visually suggestive answer. Note that the percentage of damaged pixels in both examples is exactly the same. In the presence of large gaps, more sophisticated inpainting methods, such as inpainting by coherence transport (see Section 6.2), have to be consulted (see Figure 2.3).

2.1. Variational and PDE Methods

A *variational* approach for image inpainting is a method that computes the inpainted image as a minimiser of an objective functional. The latter is the

Figure 2.3. Large gap image restoration using coherence transport inpainting (see Section 6.2).

result of careful modelling by the user. Its generic form can be derived by the following Bayesian considerations.

Remark 2.1.1 Image inpainting is part of a larger class of mathematical problems that are called *ill-posed inverse problems*. The damaged image is the result of a certain kind of degradation process (the forward operation), and our task is to 'invert' this degradation and restore the original image. This is, of course, an ill-posed problem because – without further prior information – the image information inside the damaged part is completely lost and can only be guessed. Variational (regularisation) approaches to inverse problems aim to turn an ill-posed problem into a well-posed one by inserting prior information (in terms of regularity of the solution) into the problem.

The Bayesian Approach Variational methods can be best explained from the point of view of inverse problems. In a wide range of image processing tasks one encounters the situation that the observed image g is corrupted, for example, by noise or blur. The goal is to recover the original image u from the observed datum g. In mathematical terms, this means that one has to solve an inverse problem $Tu = g$, where T models the process through which the image u went before observation. In the case of an operator T with unbounded inverse, this problem is ill posed. In such cases, one modifies the problem by introducing some additional a priori information on u, usually in terms of a regularising term involving, for example, the total variation of u. This results in a minimisation problem for the fidelity $Tu - g$ plus the a priori information modelled by the regularising term $R(u)$. The underlying formalism for such an approach is given by Bayesian statistics. In the terminology of prior image models, the regularising term is the so-called prior image model, and the fidelity term is the data model. The concept of prior models has a long tradition in the regularisation for ill-posed inverse problems (see, e.g., [EHN96]) and has been explained for images by Mumford [GG84, Mum94, ZM97, MG01b]. For a general overview on this topic, see also [CS05a, AK06, SGG$^+$08 and BL11].

Inpainting approaches also can be formulated within this framework. When starting with the Bayesian approach for image inpainting, we assume that the observed image g is

$$g = \begin{cases} u+n & \text{in } \Omega \setminus D \\ \text{unknown} & \text{in } D. \end{cases}$$

In most of the examples we are going to consider, the noise part is in fact zero. This is so because those examples are simulated – taking a perfect photograph and cutting some holes in it. However, in practice, this assumption is very reasonable because the deterioration of an image (a photograph or a painting) by missing parts is usually accompanied by a low-quality, that is, noisy, image as a whole.

To simplify our statistical considerations, we go back to the discrete setting for a moment. Given a damaged image $g \in \mathbb{R}^N \times \mathbb{R}^N$, we can formulate two components for solving a general inverse problem:

- *Data model*: $g = Tu + n$, where $u \in \mathbb{R}^N \times \mathbb{R}^N$ is the original image (to be reconstructed), T is a linear transformation ($T = \mathcal{S}$, a sampling operator for image inpainting) and n is the noise which for our purpose is assumed to be Gaussian distributed with mean 0 and standard deviation σ
- *A priori probability density*: $P(u) = e^{-p(u)}du$. This is a priori information on the original image.

Then the a posteriori probability for u knowing g given by Bayes is

$$P(u|g) = \frac{P(g|u)P(u)}{P(g)},$$

with

$$P(g|u) = e^{-(1/2\sigma^2)\sum_{i,j}|(Tu)_{i,j}-g_{i,j}|^2}, \qquad P(u) = e^{-p(u)}$$

The idea of maximum a posteriori (MAP) image reconstruction is to find the 'best' image as the one which maximises this probability or, equivalently, which solves the minimisation problem

$$\min_u \left\{ p(u) + \frac{1}{2\sigma^2} \sum_{i,j} |g_{i,j} - (Tu)_{i,j}|^2 \right\}.$$

Variational Methods Let us take our derivation from the preceding section and turn it into an inpainting model in the continuum. Let $\Omega \subset \mathbb{R}^2$ be an open and bounded domain with Lipschitz boundary, and let B_1, B_2 be two Banach spaces with $B_2 \subseteq B_1$, $g \in B_1$ denoting the given image and $D \subset \Omega$ the missing domain. A general variational approach in image inpainting is formulated mathematically as a minimisation problem for a regularised cost functional $\mathcal{J}\colon B_2 \to \mathbb{R}$,

$$\mathcal{J}(u) = R(u) + \frac{\lambda}{2} \left\| \mathbb{1}_{\Omega \setminus D}(g-u) \right\|_{B_1}^2 \to \min_{u \in B_2}, \tag{2.1}$$

where $R\colon B_2 \to \mathbb{R}$,

$$\mathbb{1}_{\Omega \setminus D}(x) = \begin{cases} 1 & x \in \Omega \setminus D \\ 0 & x \in D, \end{cases} \tag{2.2}$$

is the indicator function of $\Omega \setminus D$, and $\lambda \gg 1$ is a large constant. This constant is the tuning parameter of the approach. As before, $R(u)$ denotes the regularising term and represents a certain a priori information from the image u; that is, it determines in which space the restored image lies. In the context of image inpainting, that is, in the setting of (2.1) and (2.2), it plays the main role of filling in the image content in the missing domain D, for example, by diffusion and/or transport. The fidelity term $\left\| \mathbb{1}_{\Omega \setminus D}(g-u) \right\|_{B_1}^2$ of the inpainting approach forces the minimiser u to stay close to the given image g outside the inpainting domain (how close depends on the size of λ). In this case, the operator T from the general approach equals the indicator function of $\Omega \setminus D$. In general, we have $B_2 \subset B_1$, which signifies the smoothing effect of the regularising term on the minimiser $u \in B_2(\Omega)$.

Note that the variational approach of (2.1) and (2.2) acts on the whole image domain Ω (global inpainting model) instead of posing the problem on the missing domain D only. This has the advantage of simultaneous noise removal in the whole image and makes the approach independent of the number and shape of the holes in the image. In this global model, the boundary condition for D is superimposed by the fidelity term.

Before getting into the development of inpainting algorithms, one has to understand what an image really is. In the framework of image prior models, this knowledge is encoded in the regularising term $R(u)$. As a consequence, different image prior models, result in different inpainting methods. As pointed out in [SC02], the challenge of inpainting lies in the fact that image functions are complex and mostly lie outside the usual Sobolev spaces. Natural images, for example, are modelled by Mumford as distributions (cf. [Mum94]). Texture images contain oscillations and are modelled by Markov random fields (see, e.g., [GG84 and Bre99]) or by functions in negative Sobolev spaces (see, e.g., [Mey01, VO03, OSV03 and LV08]. Most non-texture images are modelled in the space of functions of bounded variation [ROF92, CL97] and in the Mumford-Shah object-boundary model (cf. [MS89]).

Note also that despite its similarity to usual image enhancement methods such as de-noising or de-blurring, inpainting is very different from these approaches. This is so because the missing regions are usually large, that is, larger than the type of noise treated by common image enhancement, algorithms. Additionally, in image enhancement, the pixels contain both noise and original image information, whereas in inpainting, there is no significant information inside the missing domain. Hence, reasonable energy-based approaches in denoising do not necessarily make sense for inpainting. An example of this discrepancy between inpainting approaches and existing image enhancement methods is given in the work of Chan and Shen [CS05b]. Therein the authors pointed out that the extension of existing texture modelling approaches in de-noising, de-blurring and decomposition to inpainting is not straightforward. In fact, the authors showed that the Meyer model [Mey01] modified for inpainting, where the fidelity term modeled in Meyer's norm only acts outside the missing domain, is not able to reconstruct interesting texture information inside the gap: For every minimiser pair (u,v) (where v represents the texture in the image) of the modified Meyer model, it follows that v is identically zero inside the gap D.

PDEs for Image Inpainting To segue into the PDE-based approach for image inpainting, we first go back to the general variational model in (2.1) and (2.2). Under certain regularity assumptions on a minimiser u of the functional $\mathcal{J}$, the minimiser fulfils a so-called optimality condition on (2.1), that is, the corresponding Euler-Lagrange equation. In other words, for a minimiser u, the first variation, that is, the Fréchet derivative of $\mathcal{J}$, has to be zero. In the case $B_1 = L^2(\Omega)$, in mathematical terms this reads

$$-\nabla R(u) + \lambda \mathbb{1}_{\Omega\setminus D}(g-u) = 0 \quad \text{in } \Omega, \tag{2.3}$$

which is a PDE with certain boundary conditions on $\partial\Omega$. Here ∇R denotes the Fréchet derivative of R over $B_1 = L^2(\Omega)$ or, more generally, an element from the sub-differential of $R(u)$. The dynamic version of (2.3) is the so-called steepest-descent or gradient-flow approach. More precisely, a minimiser u of (2.1) is embedded in an evolutionary process. We denote it by $u(\cdot,t)$. At time $t=0$, $u(\cdot,t=0) = g \in B_1$ is the original image. It is then transformed through a process that is characterised by

$$u_t = -\nabla R(u) + \lambda \mathbb{1}_{\Omega\setminus D}(g-u) \quad \text{in } \Omega. \tag{2.4}$$

Given a variational formulation such as (2.1) and (2.2), the steepest-descent approach is used to numerically compute a minimiser of $\mathcal{J}$, whereby (2.4) is iteratively solved until one is close enough to a minimiser of $\mathcal{J}$.

In other situations we will encounter equations that do not come from variational principles, such as curvature-driven diffusion (CDD) inpainting [CS01b] and Cahn-Hilliard and TV-H^{-1} inpainting in Section 5.3. Then the inpainting approach is directly given as an evolutionary PDE, that is,

$$u_t = F(x, u, Du, D^2u, \ldots) + \lambda \mathbb{1}_{\Omega \setminus D}(g - u), \tag{2.5}$$

where $F\colon \Omega \times \mathbb{R} \times \mathbb{R}^2 \times \mathbb{R}^4 \times \cdots \to \mathbb{R}$ and belongs to the class of PDE-based inpainting approaches.

2.1.1. A Brief Historical Account on the State of the Art

Depending on the choice of the regularising term R and the Banach spaces B_1, B_2, that is, the flow $F(x, u, Du, \ldots)$, various inpainting approaches have been developed. These methods can be divided into two categories: texture inpainting, which is mainly based on synthesising the texture and filling it in, and non-texture (or geometrical/structure) inpainting, which concentrates on recovery of the geometrical part of the image inside the missing domain. In the following we shall only concentrate on non-texture images. In fact, the usual variational/PDE approach in inpainting uses local PDEs which smooth out every statistical fluctuation; that is, they do not see a global pattern such as texture in an image. In [SC02], the authors call this kind of image restoration *low-level inpainting* because it does not take into account global features such as patterns and textures. For now, let us start with the presentation of existing non-texture inpainting models.

Imitating Art Restorers The terminology of digital inpainting first appeared in the work of Bertalmio et al. [BSCB00]. Their model is based on observations about the work of museum artists who restore old paintings. Their approach follows the principle of prolonging the image intensity in the direction of the level lines (sets of image points with constant grey values) arriving at the hole. This results in solving a discrete approximation of the PDE

$$u_t = \nabla^{\perp} u \cdot \nabla \Delta u, \tag{2.6}$$

solved within the hole D extended by a small strip around its boundary. This extension of the computational domain around the strip serves as the intact source of the image. It is implemented in order to fetch the image intensity and direction of the level lines which are to be continued. Equation (2.6) is a transport equation for the image smoothness modelled by Δu along the level lines of the image. Here $\nabla^{\perp} u$ is the perpendicular gradient of the image function u; that is, it is equal to $(-u_y, u_x)$. To avoid the crossing of level lines, the authors additionally apply intermediate steps of non-linear diffusion, which may result in the solution of a PDE such as

$$u_t = \nabla^{\perp} u \cdot \nabla \Delta u + \nu \nabla \cdot (d(|\nabla u|) \nabla u),$$

where $d(s)$ defines the diffusivity coefficient, and $\nu > 0$ is a small parameter. In [BBS01], the authors interpret a solution of the latter equation as a direct solution of the Navier-Stokes equation for an incompressible fluid, where the image intensity function plays the role of the stream function whose level lines define the stream lines of the flow. Note that the advantage of this viewpoint is that one can exploit a rich and well-developed history of fluid problems, both analytically and numerically. Also note that the model of Bertalmio et al. actually is a third-order non-linear PDE. In the next section we shall see why higher-order PDEs are needed to solve the inpainting task satisfactorily. We will discuss Bertalmio's idea of inpainting with transport equations together with inpainting by coherent transport [BM07] in more detail in Chapter 6.

In a subsequent work of Ballester et al. [BBC$^+$01] the authors adapted the ideas of [BSCB00] of simultaneous grey level and gradient continuation to define a formal variational approach to the inpainting problem. Their variational approach is solved via its steepest descent, which leads to a set of two coupled second-order PDEs, one for the grey levels and one for the gradient orientations.

An Axiomatic Approach and Elastica Curves Chronologically earlier, Caselles, Morel and Sbert [CMS98a] and Masnou and Morel [MM98] initiated the variational/PDE approach for image interpolation. In [CMS98a], the authors show that any operator that interpolates continuous data given on a set of curves can be computed as a viscosity solution of a degenerate elliptic PDE. This equation is derived via an axiomatic approach in which the basic interpolation model, that is, the PDE, results from a series of assumptions about the image function and the interpolation process. See Section 4.1 for more details.

The approach of Masnou and Morel [MM98] belongs to the class of variational approaches and is based on the work of Nitzberg et al. on segmentation [NMS93]. In [NMS93], Nitzberg et al. present a variational technique for removing occlusions of objects with the goal of image segmentation. Therein the basic idea is to connect T-junctions at the occluding boundaries of objects with Euler elastica-minimising curves. A curve is said to be *Euler's elastica* if it is the equilibrium curve of the Euler elastica energy

$$E(\gamma) = \int_\gamma (\alpha + \beta\kappa^2)\, ds,$$

where ds denotes the arc-length element, $\kappa(s)$ is the scalar curvature and α, β are two positive constants. These curves were originally obtained by Euler in 1744 (cf. [Lov13]) and were first introduced in computer vision by Mumford [Mum94]. The basic principle of the elastica curves approach is to prolongate edges by minimising their length and curvature. In [Mum94 and NMS93], it is based on a priori edge detection. Hence, this approach is only applicable to highly segmented images with few T-junctions and is not applicable to natural images. Moreover, edges alone are not reliable information because they are sensitive to noise. In [MM98], Masnou and Morel extended Mumford's idea of length and curvature minimisation from edges to all the level lines of the image function. Their approach is based on the global minimisation of a discrete version of a constrained Euler elastica energy for all level lines. This level-line approach has the additional advantage that it is contrast invariant; this is different from the edge approach of Nitzberg et al. [NMS93], which depends on the difference of grey values. The discrete version of the Euler elastica energy is connected to the human-vision approach of Gestalt theory, in particular, Kanizsa's amodal completion theory [Kan79, Kan85]. Gestalt theory tries to explain how the human visual system understands partially occluded objects. This gave the approach in [MM98] its name *disocclusion* instead of *image inpainting*. Details of the theoretical justification of the model in [MM98] and the algorithm itself were published much later by Masnou [Mas02]. Note that the Euler elastica energy was used later for inpainting by Chan and Shen in a functionalised form (cf. [CKS02] and later remarks within this section).

Non-Linear Diffusion: Total Variation Inpainting and CDD Another variational inpainting approach is presented in the work of Chan and Shen [SC02]. Their approach is chronologically in between the two works of Bertalmio et al. (i.e., [BSCB00 and BBC$^+$01]. The desire was to create a scheme which is motivated by existing de-noising/segmentation methods and is mathematically easier to understand and analyse. Their approach is based on the most famous model in image processing, the total variation (TV) model, where $R(u) = |Du|(\Omega) \approx \int_\Omega |\nabla u|\, dx$ denotes the total variation of u and $B_1 = L^2(\Omega)$ and $B_2 = BV(\Omega)$ are the space of functions of bounded variation (cf. also [SC01, SC02, RO94, ROF92]). It results in the action of non-linear diffusion inside the inpainting domain, which preserves edges and diffuses homogeneous regions and small oscillations such as noise. More precisely, the corresponding steepest-descent equation reads

$$u_t = \nabla \cdot \left(\frac{\nabla u}{|\nabla u|} \right) + \lambda \mathbb{1}_{\Omega \setminus D}(g-u).$$

The disadvantage of the total variation approach in inpainting is that the level lines are interpolated linearly (see Section 5.1). This means that the direction of the level lines is not preserved, since they are connected by a straight line across the missing domain. A straight-line connection might still be pleasant for small holes but very unpleasant in the presence of larger gaps, even for simple images. Another consequence of the linear interpolation is that level lines might not be connected across large distances (see Section 5.1). A solution for this is the use of higher-order PDEs such as the first works of Bertalmio et al. [BSCB00] and Ballester et al. [BBC+01] and the elastica approach of Masnou and Morel [MM98], as well as some PDE/variational approaches such as exemplar-based inpainting proposed later.

Within this context, the authors in [CS01b] proposed a new TV-based inpainting method. In their model, the conductivity coefficient of the non-linear diffusion depends on the curvature of the level lines, and it is possible to connect the level lines across large distances. This new approach is a third-order diffusion equation and is called inpainting with *curvature-driven diffusions* (CDDs). The CDD equation reads

$$u_t = \nabla \cdot \left(\frac{d(\kappa)}{|\nabla u|} \nabla u \right) + \lambda \mathbb{1}_{\Omega \setminus D}(g-u),$$

where $d\colon B \to [0,+\infty)$ is a continuous function, which penalises large curvatures and encourages diffusion when the curvature is small. Here B is an admissible class of functions for which the curvature κ is well defined, for example, $B = C^2(\Omega)$. It is of similar type as other diffusion-driven models in imaging such as the Perona-Malik equation [PM90, MS95], and like the latter, it does not (in general) follow a variational principle. To give a more precise motivation for the CDD inpainting model, let us recall that a problem of the usual TV model is that the diffusion strength depends only on the contrast or strength of the level lines. In other words, the non-linear diffusion of the TV model diffuses with conductivity coefficient $1/|\nabla u|$. Hence, the diffusion strength does not depend on geometrical properties of the level line, given by its curvature. In the CDD model, the conductivity coefficient is therefore changed to $d(|\kappa|)/|\nabla u|$, where d annihilates large curvatures and stabilises small ones. Interestingly enough, CDD performs completely orthogonally to the transport equation of Bertalmio et al. [BSCB00]. Their equation transports the smoothness along the level lines, whereas the CDD equation diffuses image pixel information perpendicularly to the level lines. We will discuss these inpainting mechanisms of transport and diffusion in more detail in Chapter 8.

Euler Elastica Inpainting This observation gave Chan, Kang and Shen the idea to combine both methods, which resulted in the Euler elastica inpainting model (see [CKS02, CS01a]). Their approach is based on the earlier work of Masnou and Morel [MM98], with the difference that the new approach poses a functionalised model. This means that instead of an elastica curve model for the level lines of the image, they rewrote the elastica energy in terms of the image function u. Then the regularising term reads $R(u) = \int_\Omega (\alpha + \beta(\nabla \cdot (\nabla u/|\nabla u|))^2)|\nabla u|\, dx$, with positive weights α and β, $B_1 = L^2(\Omega)$ and $B_2 = BV(\Omega)$. In fact, in [CKS02], the authors verified that Euler elastica inpainting model combines both inpainting processes [BSCB00 and CS01b]. They also presented a very useful geometrical interpretation for all three models. We shall discuss this issue in a little more detail in Chapter 8, where we compare this geometrical interpretation with those for two other higher-order inpainting schemes discussed in Sections 5.3 and 5.4.

Active Contour Models Other examples to be mentioned for (2.1) are the active contour model based on Mumford and Shah's segmentation [MS89, SC02, TYJW01, ES02] and its high-order correction, the Mumford-Shah-Euler image model [ES02]. The latter improves the former by replacing the straight-line curve model with the elastica energy. The Mumford and Shah image model reads

$$R(u,\Gamma) = \frac{\gamma}{2}\int_{\Omega\setminus\Gamma} |\nabla u|^2\, dx + \alpha\mathcal{H}_1(\Gamma), \tag{2.7}$$

where Γ denotes the edge collection and $\mathcal{H}_1$ the one-dimensional Hausdorff measure (generalisation of the length for regular curves). The corresponding inpainting approach minimises the Mumford-Shah image model plus the usual L^2 fidelity on $\Omega \setminus D$. The idea to consider this model for inpainting goes back to Chan and Shen [SC02] as an alternative to TV inpainting and to Tsai et al. [TYJW01]. The Mumford-Shah-Euler image model differs from (2.7) in the replacement of the straight-line model with Euler's elastica curve model

$$R(u,\Gamma) = \frac{\gamma}{2}\int_{\Omega\setminus\Gamma} |\nabla u|^2\, dx + \int_\Gamma (\alpha + \beta\kappa^2)\, ds,$$

where κ denotes the curvature of a level line inside the edge collection Γ, and α and β are positive constants as before.

Inpainting with Sparse Dictionaries Using wavelets, curvelets, shearlets and 'you-name-lets' constitutes a very popular framework for representing digital images [Mal99]. As argued by many authors, these functions offer a basis or frame that allows a sparse and multi-scale representation of the image. They are used in inpainting as well. We will not discuss them in this book, however, so let us refer to some other works [FSM09, DJL$^+$12, KKL13, GK14, KKZ14] for interested readers to start enquiring about them.

Inpainting in Transform Spaces Inpainting is not restricted to inpainting in the image domain only. Inpainting in the wavelet domain with TV inpainting is considered, for instance, in [CSZ06, DB08 and ZC10]. Inpainting problems in the Fourier domain arise in magnetic resonance imaging (MRI) [LLT03], and inpainting of imperfect and limited-angle tomographic data is seen in electron tomography [BBLS14] (see also Section 9.3).

Exemplar-Based Inpainting Not all inpainting methods rely on PDEs. In the context of texture inpainting, the synthesis of texture is discussed in [EL99]. Non-local PDEs for texture and pattern inpainting are considered in, for example, [GO07 and ZC10]. There also is an extensive literature on exemplar-based inpainting approaches, for example, [CPT03, BCG$^+$06, ACS09, FACS09, ALM10, AFCS11 and ACF12]. See also [Cas11] for a short review article on this topic. The idea of exemplar-based inpainting is to fill in gaps in images taking into account the image information not only in the vicinity of the inpainting domain – as it is for local PDE and variational methods – but also in all the intact image contents. Using sophisticated copy-and-paste procedures, intact parts of the image are copied into the inpainting domain weighted by their likelihood to be relevant to the reconstruction of the image.

Exemplar-based inpainting uses a copy-and-paste algorithm to fill in the missing areas of an image. If an image has a hole where a missing shirt button should be, for example, an algorithm that uses exemplar-based inpainting could insert the button by copying one from another location in the image; a PDE-based algorithm could not. This is of particular importance when inpainting repetitive image contents, such as patterns and texture.

Some More Remarks Other PDE- and variational-based inpainting methods have been introduced. Tschumperlé used diffusion in coherent directions for image inpainting [Tsc06], for instance. Bornemann and März perfected the idea of Bertalmio and proposed a fast inpainting method based on coherence transport [BM07, Mär11]. Tai et al. [TOH07] proposed an extension of TV inpainting using a Stokes flow. Bertozzi, Esedoglu and Gillette [BEG07,

BEglG07] proposed a modified Cahn-Hilliard equation for the inpainting of binary images (also cf. Section 5.3), and in a separate work, Grossauer and Scherzer [GS03] proposed a model based on the complex Ginzburg-Landau energy. A generalisation of Cahn-Hilliard inpainting for grey value images called TV-H^{-1} *inpainting* is proposed in [BHS09] (also see Section 5.3), and second-order total variation inpainting is discussed in [PS14 and PSS13] and also presented in Section 5.5. See also Sections 4.5 and 5.6 for more references along these lines.

There are, of course, a myriad of extensions and special cases of inpainting problems and methods in the literature. Let us mention landmark-based inpainting (see, e.g., [KCS02] for its application to inpainting from multiple views), inpainting via correspondence maps (as discussed in [DSC03]) and blind inpainting [DJL$^+$12], where the inpainting domain is unknown.

2.2. Structure Versus Texture Inpainting

Structure inpainting denotes the task of reconstructing the large-scale structures – which also can be thought of as the main geometry of the image. In the following chapters we will see that variational and PDE inpainting can perform extremely well for the task of structure inpainting. They pick up image information from the boundary (or a vicinity) of the inpainting domain and propagate it inwards. The rules that this propagation should follow are modelled into the PDE. These are usually driven by techniques used in art restoration [WG04] (see also Section 6.1) as well as studies of visual perception, in particular, the principle of 'good continuation' shaped by Kanizsa [Kan79] (see also Chapter 3). PDEs are a powerful machinery for the formalisation of these mechanisms. They are based on Newton's differential calculus and are often used to describe phenomena in nature where the variations of quantities are compared, for instance, in physics, biology, engineering, computer science, social sciences and economics. The brief overview of PDE inpainting methods given in the preceding section gives only a first glimpse into the wealth of PDE models for inpainting.

Texture and pattern, however, constitute fine-scale repetitive structures that a local inpainting method such as a PDE interpolator is not able to reproduce. Texture synthesis and exemplar-based inpainting methods as described earlier should be consulted in this case. Exemplar-based inpainting constitutes a very powerful inpainting approach which can restore images with visually often surprisingly pleasing results. As long as enough useful 'examples' are present in the intact part of the image as well as sufficient 'hints' for what of these examples should be copied inside the inpainting domain, this copy-and-paste methodology works very well. Note, however, that the output of these methods can be very sensitive with respect to the initialisation inside the inpainting domain, as well as the lack of appropriate examples. Moreover, PDE-based approaches are directly applicable to a wider range of fields and problems than exemplar-based approaches. One example is, of course, inpainting in transform spaces. In the medical field, for instance, PDE-based regularisation is used to reconstruct images from highly under-sampled MRI measurements. In the material sciences, electron tomography produces tomographic measurements which have whole parts of their

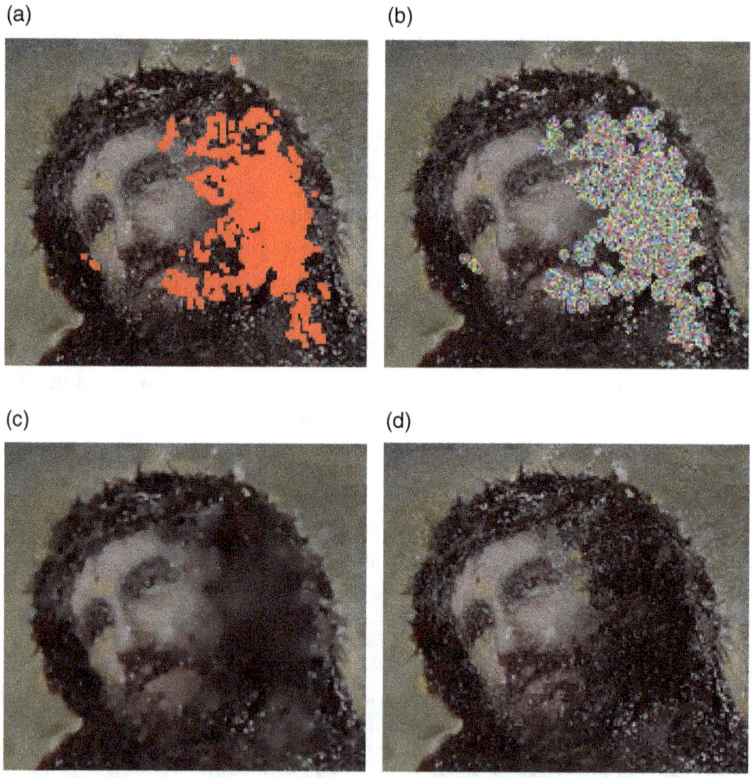

Figure 2.4. Structure and texture inpainting, two different animals. Image restoration of *Ecce Homo* in Figure 1.4 with the local PDE inpainting method TV-H^{-1} (Section 5.3) and a global exemplar-based inpainting method [ACS09, AFCS11] initialised with the result of TV-H^{-1} inpainting. (a) Mask for restoration. (b) Initialisation of the restoration algorithm with random colours. (c) Restored image with local PDE method. (d) Restored image with global inpainting.

angular dimensions missing. This problem is called *limited angle tomography* and will be discussed in Section 9.3. There is no straightforward way that such reconstructions can be done with an exemplar-based approach. Moreover, the more sophisticated exemplar-based approaches are usually computationally much more demanding than PDE-based approaches. Therefore, PDE-based approaches are usually chosen for very large-scale problems, as well for problems which need real-time solutions. Unifying approaches for simultaneous structure and texture inpainting aim to combine the strengths of both methods (see, e.g., [BVSO03, Gro04, ESQD05, CGMP11 and BBCS10]).

Figures 2.4 and 2.5 show two inpainting problems and their corresponding PDE- and exemplar-based solutions. This book discusses only local (structure) inpainting methods, in particular, PDE approaches.

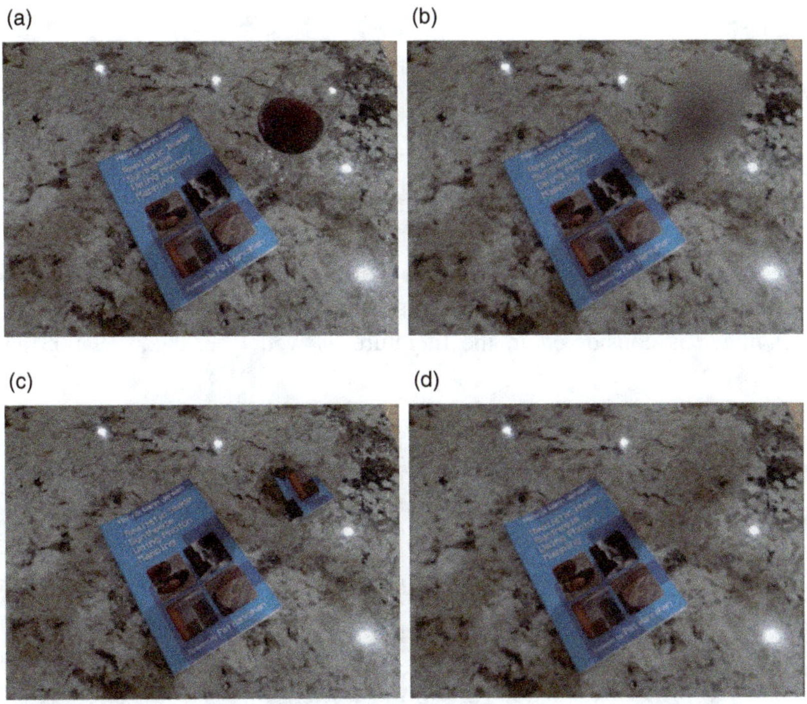

Figure 2.5. Structural inpainting, such as harmonic inpainting, can aid the exemplar-based inpainting result produced using the method from [ACS09, AFCS11]. (a) Original photo. (b) Harmonic inpainting. (c) Exemplar-based inpainting without smart initialisation. (d) Exemplar-based inpainting initialised with harmonic inpainting result from Figure 2.2.

2.3. Inpainting of Colour Images

So far we have considered grey scale images only, that is, $u \in \mathbb{R}^{n\times m}$. However, it is easy to derive a first straightforward extension of inpainting methods for coloured images if we treat each colour channel of the function separately. That is to say, when $\mathbf{u} \in (\mathbb{R}^3)^{n\times m}$, where for simplicity we consider red-green-blue (RGB) images, we apply the inpainting method to u_1, u_2, u_3 independently. Another possibility is the extension of the functional definitions or PDEs in the inpainting approaches discussed so far for vector-valued images. For instance, in the case of TV inpainting that will be presented in Section 4.3, if $\mathbf{u} = (u_1, u_2, u_3) \in \mathrm{BV}(\Omega;\mathbb{R}^3)$, then the total variation (TV) of u is defined as

$$TV(\mathbf{u}) = \sup\left\{\sum_{i=1}^{3}\int_\Omega u_i \mathrm{div}\phi_i : \phi \in \left[C_c^1(\Omega,\mathbb{R}^2)\right]^3,\ \|\phi\|_\infty \le 1\right\} \tag{2.8}$$

(see, e.g., [BC98, For06, FM07, FR08 and MBBS13]). We do not go into more detail on the vector-valued case here. We rather apply the inpainting methods that are formulated for scalar-valued functions to each colour channel separately. We use the RGB colour space to do so (compare Figure 2.6). It is questionable that this is the optimal choice. Inpainting in other colour spaces is possible by converting the RGB image. In [BSCB00], for instance, a colour model related to the LUV model is used to avoid the appearance of spurious colours. Let us also remark that inpainting colours in a grey level image is considered in the literature as well (see, e.g., [Sap05 and FM07]).

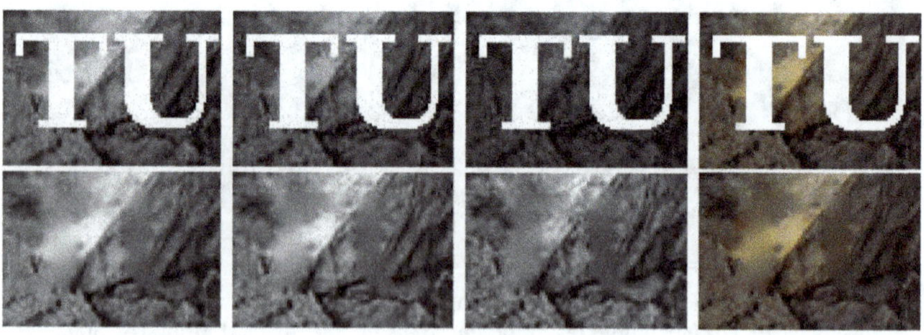

Figure 2.6. Channel-by-channel inpainting of colour images. In the first row, the red, green and blue colour channel and the full RGB image are shown. The second row contains the corresponding inpainting results.

2.4. Video Inpainting

The inpainting of videos holds, at the same time, more possibilities and more challenges than image inpainting. One of the main challenges is the high dimensionality of the data, having to deal now not only with one single image but with a sequence of image frames. This is a possibility at the same time because damaged or occluded areas in one frame might be intact in an earlier or later frames in the sequence. But how do we find the correspondence between a damaged pixel in one frame and the corresponding intact pixel in another frame and, if found, how should it be mapped to the damaged part? Much research effort is currently going into video inpainting with mixed success, the most successful approaches being exemplar-based inpainting procedures, whose main problem still is their long execution time. We do not discuss video inpainting methods in this book, but we refer readers to some recent works on the use of optical flow and exemplar-based inpainting for more information and further references [DSC03, WSI04, PSB05, PSB07, WSI07, LS11, WPUB11, GTK$^+$12, SDC14, NAF$^+$14].

3

The Principle of Good Continuation

Image inpainting is a mathematically highly ill-posed process. Once something is completely lost, we can never know for sure what has been there – except if we knew the image before it was damaged or we can travel back in time and have a look. This means that image inpainting does not have a unique solution in general. Since we are intelligent human beings, we can narrow done our search by using roughly two main strategies: our *natural visual perception and experience* and our *historical knowledge* (almost like travelling back in time) about the image and its hole. Both these strategies are crucial for modelling of the desired inpainting process and will determine the inpainting methods that we will consider in subsequent chapters.

Historical knowledge of the image could be many things and depend very much on the kind of application. In art restoration, known characteristics and techniques used by the painter, intact illustrations of a similar scene by different artists (such is the case in many religious paintings from the eighteenth century), and knowledge about general painting techniques, colour mixtures and materials used at the time the painting was created have been used. In video inpainting, knowledge from earlier frames could be used to extrapolate in time into later frames where damage appears. In medical imaging applications, anatomical knowledge of a template human body is used, and similarly, there are many other application in which geometrical knowledge about the shapes of objects one is looking for can be used (compare the application on the inpainting of satellite images of roads in Section 9.2). This part of the interpolation process depends on the context of the interpolation.

Visual perception, however, gives a context-free continuation of an incomplete scene. It constitutes our – either natural or learned – ability to automatically interpolate broken or occluded structures. This automatic continuation follows certain laws. In Figure 3.1, what is the correct interpretation of the image we see? Is it one straight line or two straight-line endings? What

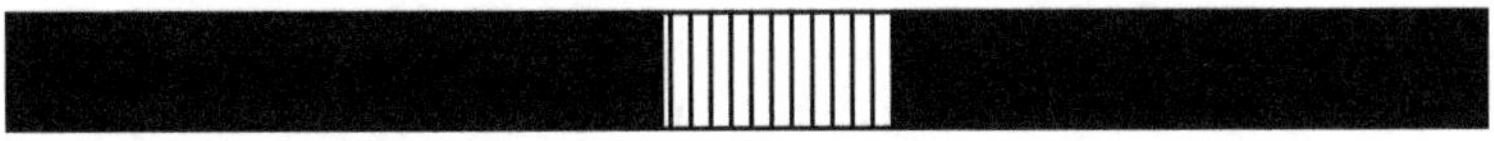

Figure 3.1. Is this a straight line with a hole? Or two straight lines?

is more probable is determined by our way of perception and is driven by our experience. Gestalt theory studies these possible scenarios, assigns probabilities to them and discusses visual paradoxes. In what follows, we give an overview of the main principles of the Gestaltist school that are relevant for image completion and interpolation. They will return to us later when those principles are being formalised and infiltrated into PDE models for image inpainting.

3.1. Gestalt Theory

In German, *Gestalt* is the perceived shape of an object. Gestalt theory studies the laws of visual reconstruction. As a working model, it starts with the representation of shapes by a collection of points with different shades of grey. These points constitute a microscopic model of the rough datum that corresponds to local retina cell stimuli. This seems very similar to the idea of a digital image, where continuous structures are also missing but objects are represented by pixels, that is, squared dots with different grey or colour values. Gestaltists (scientists who work on Gestalt theory) investigate the principles of how continuous shapes (*Gestalten* or, anglicised, *gestalts*) are reconstructed from such point clouds by our perception apparatus. It is almost like an encryption process that identifies physical objects with accumulations of points. Starting from a very localised representation of image structures, Gestalt theory aims to infer a global percept of an image.

Historically, Gestalt theory first appeared in the work of Wertheimer [Wer23], who formulated two main organising laws for gestalts: grouping based on single qualities of points or collections of points following defined laws and collaboration and conflicts of these laws. The 'bible' of Gestaltists is the book series *Gesetze des Sehens* by Metzger [Met36]. Moreover, many important concepts used in digital image processing and computer vision nowadays are based on the work of Kanizsa [Kan79], who gave a convincing list of gestalt/grouping laws, among them his amodal completion theory, which is discussed in more detail in the next section. A first systematic formalisation of Gestalt principles into mathematically computable gestalt laws that can be used to create image analysis and processing algorithms is beautifully presented in the book by Desolneux, Moisan and Morel [DMM07]. Earlier works in which gestalt laws are taken and put into algorithmic language are, for instance, the works of Shaáhua and Ullmann [US88] and Guy and Medioni [GM93], where important global perceptual image structures were computed following the laws of Gestalt theory.

We start with the definition of what a gestalt is and cite Loewe [Low85], who described a gestalt as a set of points whose spatial arrangement could not occur in noise. Its specific turnout depends on the type of gestalt, that is, the type of image structure or shape, that is considered. A perceptual boundary, for instance, is defined in [DMM07] as a level line whose gradient is large enough that no such line is likely to occur in white noise. Following the Gestaltist idea, every gestalt, every perception, is the result of a construction that follows gestalt laws. The first ones are the *grouping laws*, where a *group* is a collection of points with common characteristics. If the group's characteristics are significant in the image scene, that is, in quantity, in contrast with the remainder of the scene and not in conflict with other possible groups, such a group becomes a gestalt. Kanizsa [Kan79] provided the following list of elementary grouping laws: *vicinity, similarity, continuity of direction, amodal completion, closure, constant width, tendency to convexity, symmetry, common motion* and *past experience.*

Amodal completion is the grouping law that will be of most interest for the problem of image inpainting. *Amodal* refers to a completion without the chromatic attributes of the visual modality that is completed. It is in action when two curves meet in the form of a T-junction (compare Figure 3.2). The leg of the T-junction is perceived as an interrupted curve that belongs to the boundary of the occluded object.

Whenever possible, our perception extrapolates this leg and connects it with another leg. This process is called *amodal completion.* The connection of the two legs follows the good continuation principle, which says that the interpolated curve is as similar as possible to the two given pieces. In the image inpainting methods that we will discuss in what follows, this good continuation

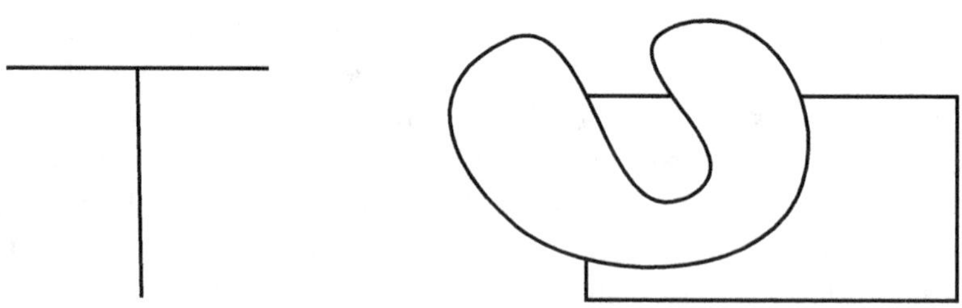

Figure 3.2. The vertical curve meets the horizontal curve and forms a T-junction where they meet. A T-junction indicates a possible occlusion.

principle will lead us to interpolation laws that control the curvature of interpolated level lines of the image (cf. Section 5.2.3). Other grouping laws that should guide the inpainting process are the colour-constancy law (connected regions with similar colours should be unified), the vicinity law (close-by points or point clouds are likely to be considered as one large object) and the similarity law (similar structures should be grouped together).

As in image restoration, the grouping of points does not have to be unique. Point constellations may result in perceptually ambiguous situations. Indeed, groupings of points might compete or be in conflict with each other. In most situations, however, one grouping dominates (i.e., masks) the others and results into the perceived gestalt. The question of how to formalise (automate our evaluation on) what is likely to be perceived in a digital image boils down to the how different groupings and their likelihoods are quantified. We refer the reader to [DMM07] for a derivation of a computational Gestalt theory. In our context of image inpainting it suffices to mention that amodal completion is a very strong grouping law and mostly wins over the others. Hence, it will be one of our main driving principles for the discussion of image inpainting methods.

3.2. Kanizsa's Amodal Completion

Investigating the process of visual perception, Kanizsa got very interested in the concept and underlying laws of amodal completion. In his article 'Seeing and Thinking' [Kan85], he questioned the relation between visual completion as an automatic process (that follows certain rules) and its interpretation (that arises from thinking about it and should be considered as a secondary step in perception). He said:

> If coding, recognition and interpretation necessarily presuppose the visual data that are the object of these operations, and if these data already possess an organization of their own, the problem which arises at this point is that of their formation. In other words: how do we form the cues which must be interpreted? ... what we see would not only be utilized by inferential processes in the interpretation phase but would also be the product of unconscious inferences drawn during the primary process. ... The nature of these principles must be discovered and not given for granted beforehand. ... Moreover, a particular class of phenomena is regarded as the result of processes similar to those which occur in standard problem-solving situations. When the proximal stimulation is plurivocal – that is, when it can be 'read' in more than one way – the visual system is obliged to choose from a certain number of possible solutions. The solution eventually chosen would be the most logical, self-consistent, parsimonious and elegant out of those which are theoretically possible.

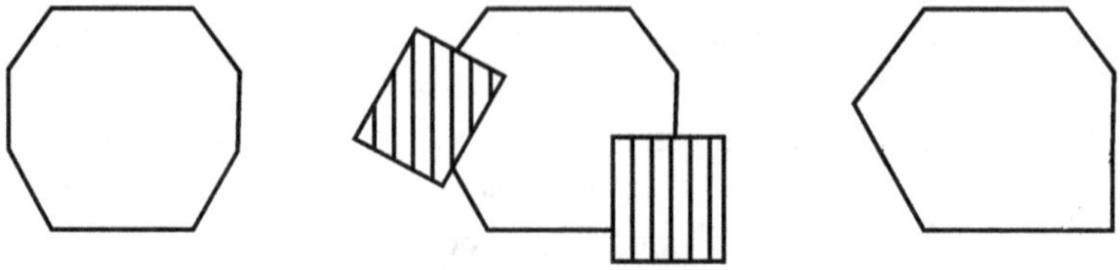

Figure 3.3. Visual completion with and without contextual information. From left to right: An octagon, an occlusion and its visual completion.

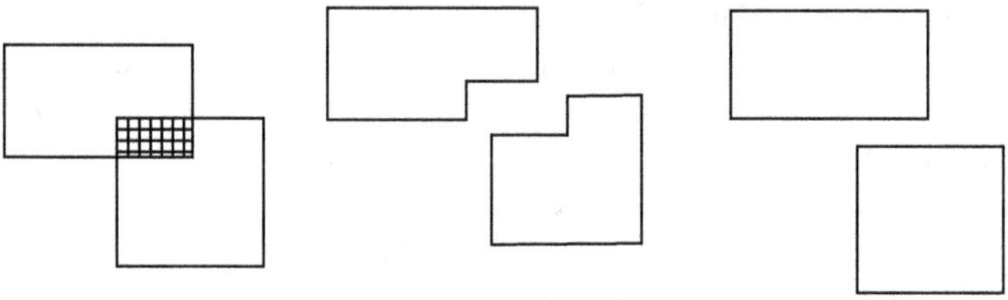

Figure 3.4. T-junctions are more likely to be connected than to remain endings.

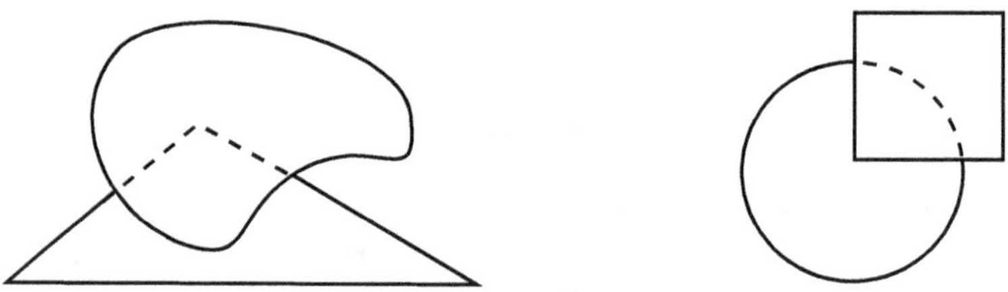

Figure 3.5. Quality of the continuation: preservation of smoothness of the T-legs.

In his monograph Kanizsa discussed a series of examples. Some of them are featured in Figures 3.3 through 3.5. He concentrated on visual completion that does not depend on contextual knowledge of the perceiver (compare Figure 3.3). The driving principle is 'good continuation' that prefers connection of apparently broken-off (or occluded) structures over their disconnection (compare Figure 3.4), as well as smooth continuation over other possible connections (also called *co-curvilinearity* [Low85]; compare Figure 3.5). In [GM93], the authors also presented the law of constant curvature, which says that we intend to extend a curve with constant curvature with the same curvature (compare Figure 3.6). In [Kan85], Kanizsa concluded that

> the visual system, in the cases in which it is free to do so, does not always choose the solution that is most coherent with the context, as normal reasoning would require. This means that seeing follows a different logic – or, still better, that it does

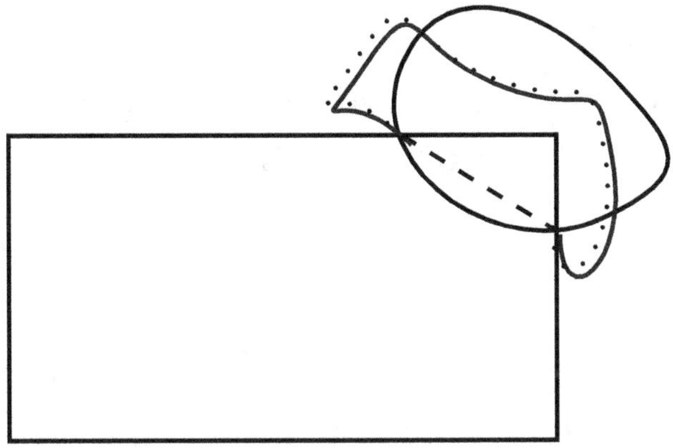

Figure 3.6. Constant curvature reconstruction is preferred over other possible scenarios (without other clues). The two solid straight lines constitute the most probable continuation scenario with respect to our perception.

> not perform any reasoning at all but simply works according to autonomous principles of organization which are not the same principles which regulate thinking.

The discovery and formalisation of those principles is one of the main topics in the design of local image inpainting methods. Kanizsa's study of amodal completion and his principles of 'good continuation' have influenced many state-of-the-art inpainting methods. The integration of contextual information into the inpainting process is described in Kanizsa's secondary step, which involves thinking and interpretation. State-of-the-art machine learning methods and exemplar-based inpainting methods are designed to mimic this secondary step by learning the correct filling of the gap from examples.

4

Second-Order Diffusion Equations for Inpainting

Before we start, let us emphasise here once more that in this and most subsequent chapters we focus on local structural inpainting methods. Structural image inpainting means that we fill in missing parts in images by using local structural information only. To do so, we formulate a partial differential equation (PDE) or a variational approach that picks up this information in terms of colour/grey values and image edges and propagates the information into the missing domain by means of transport and diffusion.

In this chapter we commence our presentation of variational and PDE methods for image inpainting with an axiomatic derivation of a PDE interpolator proposed by [CMS98a]. The resulting generic second-order PDE gives rise to our first three PDE inpainters: harmonic inpainting, total variation (TV) inpainting and absolutely minimising Lipschitz extensions (AMLE) inpainting. In the last section we focus on an extension of TV inpainting that leads us to the discussion of higher-order, in particular, curvature-based, PDEs for image inpainting in Chapter 5.

Throughout this chapter we use the following definitions and assumptions:

- The image domain $\Omega \subset \mathbb{R}^2$ is bounded and open with Lipschitz boundary $\partial\Omega$.
- The inpainting domain $D \subset \Omega$ lies in the interior of Ω, that is, $\partial D \cap \partial\Omega = \emptyset$.

4.1. An Axiomatic Approach to Image Inpainting

One of the most pioneering contributions in PDE-based image processing methods certainly is the work of Alvarez et al. [AGLM93]. There the authors derive a general second-order PDE whose evolution describes a multi-scale analysis of an image, that is, a family of transforms which when applied to a given image produces a sequences of new images. A typical example of a multi-scale analysis of evolution equations is the scale space [Wei98] produced by the heat equation. In this case, the transform is a Gaussian kernel whose standard deviation grows like the square root of time. In [AGLM93], the authors showed that all sequences of images derived from a multi-scale analysis of a given image are solutions of this PDE, given that certain reasonable assumptions on the nature of the analysis are satisfied. These assumptions constitute formal requirements, that is, axioms on image processing transforms classified into three categories: architectural requirements, stability requirements and morphological requirements. Based on these axioms, different models can be characterised, and a new multi-scale approach is proposed that subsumes all these axioms into one single equation. Inspired by this axiomatic description, Caselles, Morel and Sbert [CMS98a] formalised requirements that an image interpolation operator should satisfy and derived a generic second-order PDE for image inpainting which models the desired interpolation operator.

4.1.1. The General Form of the Interpolation Operator

Let us start by introducing a specific notation that is convenient to formulate the general form of the interpolation operator.

Definition 4.1.1 Directional Hessian For $u\colon \Omega \to \mathbb{R}$, let D^2u be the Hessian of u. Then we can specify second-order derivatives of u in a specific direction by assigning the following quadratic form

$$Au(v,w) := \sum_{i,j} A_{i,j} v_i w_j, \quad A \in \mathbb{R}^{2\times 2}, v, w \in \mathbb{R}^2,$$

to $A = D^2$. Then the second partial derivatives in the direction of the gradient of u read $u_{\eta\eta} = D^2u(\nabla u/|\nabla u|, \nabla u/|\nabla u|)$, and the second partial derivatives in the direction tangent to the level lines are given by $u_{\xi\xi} = D^2u\left(\nabla^{\perp}u/|\nabla u|, \nabla^{\perp}u/|\nabla u|\right)$. Of course, we can have the mixed case as well, which gives $u_{\xi\eta} = D^2u\left(\nabla^{\perp}u/|\nabla u|, \nabla u/|\nabla u|\right)$.

Before formalising the axioms, the interpolation operator should fulfil, let us discuss three possible choices that will make the formalism more intuitive and will be considered again later as examples of the general PDE are derived. We consider the following simple situation: let u be an image function considered on a finite grid imposed on a continuous domain $\Omega \subset \mathbb{R}^2$ except in one point x_0 of the grid. In [CMS98a], the following choices for interpolating the value of u in $x_0 = (x_0^1, x_0^2)$ are proposed:

- Compute $u(x_0)$ as the *mean value of neighbouring pixels*, that is,

$$u(x_0) = \frac{1}{4}\left(u(x_0-(h,0))+u(x_0+(h,0)) + u(x_0-(0,h))+u(x_0+(0,h))\right),$$

  where h is the step size of the equidistant grid in Ω. Looking at this formula, we may realise that this is nothing other than the five-point formula which approximates the Laplacian of u in the continuous setting. In fact, by Taylor expansion and by letting $h \to 0$, we get that – in the continuum – u is a solution of the Laplace equation in x_0, that is,

$$\Delta u(x_0) = 0. \tag{4.1}$$

  We will consider the Laplace equation for image inpainting in more detail in Section 4.2 as the method of *harmonic inpainting*.
- Compute $u(x_0)$ as the *median value of neighbouring pixels*. It has been shown in [BG95 and Eva93] that iterating a median filter n times based on a grid with step size h and letting $h \to 0$, $n \to \infty$, the dynamics of this filtering technique converge to a mean curvature flow [Eva98]

$$|\nabla u| \cdot \operatorname{div}\left(\frac{\nabla u}{|\nabla u|}\right)(x_0) = D^2 u\left(\frac{\nabla^\perp u}{|\nabla u|}, \frac{\nabla^\perp u}{|\nabla u|}\right)(x_0) = 0. \tag{4.2}$$

  Inpainting by mean-curvature flow will appear again later when we look at *total variation inpainting* in Section 4.3, Remark 4.3.2.
- Compute $u(x_0)$ by prolonging the image information from neighbouring pixels which lie along the gradient of the image function. In [CT96], Casas and Torres considered this idea and proposed an interpolation algorithm that – in the simple situation here and for an image function $u \in C^2(\Omega)$ – boils the interpolation down to computing u in x_0 as

$$u(x_0) = \frac{1}{2}\left(u(x_0+h\nabla u)+u(x_0-h\nabla u)\right)+\mathcal{O}(h^2),$$

  where the gradient $\nabla u(x_0)$ can be approximated by central differences evaluated in the neighbouring (known) grid points. Similarly as before, by Taylor expansion and by letting $h \to 0$, we get

$$D^2 u\,(\nabla u, \nabla u)\,(x_0) = 0. \tag{4.3}$$

  This equation gives rise to the inpainting model of *absolutely minimising Lipschitz extensions (AMLE)* [Aro67, CMS98a, Jen93], which we will describe in more details later on in this chapter.

Now let us continue with the formalisation of the inpainting axioms. As we will see, the preceding interpolation equations (4.1)–(4.3) represent all possible choices within the axiomatic setup of [CMS98a].

Axiomatic Description

The interpolation axioms are defined for the generic situation of one inpainting domain D whose boundary is a continuous and simple Jordan curve Γ (see Definition B.13.1 in Appendix B). Moreover, we set $\mathcal{C}$ to be the set of all continuous and simple Jordan curves Γ and $\mathcal{F}(\Gamma)$ to be the set of continuous functions defined on the curve Γ taking values in $\mathbb{R}$. An interpolation operator E is defined as a mapping from $\mathcal{F}(\Gamma) \times \mathcal{C}$ to a function on the inpainting domain $D(\Gamma)$. It associates an interpolant $E(\phi, \Gamma)$ to a continuous function $\phi \in \mathcal{F}(\Gamma)$ and a curve $\Gamma \in \mathcal{C}$ in every point in $D(\Omega)$. From this interpolation operator we demand the following axioms to be fulfiled:

A1 Comparison Principle For any $\Gamma \in \mathcal{C}$ and for any two functions $\phi, \psi \in \mathcal{F}(\Gamma)$, we have

$$\text{If} \quad \phi \leq \psi, \quad \text{then} \quad E(\phi, \Gamma) \leq E(\psi, \Gamma).$$

A2 Stability Principle For any $\Gamma \in \mathcal{C}$ and any $\phi \in \mathcal{F}(\Gamma)$, we have

$$\text{For any} \quad \Gamma' \in \mathcal{C} \quad \text{with} \quad D(\Gamma') \subset D(\Gamma),$$
$$\text{it follows that } E(E(\phi, \Gamma)|_{\Gamma'}, \Gamma') = E(\phi, \Gamma)|_{D(\Gamma')}.$$

A3 Regularity Principle Let $A \in SM(2)$, the space of all symmetric two-dimensional matrices, $p \in \mathbb{R}^2 \setminus \{0\}$, $c \in \mathbb{R}$ and

$$Q(y) = \frac{A(w-v, w-v)}{2} + \langle p, w-v \rangle + c,$$

where here and in what follows $\langle v, w \rangle = \sum_{i=1}^{2} v_i w_i$, and A is defined as in Definition 4.1.1. Further,

$$\frac{E(Q|_{\partial B(x,r)}, \partial B(x,r))(x) - Q(x)}{r^2/2} \to F(A, p, c, x), \qquad \text{as } r \to 0^+,$$

where F: $SM(2) \times \mathbb{R}^2 \setminus \{0\} \times \mathbb{R} \times \mathbb{R}^2 \to \mathbb{R}$ is a continuous function, and $B(x,r)$ is the two-dimensional ball with radius r and centre x.

Indeed, if we pause here for a brief reflection, we see that these axioms are reasonable assumptions for an inpainting operator. The comparison principle (A1) asks the inpainting process to be monotonic, whereas the stability principle (A2) guarantees that no additional information can be gained by iterating the interpolation procedure. A principle which could be added to this list is a maximum (minimum) principle, saying that the supremum (infimum) of the interpolant $E(\phi, \Gamma)$ in D is always smaller (greater) than or equal to the supremum (infimum) of the given boundary data ϕ. Note that not all inpainting models proposed in the literature necessarily fulfil these criteria. Higher-order variational and PDE models, which we will encounter in later sections of this book, in general do not have a comparison or maximum principle. The third principle (A3) is a natural assumption on the regularity of the interpolation operator. In what follows we will see that in our setting in particular, $A = D^2u$, $p = \nabla u$, and $c = u$, and one can find an interpolant $u \in C^2$ that fulfils the regularity principle (A3).

For any $\Gamma \in \mathcal{C}$, $\phi \in \mathcal{F}(\Gamma)$, we also demand the following morphological invariances:

A4 Translation Invariance

$$E(\tau_h\phi, \Gamma - h) = \tau_h E(\phi, \Gamma),$$

where τ_h is a translation by h, that is, $\tau_h\phi(x) = \phi(x+h)$, $x \in \Gamma$ and $h \in \mathbb{R}^2$. This axiom requires the inpainting operator to be invariant under translation of the boundary data ϕ by a vector h.

A5 Rotation Invariance

$$E(R\phi, R\Gamma) = RE(\phi, \Gamma),$$

where R: $\mathbb{R}^2 \to \mathbb{R}^2$ is an orthogonal map in $\mathbb{R}^2$. We denote by $R\phi(x) := \phi(R^t x)$ and $R\Gamma \in \mathcal{C}$. This means that the inpainting operator shall be invariant under rotation.

A6 Grey Scale Shift Invariance

$$E(\phi + c, \Gamma) = E(\phi, \Gamma) + c, \quad c \in \mathbb{R}.$$

A7 Linear Grey Scale Invariance

$$E(\lambda\phi, \Gamma) = \lambda E(\phi, \Gamma), \quad \lambda \in \mathbb{R}.$$

A8 Zoom Invariance

$$E(\delta_\lambda\phi, \lambda^{-1}\Gamma) = \delta_\lambda E(\phi, \Gamma),$$

where $\delta_\lambda\phi(x) = \phi(\lambda x)$, $\lambda > 0$, which means that the inpainting operator shall be invariant under zooming.

Assuming that (A1)–(A8) are valid, an interpolant u for smooth boundary data ϕ can be characterised as a so-called viscosity solution of a second-order PDE as follows (see Section B.12 of Appendix B for a definition of viscosity solutions):

Theorem 4.1.2 [CMS98a] *Let E be an interpolation operator that satisfies axioms (A1)–(A8). Let ϕ be a continuous function defined on the boundary of* *the inpainting domain ∂D and $u = E(\phi, \partial D)$ its interpolant in D. Then $u \in C(D)$ is a viscosity solution of the equation*

$$\begin{aligned} G(A) = G(a,b,c) = 0, &\quad \text{in } D, \\ u|_{\partial D} = \phi, & \end{aligned} \tag{4.4}$$

where A is defined as in Definition 4.1.1, that is,

$$A = \begin{pmatrix} u_{\eta\eta} & u_{\xi\eta} \\ u_{\xi\eta} & u_{\xi\xi} \end{pmatrix}, \tag{4.5}$$

and $a = u_{\eta\eta}, b = u_{\xi\eta}$, and $c = u_{\xi\xi}$, where the derivatives are meant in the weak sense as specified in Definition B.12.1. Moreover, $G(A)$ is a non-decreasing function of A satisfying $G(\lambda A) = \lambda G(A)$ for all $\lambda \in \mathbb{R}$.

Proof ([CMS98b]) The proof of Theorem 4.1.2 can be found in an extended version [CMS98b] of the published manuscript [CMS98a]. It is divided into a series of theorems and propositions. Starting from a characterisation of the interpolation operator that just fulfils axioms (A1)–(A3), the authors gradually add the other assumptions and eventually end up with (4.4). In the following, we will sketch their proof in three steps:

Step 1: We show that the validity of the regularity principle (A3) extends to the class of smooth functions in $\mathbb{R}^2$. In particular, we prove that for any smooth function $\phi \in C_b^2(\mathbb{R}^2)$ and any $x \in \mathbb{R}^2$ such that $\nabla\phi(x) \neq 0$, we have

$$\frac{E(\phi|_{\partial B(x,r)}, \partial B(x,r))(x) - \phi(x)}{r^2/2} \to F(D^2\phi(x), \nabla\phi(x), \phi(x), x), \quad \text{as } r \to 0^+, \tag{4.6}$$

where F is continuous and non-decreasing in A.

Without loss of generality, let $x = 0$ and $\phi \in C_b^2(\mathbb{R}^2)$ such that $\nabla\phi(0) \neq 0$. Then we define for $\eta \in \mathbb{R}$ the function

$$Q_\eta(z) := \phi(0) + \nabla\phi(0)z + \frac{1}{2}D^2\phi(0)(z,z) + \frac{\eta}{2}\langle z,z\rangle, \quad z \in \mathbb{R}^2.$$

As ϕ is continuous, in a neighbourhood of 0 we have

$$Q_{-\epsilon}(x) \le \phi(x) \le Q_\epsilon(x), \quad \epsilon > 0,$$

and, in particular, $Q_{-\epsilon}(0) = Q_\epsilon(0) = \phi(0)$. Because of the comparison and stability principles (A1 and A2), the preceding inequality also holds for the interpolation of the involved functions on $\partial B(0,r)$. The assertion (4.6) follows by letting $r \to 0$, using the regularity principle (A3) and eventually taking the limit $\epsilon \to 0$. The continuity of F follows from the regularity principle and the monotonicity in A.

Step 2: Let $\phi \in \mathcal{F}(\partial D)$ and $u = E(\phi, \partial D)$. We continue by proving that u is a viscosity solution of

$$\begin{aligned} F(D^2 u, \nabla u) &= 0, \quad \text{in } D, \\ u|_{\partial D} &= \phi. \end{aligned} \tag{4.7}$$

To do so, let $\psi \in C_b^\infty(D)$, and suppose that $u - \psi$ has a local maximum at $x = x_0$ and $\nabla \psi(x_0) \neq 0$. Then, for some $r > 0$, we have

$$u(x) - \psi(x) \leq u(x_0) - \psi(x_0), \qquad \text{on } \partial B(x_0, r).$$

Again, by using the comparison and stability principles, we have

$$0 \leq E(\psi + u(x_0) - \psi(x_0), \partial B(x_0, r))(x_0) - u(x_0).$$

By consecutively using the grey scale shift invariance (A6) and (4.6) from step 1 of the proof for the limit $r \to 0$, we get

$$0 \leq F(D^2 \psi(x_0), \nabla \psi(x_0), \psi(x_0), x_0).$$

Analogously, we get the reverse inequality if $u - \psi$ has a local minimum in x_0. Hence, u is a viscosity solution of

$$\begin{aligned} F(D^2 u, \nabla u, u, x) &= 0, \quad \text{in } D, \\ u|_{\partial D} &= \phi. \end{aligned}$$

It remains to show that F is independent of u and x. This is a consequence of the translation invariance (A4) and the grey scale shift invariance (A6), respectively. Indeed, defining for $A \in SM(2)$, $x, x_0, p, h \in \mathbb{R}^2$, $c \in \mathbb{R}$, and

$$Q(x) := \frac{A(x - x_0, x - x_0)}{2} + \langle p, x - x_0 \rangle + c,$$

we have for $r > 0$

$$\begin{aligned} &E(\tau_h Q|_{\partial B(x_0,r)}, \partial B(x_0, r) - h)(x_0 - h) - \tau_h Q(x_0 - h) \\ &= E(Q|_{\partial B(x_0,r)}, \partial B(x_0, r))(x_0) - Q(x_0). \end{aligned}$$

Letting $r \to 0$ and by steps 1 and 2 of this proof, we get

$$F(A, p, c, x_0 - h) = F(A, p, c, x_0), \quad \forall h \in \mathbb{R}^2$$

and hence that F is independent of x. Similarly, using (A6), we show that F is independent of c.

Step 3: We conclude the proof by showing that $F(D^2 u, \nabla u)$ can be simplified to $G(A)$, with A given by (4.5).

For $p \in \mathbb{R}^2$, let R_p be a rotation matrix in $\mathbb{R}^2$ with $R_p' p = |p| e_1$, $e_1 = (1,0)$. We will first show that if $p = \nabla u$, then $G = G(R_{\nabla u}' D^2 u R_{\nabla u})$. To do this, we need some additionaly linearity and invariance properties of F, which are immediate consequences of axioms (A1)–(A8):

(P1) $F(\lambda A, \lambda p) = \lambda F(A,p), \quad \lambda \in \mathbb{R}$
(P2) $F(\lambda^2 A, \lambda p) = \lambda^2 F(A,p), \quad \lambda > 0$
(P3) For an orthogonal matrix R in $\mathbb{R}^2$, we have $F(R'AR, R'p) = F(A,p)$.

Then, if $p \neq 0$, we have

$$F(A,p) = F(R_p' A R_p, R_p' p) = F(R_p' A R_p, |p| e_1) = F(R_p' A R_p, e_1) = G(R_p' A R_p).$$

Then, with $\vec{n} = p/|p|$, we may write

$$R_p' D^2 u R_p = \begin{pmatrix} D^2 u(\vec{n},\vec{n}) & D^2 u(\vec{n},\vec{n}^\perp) \\ D^2 u(\vec{n}^\perp,\vec{n}) & D^2 u(\vec{n}^\perp,\vec{n}^\perp) \end{pmatrix}.$$

This finishes the proof. ■

If G is differentiable at zero, we can get a more explicit formula for equation (4.4).

Proposition 4.1.3 [CMS98a] *If in addition to the assumptions in Theorem 4.1.2 the function G is differentiable at* $(0,0,0)$*, then* (4.4) *more explicitly reads*

$$\alpha D^2 u\left(\frac{\nabla u}{|\nabla u|}, \frac{\nabla u}{|\nabla u|}\right) + 2\beta D^2 u\left(\frac{\nabla u}{|\nabla u|}, \frac{\nabla^\perp u}{|\nabla u|}\right)$$
$$+ \gamma D^2 u\left(\frac{\nabla^\perp u}{|\nabla u|}, \frac{\nabla^\perp u}{|\nabla u|}\right) = 0, \quad \text{in } D,$$
$$u|_{\partial D} = \phi, \tag{4.8}$$

where $\alpha, \gamma \geq 0$ *and* $\alpha\gamma - \beta^2 \geq 0$.

The proof of Proposition 4.1.3 follows from a simple Taylor expansion and can be found in [CMS98b]. But what is the right choice for α, β, γ, that is, what is the right choice for a sound interpolation operator? Let us discuss some examples.

Example 4.1.4 Harmonic inpainting For $\beta = 0$ and $\alpha = \gamma$, we obtain the introductory example (4.1), the Laplace equation

$$\Delta u = 0, \quad \text{in } D,$$
$$u|_{\partial D} = \phi, \tag{4.9}$$

where we have used

$$D^2u\left(\frac{\nabla u}{|\nabla u|},\frac{\nabla u}{|\nabla u|}\right)+D^2u\left(\frac{\nabla^\perp u}{|\nabla u|},\frac{\nabla^\perp u}{|\nabla u|}\right)=|\nabla u|^2\Delta u.$$

We will discuss this interpolation operator again in Section 4.2. It constitutes what we will call *harmonic inpainting* and boils down to averaging the boundary data inside D.

More generally, from $\beta=0$ and $\alpha/\gamma=(p-1)$, $p\geq 1$, one recovers the p-Laplacian equation

$$\begin{aligned}\operatorname{div}\left(|\nabla u|^{p-2}\nabla u\right)&=0, \quad \text{in } D,\\ u|_{\partial D}&=\phi.\end{aligned} \tag{4.10}$$

For $\infty>p>2$, this inpainting equation is able to interpolate isolated points, but the gradient of its solution might be unbounded (see [CMS98a]).

Example 4.1.5 Total Variation Inpainting For $\alpha=\beta=0$ and $\gamma\neq 0$, we obtain the mean-curvature flow that we have encountered already in (4.2), that is,

$$\begin{aligned}D^2u\left(\frac{\nabla^\perp u}{|\nabla u|},\frac{\nabla^\perp u}{|\nabla u|}\right)&=0, \quad \text{in } D,\\ u|_{\partial D}&=\phi.\end{aligned} \tag{4.11}$$

This model has several issues. Equation (4.11) does not obtain a unique viscosity solution or might not even have a viscosity solution at all (even for smooth boundary data) (see [CMS98a]). We refer to Exercise 10 in Appendix A for an example of non-uniqueness, as well as to Section 4.3 for a more detailed discussion of the appropriateness of a unique versus a non-unique interpolation – remember that inpainting problems may present ambiguous situations (see Chapter 3). In Section 4.3 we will revisit this model in its variational formulation (compare in particular Remark 4.3.2). The inpainting model that arises is called *total variation (TV) inpainting*. Moreover, in Section 5.2 we will discuss an extended version of this approach, which adds higher-order differential information to the preceding equation.

Example 4.1.6 Absolutely Minimising Lipschitz Extension Inpainting If $\beta=\gamma=0$ and $\alpha\neq 0$ or, in other words, by letting $p=\infty$ in (4.10), we get

$$\begin{aligned}D^2u\left(\frac{\nabla u}{|\nabla u|},\frac{\nabla u}{|\nabla u|}\right)&=0, \quad \text{in } D,\\ u|_{\partial D}&=\phi,\end{aligned} \tag{4.12}$$

[compare also (4.3)]. We will devote Section 4.4 to this equation, so an extensive discussion of this case will be postponed.

4.2. Harmonic Image Inpainting

Harmonic image inpainting is one of the most basic, simple and best understood methods for image interpolation. The inpainted image is computed as a solution of the Laplace equation or as a minimiser of the Dirichlet energy over the inpainting domain. It constitutes a smooth and linear interpolation process that roughly fills in missing grey values by averaging the given grey values on the boundary of the inpainting domain. Let us formulate the inpainting procedure in two ways:

Local PDE Formulation Let $g \in L^2(\Omega)$ be the given image, whose grey values got lost inside the inpainting region $D \subset \Omega$. As in the preceding section, the boundary of $D \subset \Omega$ is assumed to be a continuous and simple Jordan curve. Then the inpainted image $u \in H^1(D)$ is computed as the weak solution [Eva98] of the Laplace equation

$$\begin{aligned} \Delta u &= 0, \quad \text{in } D, \\ u &= g, \quad \text{on } \partial D, \end{aligned} \tag{4.13}$$

that is, u fulfils

$$\begin{aligned} \int_D \nabla u \cdot \nabla \psi \, dx &= 0, \quad \forall \psi \in H_0^1(D), \\ u &= g, \quad \text{a.e. on } \partial D. \end{aligned}$$

We call the weak solution of (4.13) the *harmonic extension* u of f from $\Omega \setminus D$ to D. The inpainted image is a smooth extension of g into D. The solution u of (4.13) can be understood as the solution of the minimisation problem

$$\min_{u \in H^1(\Omega)} \left\{ \int_D |\nabla u|^2 \, dx \quad \text{such that} \quad u = g \text{ in } \Omega \setminus D \right\}.$$

The quality of harmonic extensions has to be studied for the class of smooth functions $g \in C^2(\Omega)$. To do so, Chan and Shen [SC02] first described a smooth interpolant more generally as the sum of a harmonic and an anti-harmonic term by applying Green's second formula. That is, let $x = (x^1, x^2) \in D$ and $G(x_0, x)$ the Green's function for the grounded Poisson equation for a fixed $x_0 \in D$, that is,

$$-\Delta G = \delta_{x-x_0} \quad \text{in } D, \quad G = 0, \quad \text{on } \partial D. \tag{4.14}$$

Then, by applying Green's second formula to a given smooth image g and to the Green's function G, we get a definition for g in every point $x_0 \in D$ which reads

$$g(x_0) = \underbrace{\int_{\partial D} g(x(s)) \frac{-\partial G(x_0, x(s))}{\partial \vec{n}}\, ds}_{u(x_0)=g^h(x_0)} + \underbrace{\int_D G(x_0, x)(-\Delta g)\, dx}_{g^a(x_0)},$$

Here the first term of the expression on the right, denoted by g^h, is the harmonic part of the smooth function g, whereas the second term, denoted by g^a, is the anti-harmonic part of g. More precisely, $g^h(x_0)$ is the extension of $g|_{\partial D}$ with respect to the harmonic measure $(-\partial G(x_0, x(s))/\partial \vec{n})\, ds$ sourced at $x_0 \in D$. The remainder $g^a(x_0)$ is the error that the harmonic extension commits with respect to the true smooth image g. Hence, in order to make statements about the accuracy of harmonic extensions for the interpolation of smooth functions, one needs to get a better grip on the anti-harmonic term. The latter is done by estimates on Green's function G. More precisely, we have the following theorem (see also [SC02]):

Theorem 4.2.1 *Let d be the diameter of the inpainting domain D, that is, $d = \max_{x,y \in \partial D} |x-y|$. Let G be the Green's function solving* (4.14). *Then*

$$\int_D G(x_0, x)\, dx \leq \frac{d^2}{4}.$$

Proof The proof of the assertion is done in the following steps: first, we take a disc B_d of radius d and arbitrary centre $m \in D$. Such a disc automatically covers the domain D, that is, $D \subset B_d$. Using the comparison property of the Green's function proved in Exercise 2 in Appendix A, we have that $G_D(x_0, x) \leq G_{B_d}(x_0, x)$ for all $x, x_0 \in D$. The integral of the Green's function G_{B_d} can be computed explicitly by using Exercise 3 in Appendix A. To do so, without loss of generality (w.l.o.g.) let m be the origin. Then the Green's function on B_d can be mapped to the Green's function on B_1 by the simple rescaling $G_{B_d}(x_0, x) = G_{B_1}(x_0/d, x/d)$, and we have

$$\begin{aligned}\int_D G(x_0, x)\, dx &\leq \int_D G_{B_d}(x_0, x)\, dx \leq \int_{B_d} G_{B_d}(x_0, x)\, dx \\ &= \int_{B_d} G_{B_1}(x_0/d, x/d)\, dx = d^2 \int_{B_1} G_{B_1}(x_0/d, \tilde{x}) d\tilde{x} \\ &= d^2 \frac{1 - |x_0/d|^2}{4} \leq \frac{d^2}{4}.\end{aligned}$$

■

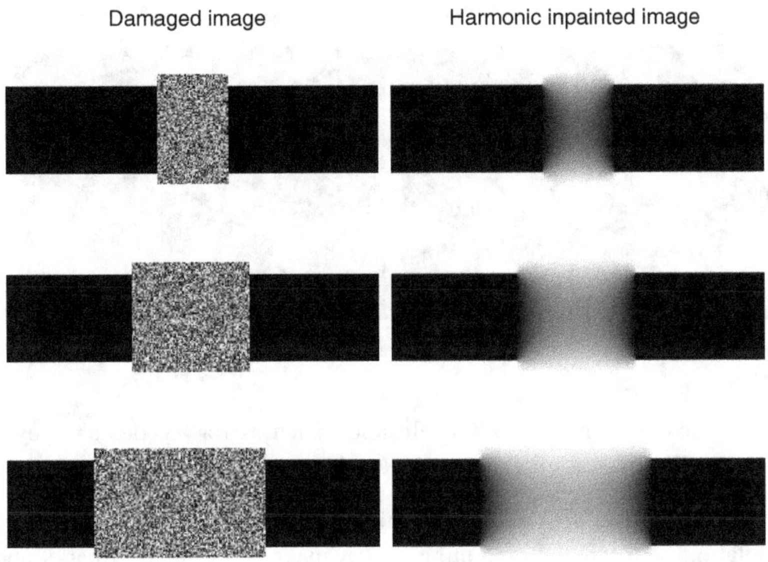

Figure 4.1. Evaluation of harmonic inpainting for a simple example of a broken line. Accepting the connected line as the desired solution to the inpainting problem, the quality of harmonic inpainting suffers from its inability to interpolate discontinuities (edges) and the increase in error when the hole becomes larger.

Using Theorem 4.2.1, the accuracy of the harmonic extension $u = g^h$ of a smooth function g in D is derived. In fact, for any smooth function g on D, one can find a uniform bound M on its Laplacian, that is, $|\Delta g| \leq M$ in D. Hence,

$$|g(x_0) - g^h(x_0)| = |g^a(x_0)| \leq M \int_D G(x_0, x)\, dx \leq \frac{Md^2}{4}$$

for all $x_0 \in D$. This gives $\|g - u\|_\infty = \|g - g^h\|_\infty = \mathcal{O}(d^2)$, which is the maximal error harmonic inpainting (4.13) commits under the assumption that a smooth function g is the ground truth. In [CK06], Chan and Kang added a more detailed analysis for error estimation in image inpainting to the preceding assertion. In particular, they sharpened the preceding error estimate for harmonic inpainting by replacing the diameter d of the inpainting region by the smaller axis b of an ellipse that covers D. This supports our discussion in Chapter 1, where we conjectured that inpainting of narrow domains gives better results than inpainting of thicker domains with the same area.

Figure 4.1 shows the inpainting of a black stripe on a white background. The stripe has increasingly larger holes which break it into two pieces. Harmonic inpainting clearly fails in producing a desirable interpolation result – especially when having the principle of good continuation in mind. For the example of

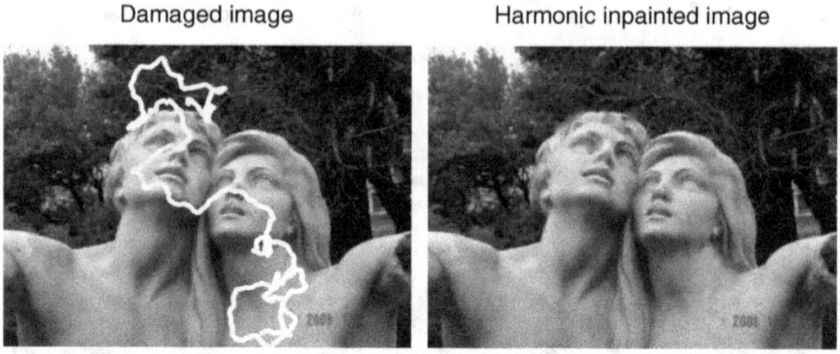

Figure 4.2. Harmonic inpainting for a natural image.

inpainting natural images with small holes, such as for the damaged eye in Figure 4.2, however, it produces an acceptable inpainting result.

Remark 4.2.2 Before we finish this section with a slightly different formulation of harmonic inpainting, let us make a few final remarks about its current form. For a reasonable definition of the harmonic extension of an image function g, we had to assume that the g we are aiming to interpolate is sufficiently smooth, more precisely is at least in H^1. Such an g cannot feature image edges exactly which would turn up as discontinuities in the function g (compare Exercise 4 in Appendix A). In general, such an assumption is rather unrealistic. Image interpolants should be able to accommodate discontinuities and indeed propagate them into the inpainting domain. When inspecting the inpainting result in Figure 4.2, it looks mostly acceptable on first glance but clearly suffers from the smoothing of discontinuities by giving a rather blurred impression of the contents we expect to be interpolated. We will return to this thought when motivating the TV inpainting model in Section 4.3. Moreover, note that harmonic inpainting does not permit the interpolation of isolated points in two space dimensions. Indeed, the problem $\Delta u = 0$ in $B(0,r) \setminus \{0\}$, $u = 0$ on $\partial B(0,r)$ and $u(0) = 1$ does not have a solution (neither strong nor weak) (see Exercise 5 in Appendix A). The desire to do so in certain applications gave rise to the AMLE inpainting model presented in Section 4.4.

We continue with relating harmonic inpainting with the minimisation of a suitable energy functional defined over the whole image domain Ω. As discussed earlier, such a global approach might be interesting in particular when g in $\Omega \setminus D$ is not perfect, that is, noisy.

Global Variational Formulation Let $g \in L^2(\Omega)$ be the given image with inpainting domain $D \subset \Omega$. Then the inpainted image u can be defined as the minimiser of the energy

$$\int_\Omega |\nabla u|^2 \, dx + \lambda \| \mathbb{1}_{\Omega \setminus D}(u-g) \|^2_{L^2(\Omega)}, \tag{4.15}$$

where $\lambda \gg 1$ is a large constant. Minimising (4.15) over $H^1_0(\Omega)$, a minimiser is equivalently characterised as the solution to the corresponding Euler-Lagrange equation

$$\begin{aligned} -\Delta u &= 0, && \text{in } D, \\ -\Delta u &= \lambda(g-u), && \text{in } \Omega \setminus D, \\ u &= 0, && \text{on } \partial\Omega. \end{aligned}$$

This formulation is in flavour of many existing variational approaches in image processing. It is defined as a minimisation problem over the whole image domain Ω, not enforcing the boundary values $u = g$ on ∂D exactly but approximately only. As such, it also can handle slightly degraded, that is, noisy, and not completely trusted image information inside the intact part $\Omega \setminus D$. The latter gets enhanced as well during the inpainting process. Moreover, no regularity assumptions on the boundary of the inpainting domain D are needed for the sake of mathematical well-posedness of the Euler-Lagrange equation (compare, e.g., Figure 2.2).

4.3. Total Variation Inpainting

The performance of harmonic image inpainting is limited. This is mainly because of its linear nature, which cannot pick up structures such as edges from the intact part $\Omega \setminus D$ of the image domain Ω. Rather, the harmonic extension is done by averaging (i.e., diffusing) grey values from the boundary ∂D into the inpainting domain, leading to smooth interpolations in which image edges get blurred. This is a general issue with linear approaches for image enhancement. Image de-noising with Gaussian filtering [equivalent to de-noising with harmonic functions as in (4.13); cf. Exercise 1 in Appendix A], for instance, successfully eliminates noise in an image by attenuating high-frequency oscillations but at the same time diminishes intrinsic image structures. One way out of this is to switch to non-linear imaging approaches, which eliminate noise while keeping the significant edges in an image. One of the standard models for non-linear imaging is total variation (TV) regularisation. Since the pioneering work of Rudin, Osher and Fatemi [ROF92], TV regularisation is well established in many areas of image processing, for example, image de-noising, image de-blurring and image segmentation. Hence, it makes sense to explore its applicability in the context

Figure 4.3. TV inpainting for the removal of writing. Photo courtesy of Kostas Papafitsoros.

of image inpainting as a first step towards improving on the harmonic inpainting approach (4.13).

TV inpainting was proposed by Chan and Shen [SC02] within a global variational setup similar to (4.15). As before, let $g \in L^2(\Omega)$ be the given image, whose image contents got lost inside the inpainting domain $D \subset \Omega$. Then the inpainted image u is recovered as a minimiser of

$$\mathcal{J}(u) = |Du|(\Omega) + \frac{\lambda}{2}\|\mathbb{1}_{\Omega\setminus D}(u-g)\|^2_{L^2(\Omega)}, \tag{4.16}$$

Figure 4.3 shows a first inpainting example using (4.16) that suggests that TV inpainting might be a promising method for image restoration.

Let us start with a few words on the notion of total variation, its mathematical properties and its very important role in image processing.

4.3.1. The Total Variation as an Image Interpolator

As we have seen in our discussion on harmonic inpainting in Section 4.2, the problem with forcing an image to be continuous is that edges in an image (which are jumps in the image function) cannot be represented (see Remark 4.2.2 and Exercise 4 in Appendix A). However, edges constitute one of the most important features in images, and whenever we are processing an image – say, we de-noise, de-blur or interpolate it – we seek methods that can represent and preserve edges in the process. These considerations call for a less smooth space than Sobolev spaces such as H^1. This less smoother space turns out to be the *space of functions of bounded variation.*

But let us first come back to the motivations behind the use of total variation in image processing. Total variation had its debut in the image processing literature with the seminal papers of Rudin, Osher and Fatemi [ROF92] and Chambolle and Lions [CL97]. It was proposed based on the desire to find a smoothing procedure for images that is able to preserve sharp image structures such as edges. Let us repeat here the definition of the total variation, which, together with some of its main properties, will be discussed in more detail in Section B.11 in Appendix B. The total variation can be defined for absolutely integrable functions $u \in L^1(\Omega)$.

Definition 4.3.1 Total variation The *total variation* of an image is defined by duality: for $u \in L^1_{loc}(\Omega)$, it is given by

$$TV(u) = |Du|(\Omega) = \sup\left\{-\int_\Omega u \operatorname{div}\varphi \, dx : \varphi \in C_c^\infty(\Omega;\mathbb{R}^2), \quad |\varphi(x)| \leq 1 \forall x \in \Omega\right\}. \tag{4.17}$$

It is nothing but the TV measure of the Radon measure Du as the distributional derivative of u given by the Riesz representation theorem [AFP00] (compare Exercise 6 in Appendix A).

The space $BV(\Omega)$ of functions with bounded variation (BV) is the set of functions $u \in L^1(\Omega)$ such that $TV(u) < \infty$. Endowed with the norm

$$\|u\|_{BV} = \|u\|_1 + TV(u),$$

the space $BV(\Omega)$ is a Banach space.

Using this definition, it can be easily seen (at least formally) that for $u \in W^{1,1}(\Omega)$, the total variation $|Du|(\Omega)$ coincides with $\|\nabla u\|_1$ (compare Exercise 7 in Appendix A). Moreover, for $u = \chi_C$ for a subset $C \subset \Omega$ with smooth boundary, we have $|Du|(\Omega) = |Du|(C) = \mathcal{H}^1(\partial C)$, the perimeter of C in Ω, where $\mathcal{H}^1$ is the one-dimensional Hausdorff measure. If you want to learn more about total variation and its role in image processing, refer to [CL97, AV97, CS05a, AK06, CCC+10, CCN11 and BL11] for excellent monographs on the topic. For our purpose – for now – it is enough to know that the total variation allows the image interpolant u to feature discontinuities (see Figure 4.4). The quality of the interpolation, that is, how well it indeed picks up edge information from the given f and propagates it inwards, will be discussed in more detail in Section 4.3.3 when we introduce higher-order PDEs for image inpainting.

In one space dimension the total variation of a signal u on the interval $(-1,1)$ is just the maximal sum over the absolute differences of function values for partitions of $(-1,1)$ [Zie89]. Hence, in one space dimension it is easily demonstrated in Figure 4.5 that the total variation of two different image functions can be the same. This means that in contrast with the unique harmonic extension in Section 4.2, the TV inpainted image does not have to be unique (compare Figure 4.6 for an illustration). Let us make a final remark

Figure 4.4. TV inpainting versus harmonic image inpainting. The TV interpolator is able to reconstruct sharp image edges. From left to right: An image with stripes and a white inpainting domain, the inpainted image using harmonic inpainting and the TV inpainted image.

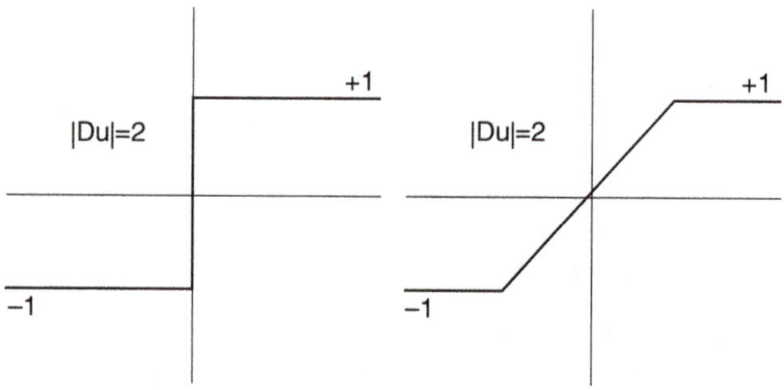

Figure 4.5. The total variation of u on the left and of u on the right is the same.

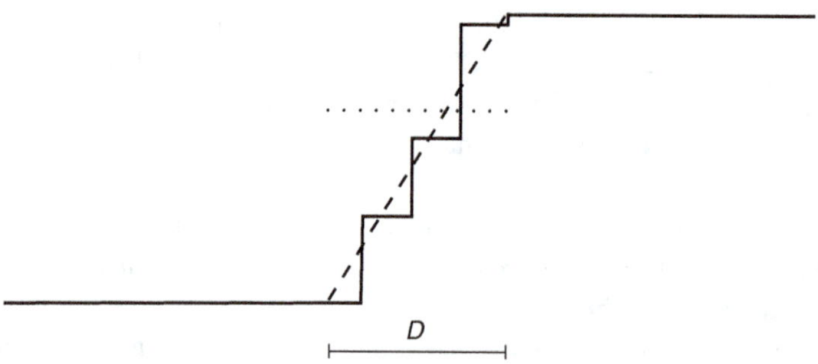

Figure 4.6. The TV inpainted image is not unique. Inside the inpainting domain D, all three functions defined by the solid, slashed and dotted lines are possible solutions of (4.16).

on the connection of TV inpainting with the interpolator (4.11) derived in Section 4.1 and then proceed with an existence proof for minimisers of (4.16) in the next section.

Remark 4.3.2 Total Variation Inpainting and the Mean-Curvature Flow Minimising the total variation is equivalent to minimising over the length of the level lines of the function (cf. Theorem 4.3.6). This already reminds us of movement by mean-curvature flow in (4.2). Let us replace for a moment the total variation in Definition 4.3.1 by the smoothed differentiable TV functional $\int_D \sqrt{|\nabla u|^2 + \epsilon}$ for a small $0 < \epsilon \ll 1$. Then, we can differentiate this functional and derive the corresponding Euler-Lagrange equation for a minimiser u to be

$$-\text{div}\left(\frac{\nabla u}{\sqrt{|\nabla u|^2 + \epsilon}}\right) = 0,$$

which is the minimal-surface equation [Eva98]. In the non-smooth case, a first-order optimality condition for a minimiser still can be derived by devising the more general terminology of the sub-differential of a convex functional (see Section B.2 in Appendix B). We refer readers to, for example, [AV97, BCN02 and CCC+10] for a detailed discussion on this and for a thorough derivation of the optimality conditions for (4.17). Let us only mention here that for smooth u, a minimiser u of (4.17) on D can be interpreted as a viscosity solution of

$$-\text{div}\left(\frac{\nabla u}{|\nabla u|}\right) = 0, \quad \text{in } D,$$
$$u = g, \quad \text{in } \Omega \setminus D,$$

which (multiplying the equation by $|\nabla u|$) brings us back to the interpolation operator (4.11) in Section 4.1. Moreover, the corresponding gradient flow of (4.17) can be formally written as

$$u_t = \text{div}\left(\frac{\nabla u}{|\nabla u|}\right). \tag{4.18}$$

Multiplying the right-hand side of (4.18) by the factor $|\nabla u|$, we get the mean-curvature flow

$$u_t = |\nabla u| \text{div}\left(\frac{\nabla u}{|\nabla u|}\right), \tag{4.19}$$

which was mentioned in [CS01a] as a morphologically invariant version of (4.18) and is used in [MO00] as an accelerated numerical method for solving (4.18). A study on the qualitative properties of solutions of (4.18) can be found in [CCN11].

4.3.2. Existence of a TV Inpainted Image

We start the analysis of (4.16) by proving the following existence result.

Theorem 4.3.3 *For a given $g \in L^2(\Omega)$, there exists a TV inpainted image u that is a solution $u \in BV(\Omega)$ for* (4.16).

For the proof, we essentially follow [CCC⁺10 and Ves01].

Proof We put the cart before the horse and start with proving that the total variation is lower semi-continuous (l.s.c.) with respect to weak convergence in L^p for $p \in [1, +\infty)$. The idea is that the total variation is the supremum of continuous functions and as such is l.s.c. More precisely, for $\varphi \in C_c^\infty(\Omega; \mathbb{R}^2)$, let us consider the functional

$$L_\varphi : u \mapsto -\int_\Omega u(x)\mathrm{div}\varphi(x)\, dx.$$

If $u_n \rightharpoonup u$ in $L^p(\Omega)$, then $L_\varphi u_n \to L_\varphi u$ (this is due to the continuity of L_φ even w.r.t. weak topologies). But then

$$L_\varphi u = \lim_{n\to\infty} L_\varphi u_n \leq \liminf_{n\to\infty} TV(u_n).$$

Taking the supremum over all $\varphi \in C_c^\infty(\Omega; \mathbb{R}^2)$ with $|\varphi(x)| \leq 1$ for all x, we deduce

$$TV(u) \leq \liminf_{n\to\infty} TV(u_n),$$

that is, TV is (sequentially) l.s.c. with respect to all the above-mentioned topologies.

Now let (u_n) be a minimising sequence for (4.16); then there exists a constant $C > 0$ such that

$$|Du_n|(\Omega) \leq C, \quad \forall n \geq 1.$$

We additionally want to prove that u_n is uniformly bounded in L^2. Once this is proved, since L^2 is reflexive, we can then extract a subsequence that weakly converges in L^2 which – together with the earlier l.s.c. result – is all we need to conclude the existence of a minimiser. We start by noting that

$$\|u_n\|_2 = \left\| u_n - \frac{\int_\Omega u_n}{|\Omega|} + \frac{\int_\Omega u_n}{|\Omega|} \right\|_2 \leq \left\| u_n - \frac{\int_\Omega u_n}{|\Omega|} \right\|_2 + \frac{1}{|\Omega|} \left| \int_\Omega u_n \right| \tag{4.20}$$

By the Poincaré-Wirtinger inequality in Theorem B.11.8, we have

$$\left\| u_n - \frac{\int_\Omega u_n}{|\Omega|} \right\|_2 \leq K_\Omega \, |Du_n|(\Omega) \leq K_\Omega \, C. \tag{4.21}$$

To control the second term of the right-hand side of (4.20), we prove that $\left|\int_\Omega u_n\right| \leq C$ for all $n \geq 1$. To do this, let us consider

$$w_n = \frac{\int_\Omega u_n}{|\Omega|} \chi_\Omega \quad \text{and} \quad v_n = u_n - w_n.$$

Then $\int_\Omega v_n = 0$ and $Dv_n = Du_n$. Hence, $|Dv_n|(\Omega) \leq C$. Using the Poincaré-Wirtinger inequality again for v_n, we obtain

$$\|v_n\|_2 \leq C.$$

We also have

$$\begin{aligned} C &\geq \|\mathbb{1}_{\Omega\setminus D} u_n - g\|_2^2 = \|\mathbb{1}_{\Omega\setminus D} v_n + \mathbb{1}_{\Omega\setminus D} w_n - g\|_2^2 \\ &\geq (\|\mathbb{1}_{\Omega\setminus D} v_n - g\|_2 - \|\mathbb{1}_{\Omega\setminus D} w_n\|_2)^2 \\ &\geq \|\mathbb{1}_{\Omega\setminus D} w_n\|_2 (\|\mathbb{1}_{\Omega\setminus D} w_n\|_2 - 2\|\mathbb{1}_{\Omega\setminus D} v_n - g\|_2) \\ &\geq \|\mathbb{1}_{\Omega\setminus D} w_n\|_2 \left[\|\mathbb{1}_{\Omega\setminus D} w_n\|_2 - 2(\|\mathbb{1}_{\Omega\setminus D}\| \cdot \|v_n\|_2 + \|g\|_2)\right]. \end{aligned}$$

Let $x_n = \|\mathbb{1}_{\Omega\setminus D} w_n\|_2$ and $a_n = \|\mathbb{1}_{\Omega\setminus D}\| \cdot \|v_n\|_2 + \|g\|_2$. Then

$$x_n(x_n - 2a_n) \leq C, \quad \text{with} \quad 0 \leq a_n \leq \|\mathbb{1}_{\Omega\setminus D}\| \cdot C + \|g\|_2 = C', \ \forall n.$$

Hence, we obtain

$$0 \leq x_n \leq a_n + \sqrt{a_n^2 + C} \leq C''.$$

This implies

$$\|\mathbb{1}_{\Omega\setminus D} w_n\|_2 = \left|\int_\Omega u_n\right| \cdot \frac{\|\mathbb{1}_{\Omega\setminus D}\|_2}{|\Omega|} \leq C'', \quad \forall n, \tag{4.22}$$

which entails that $\left|\int_\Omega u_n\right|$ is uniformly bounded. Because of (4.20)–(4.22), the sequence u_n is bounded in L^2 and in particular in L^1. Then u_n is bounded in $BV(\Omega)$, and there is a subsequence u_{n_k} and a $u \in BV(\Omega)$ such that $u_{n_k} \rightharpoonup u$ in L^2. Moreover, $\mathbb{1}_{\Omega\setminus D} u_{n_k}$ converges weakly to $\mathbb{1}_{\Omega\setminus D} u$ in L^2 because $\mathbb{1}_{\Omega\setminus D}$ is an indicator function. With the l.s.c. properties of $|Du|(\Omega)$ and $\|\mathbb{1}_{\Omega\setminus D} u - g\|_2$ w.r.t. weak convergence in L^2, we get

$$\mathcal{J}(u) \leq \liminf_{k\to\infty} \mathcal{J}(u_{n_k}) = \inf \mathcal{J},$$

so u is a minimiser of $\mathcal{J}$. ∎

In the noise-free case, that is, if we assume that $g|_{\Omega\setminus D}$ is completely intact, we can also formulate the following variational approach: assume that $g \in BV(\Omega)$, and seek the inpainted image u^* that solves

$$\min_{\{u\in L^2(\Omega):\ u|_{\Omega\setminus D}=g|_{\Omega\setminus D}\}} \{|Du|(\Omega)\}. \tag{4.23}$$

Theorem 4.3.4 *For an original image $g \in BV(\Omega)$, the minimisation problem (4.23) has a minimiser $u^* \in BV(\Omega)$*

Proof We can rewrite the constrained problem (4.23) as the following unconstrained problem:

$$u^* = \text{argmin}_u \left\{ |Du|(\Omega) + \mathbb{1}_{\{v\in L^2(\Omega):\ v|_{\Omega\setminus D}=g|_{\Omega\setminus D}\}}(u) \right\},$$

where

$$\chi_S(u) = \begin{cases} 0, & u \in S, \\ +\infty, & \text{otherwise.} \end{cases}$$

Then we can apply the direct method of calculus of variations, noting that the characteristic function is l.s.c. (see Exercise 8 in Appendix A) and using compactness properties in L^2 as in the preceding proof. ■

The inpainting approach (4.23) is preferable when g outside the inpainting domain D is perfect. The global inpainting procedure (4.16), however, is more suitable when g is corrupted outside D as well and hence cannot be totally trusted. The latter is therefore usually the approach used in practice when a global smoothing of the image is desired.

4.3.3. Insufficiency for Good Continuation

The disadvantage of the TV approach in inpainting is that the level lines are interpolated linearly. This means that the direction of the level lines is not preserved because they are connected by a straight line across the missing domain. This is due to the penalisation of the length of the level lines within the minimising process with a TV regulariser, which results in connecting level lines from the boundary of the inpainting domain via the shortest distance (linear interpolation). Please also see Sections B.9 and B.10 in Appendix B for a brief reminder on level sets and the length of a planar curve.

To see this, we take the level-line point of view. We want to derive another characterisation for the total variation in terms of the level sets of the image function u.

Definition 4.3.5 Let $E \subset \Omega$ be a measurable set in $\mathbb{R}^2$. The set E is called a *set of finite perimeter* iff its characteristic function $\chi_E \in BV(\Omega)$. We write

$$\mathrm{Per}(E;\Omega) := |D\chi_E|(\Omega),$$

for the perimeter of E in Ω .

With the notion of sets of finite perimeter, we have the following theorem:

Theorem 4.3.6 Coarea formula *Let $u \in BV(\Omega)$, and for $s \in \mathbb{R}$, we define the set $\{u > s\} := \{x \in \Omega : u(x) > s\}$ the s-sup level set of u (see also Section B.1 in the Appendix B). Then one has*

$$|Du|(\Omega) = \int_{-\infty}^{\infty} \mathrm{Per}(\{u > s\};\Omega)\, ds.$$

Proof Following [CCC+10], we will only present a sketch of the proof because the full version is very complicated and technical. The sketch is discussed in three steps:

- We start with considering affine functions $u(x) = p\,x$, $p \in \mathbb{R}$, on a simplex $\Omega = T$. Then the total variation of u in T can be easily computed to be

$$\begin{aligned} |Du|(T) &= \sup\left\{-\int_{\mathrm{T}} p\, x\mathrm{div}\varphi(x)\, dx : \varphi \in C_c^{\infty}(T;\mathbb{R}^2),\ |\varphi(x)| \le 1\ \forall x \in T\right\} \\ &= |T||p|. \end{aligned}$$

  However, the hypersurfaces $\partial\{u > s\}$ are $\{px = s\}$, and hence,

$$\int_{-\infty}^{\infty} \mathrm{Per}(\{u > s\};\Omega)\, ds = \int_{-\infty}^{\infty} d\mathcal{H}^1(\{px = s\}) = |T||p|.$$

- Now, having proved the result for affine functions on simplexes T, the idea is to triangulate Ω with simplexes T and approximate a general function $u \in BV$ by piecewise affine functions u_n on these simplexes such that $\int_\Omega |Du_n|\, dx \to |Du|(\Omega)$. By Theorem B.11.9 in Appendix B, we can approximate u with smooth functions and those (similar to finite-element theory) with piecewise affine functions. Using the l.s.c. of the total variation and Fatou's lemma (Lemma B.5.2), we get

$$\int_{\mathbb{R}} \mathrm{Per}(\{u > s\};\Omega)\, ds \le |Du|(\Omega).$$

- The proof of the reverse inequality is straightforward by the the following calculation. Again, for $u \in BV(\Omega)$, we have

$$
\begin{aligned}
\int_\Omega u(x)\operatorname{div}\varphi(x)\,dx &= \int_{\{u>0\}}\int_0^{u(x)} ds\,\operatorname{div}\varphi(x)\,dx \\
&\quad - \int_{\{u<0\}}\int_{u(x)}^{0} ds\,\operatorname{div}\varphi(x)\,dx \\
&\underbrace{=}_{\text{Fubini}} \int_0^\infty \int_\Omega \chi_{\{u>s\}}(x)\operatorname{div}\varphi(x)\,dxds \\
&\quad - \int_{-\infty}^0 \int_\Omega (1-\chi_{\{u>s\}}(x))\operatorname{div}\varphi(x)\,dxds \\
&\underbrace{=}_{\int \operatorname{div}\varphi=0} \int_{-\infty}^\infty \int_{\{u>s\}} \operatorname{div}\varphi\,dxds \\
&\leq \int_{-\infty}^\infty \operatorname{Per}(\{u>s\};\Omega)\,ds.
\end{aligned}
$$

Taking the sup over all admissible φ on the left finishes the proof. ■

Now assume for a moment that the original image g is smooth and that the missing $g|_D$ is non-constant; that is, there exists a point x_0 close to D where $\nabla g(x_0) \neq \mathbf{0}$. Then (by the inverse function theorem) level lines of g in D are well defined and distinguishable, and in particular, we have that the s-level line

$$g_s^{-1} = \Gamma_s := \{x \in \mathbb{R}^2 : g \equiv s\}$$

is uniquely labeled by the level s. Combining this with the co-area formula, we can derive the TV inpainting problem for one level line Γ_s which should be interpolated in D across the two points $p_1 \in \partial D$ and $p_2 \in \partial D$ as

$$\min_{\substack{\{\gamma_s:\ \gamma_s(p_1)=\Gamma_s(p_1),\\ \gamma_s(p_2)=\Gamma_s(p_2)\}}} \int_{\gamma_s} ds = \min_{\substack{\{\gamma_s:\ \gamma_s(p_1)=\Gamma_s(p_1),\\ \gamma_s(p_2)=\Gamma_s(p_2)\}}} \operatorname{length}(\gamma_s).$$

This means that level lines are interpolated with straight lines. Whereas a straight-line connection still might be pleasant for small holes, it is very unpleasant in the presence of larger gaps, even for simple images, as illustrated in Figure 4.7. Another consequence of the linear interpolation is that level lines may not be connected across large distances as the minimising straight-line connecting points with the shortest distance might not be the line which produces a connectedness as desired (compare Figure 4.8 and also remember our discussion on Kanisza‘s good continuation principle in Chapter 3). A

Figure 4.7. TV inpainting interpolates level lines linearly, which does not always match our visual perception.

Figure 4.8. TV inpainting for a stripe that is broken in the middle. If the gap is small enough, the stripe is nicely reconnected, and the TV interpolant preserves the sharp boundaries of the stripe. If the gap is too large, the TV interpolant breaks the line into two pieces.

solution for this is the use of higher-order derivatives for the image interpolator which encode additional information, for example, not only the length of level lines but also their curvature. A prototype extension of TV inpainting that involves higher-order derivatives is Euler's elastica inpainting model, which will be discussed in Section 5.2.

The examples in Figures 4.7 and 4.8 have been computed by the split Bregman method [GO09] using the code published in IPOL [PSS13].

4.4. Absolutely Minimising Lipschitz Extensions

The interpolation operator defined by (4.12) has an interesting connection to *absolutely minimising Lipschitz extensions* (AMLEs) [Aro67, Jen93].

Definition 4.4.1 Let $D \subset \mathbb{R}^2$ open and bounded with Lipschitz boundary. An AMLE of a function $\phi|_{\partial D}$ in D is a Lipschitz function u in D such that

$$\sup_{x\in\tilde{D}} |\nabla u(x)| \leq \sup_{x\in\tilde{D}} |\nabla \phi(x)|,$$

for all $\tilde{D} \subset D$ and ϕ such that $u - \phi$ is Lipschitz in $\tilde{D}$ and $u = \phi$ on $\partial\tilde{D}$.

This means that an AMLE u of $\phi|_{\partial D}$ is an extension of ϕ in D that has a minimal Lipschitz constant in all subdomains of D. In [Jen93], the author proves that if u is AMLE, then u is the unique viscosity solution (see Definition B.12.1 in Appendix B) of

$$D^2u(\nabla u, \nabla u) = 0, \quad \text{in } D, \tag{4.24}$$

with suitable boundary conditions specified in the following theorem. More precisely, Jensen proved the following:

Theorem 4.4.2 [Jen93] *Let ϕ be Lipschitz continuous with respect to the geodesic distance $d_D(x,y)$ between x and y. Then u is the AMLE of ϕ into D if and only if u is a viscosity solution of* (4.24) *with boundary data $u|_{\partial D} = \phi$.*

Since the numerical solution of (4.24) is usually tackled iteratively by embedding the stationary equation into an evolutionary process, it is also worth mentioning that in [CMS98b] the authors proved the existence and uniqueness of viscosity solutions $u(\cdot,t)$ for the corresponding evolutionary problem. Moreover, they showed that the solution $u(\cdot,t)$ asymptotically converges to the unique viscosity solution of the stationary equation (4.24) as $t \to \infty$.

Interpolation of Datum on Curves and Points Let us now discuss the ability of AMLE inpainting to interpolate isolated points and curves as presented in [CMS98b]. This property in particular sets AMLE inpainting apart from other inpainting approaches we have discussed so far. In particular, harmonic inpainting turned out to be insufficient for such a task (compare Exercise 5 in Appendix A). Let $D \subset \mathbb{R}^2$ be the domain to be inpainted with boundary ∂D. We assume that ∂D consists of three parts, each one representing a different morphological characteristic. That is, we have $\partial D = \partial_1 D \cup \partial_2 D \cup \partial_3 D$, where

- $\partial_1 D$ is a finite union of rectifiable Jordan curves,
- $\partial_2 D = \bigcup_{i=1}^m C_i$, where C_i are rectifiable curves, which are homeomorphic to a closed interval, and
- $\partial_3 D = \{x_i | i = 1, \ldots, N\}$ is a finite number of isolated points.

Also, the boundary data defined on ∂D are split in three parts given by

- A Lipschitz function ϕ_1 on $\partial_1 D$.
- Two Lipschitz functions ϕ_{2+}^i and ϕ_{2-}^i corresponding to the left and right limits of the interpolate on each curve C_i. More precisely, $u|_{C_i} = \phi_{2+}^i$ if $u(y) \to \phi_{2+}^i(x)$ as $y \to x$ and $\langle y, \nu_i^+(x)\rangle < 0$, where ν_i^+ is the normal on C_i pointing in an assigned positive direction (i.e., $\nu_i^- = -\nu_i^+$) and analogously for ϕ_{2-}^i. The boundary data coincide at the endpoints of each curve C_i.

- The boundary data on the isolated points $x_i \in \partial_3 D$ are a constant $u_i \in \mathbb{R}$, $i = 1, \ldots, N$.

For this setting, we have the following result:

Theorem 4.4.3 [CMS98a] *Given D with boundary ∂D and boundary data ϕ_1, $\{(\phi^i_{2+}, \phi^i_{2-})\}^m_{i=1}$, and $\{u_i\}^N_{i=1}$ as earlier, there exists a unique Lipschitz continuous viscosity solution u of*

$$
\begin{aligned}
D^2 u\left(\frac{\nabla u}{|\nabla u|}, \frac{\nabla u}{|\nabla u|}\right) &= 0, \quad \text{in } D, \\
u|_{\partial_1 D} &= \phi_1, \\
u|_{C_i} &= \phi^i_{2+} \text{ or } \phi^i_{2-}, \quad i = 1, \ldots m, \\
u(x_j) &= u_j, \qquad j = 1, \ldots, N.
\end{aligned}
$$

Proof [CMS98b] Theorem 4.4.3 is a special case of Jensen's general result from Theorem 4.4.2. To see this, we sketch the proof from [CMS98b]. Let $r > 0$ be small enough and

$$
D_r = D - \bigcup_{i=1}^{m} C_i^r - \bigcup_{j=1}^{N} B(x_j, r),
$$

where C_i^r is the curve C_i blown up by balls $B(0, r)$ of radius r (see Figure 4.9).

Blowing up the rectifiable curves C_i of $\partial_2 D$ and the points x_j of $\partial_3 D$ by balls of radius r, we get a new domain D_r whose boundary consists just of a finite union of rectifiable simple Jordan curves. Cleverly parametrising the new boundary ∂D_r, we can define the respective boundary data ϕ_r. In this situation, Theorem 4.4.2 applies, and there exists a unique viscosity solution $u_r \in W^{1,\infty}(D_r)$ of (4.24) with $u|_{\partial D_r} = \phi_r$. By a compactness argument combined with the stability of viscosity solutions, we can conclude the proof by letting $r \to 0$. ■

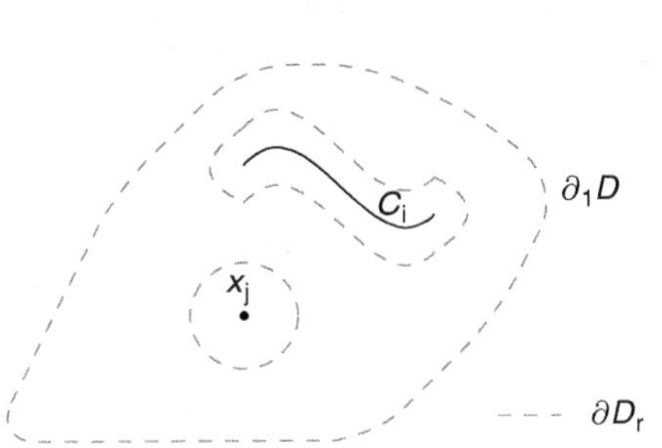

Figure 4.9. Visualisation of the boundary ∂D_r.

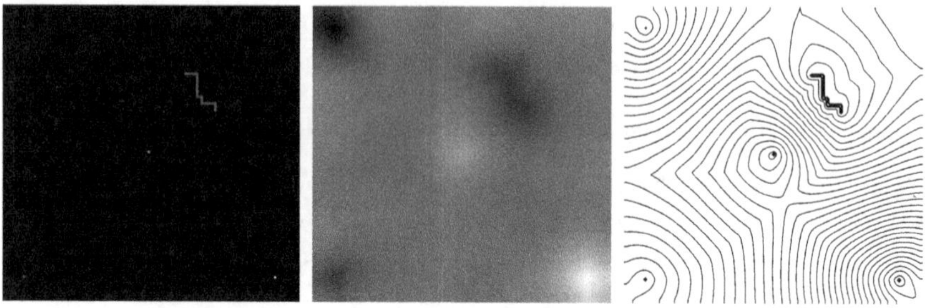

Figure 4.10. AMLE inpainting for image data g that are only given in a few points and curves. From left to right: Given image data g, AMLE inpainted image and contour plot of AMLE inpainted image.

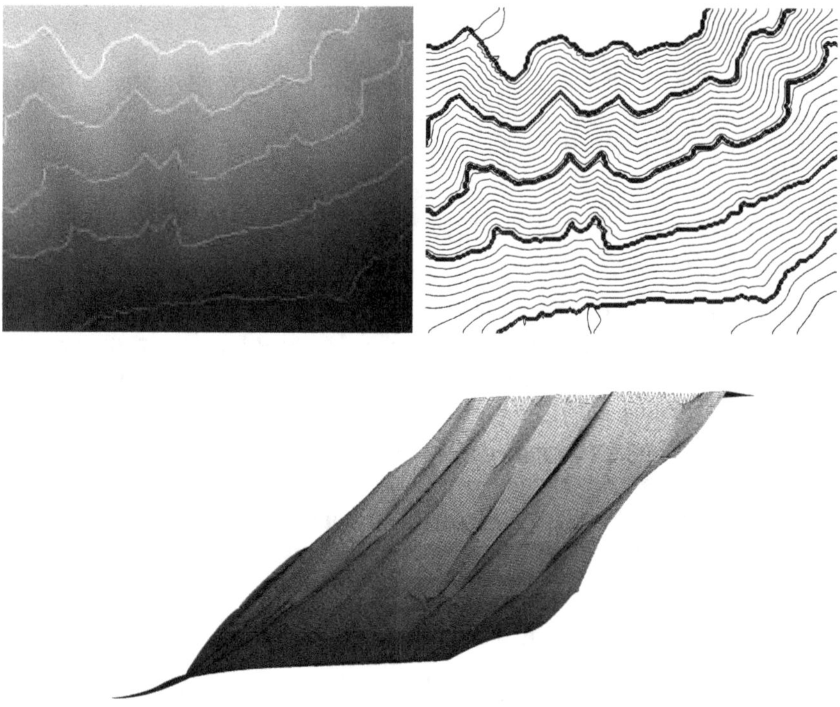

Figure 4.11. AMLE inpainting for DEM interpolation. AMLE inpainting for image data g that are only given in a few points and curves. Top: AMLE inpainting result for five given contours (highlighted contours in the plot) and contour plot of AMLE interpolated surface with given contours highlighted. Bottom: AMLE inpainted surface.

A special case of the preceding result for $D = B(0,r) \setminus \{0\}$ is the topic of Exercise 5 in Appendix A. Moreover, Figure 4.10 presents an illustration of this property of AMLE. Here we consider a global AMLE inpainting approach in which the inpainted image u is computed as a stationary state of

$$\begin{aligned} u_t &= D^2u\left(\frac{\nabla u}{|\nabla u|}, \frac{\nabla u}{|\nabla u|}\right) + \lambda \mathbb{1}_{\Omega\setminus D}(g-u) \\ &= \frac{\nabla u^t}{|\nabla u|} D^2 u \frac{\nabla u}{|\nabla u|} + \lambda \mathbb{1}_{\Omega\setminus D}(g-u). \end{aligned}$$

The validity of Theorem 4.4.3 is of special importance in certain applications of image inpainting. One example is the interpolation of natural surfaces (landscapes) which arises in the digitisation and compression of elevation maps. In [ACGR02], the authors looked at the application of AMLE inpainting for interpolating digital elevation maps (DEMs). Therein they also showed the relation between AMLE and geodesic distance transformations as well as thin-plate model. Figure 4.11 provides an example of the interpolation of a DEM.

4.5. Further Reading and Some Extensions

For more details on the axiomatic derivation presented in Section 4.1, we refer readers to [AGLM93] and to the paper by Caselles, Morel and Sbert [CMS98a] and its extended version that features more details on the derivation [CMS98b]. For harmonic inpainting, not much literature exists. We refer readers to [BW89, BW90 and BW96], where the Laplace equation (and the higher-order bi-harmonic equation) has been used for image (surface) compression. Moreover, a very interesting error analysis for the harmonic extension as well as TV inpainting can be found in [CK06]. TV inpainting was introduced by Chan and Shen in a series of papers [SC02,CS01a,SC01]. These papers are summarised in a review paper of the authors from 2005 [CS05b] as well as in their book [CS05a]. For more details on total variation and its role in image processing, we refer readers to [CCC+10]. Regarding AMLE inpainting, Almansa has dedicated part of his Ph.D. thesis to a closer investigation of this method and in particular discusses its application to surface interpolation (see [Alm02 and ACGR02].)

Anisotropic Diffusion There exist many alternatives and extensions to the second-order diffusion methods presented in the preceding sections. Our methods so far constitute linear and non-linear diffusion approaches that

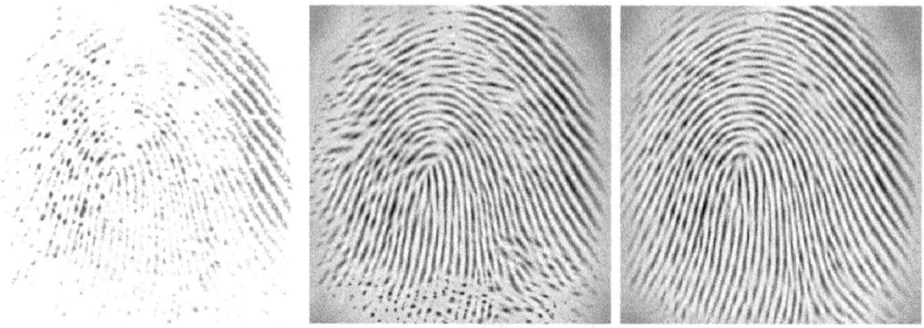

Figure 4.12. Directed diffusion and coherence-enhancing diffusion for fingerprint enhancement. From left to right: The original image (left, finger 84, impression 6, FVC2004 database 1); finger after enhancement by coherence-enhancing anisotropic diffusion filtering [Wei98] and diffusion filtering with V being an a priori estimated orientation field, as described in [GS12].

inpaint the image by diffusion in a specific direction. AMLE inpainting diffuses along the gradient of the image, TV inpainting diffuses orthogonal to the gradient – that is, tangential to level lines of the image – whereas harmonic inpainting diffuses in all directions homogeneously. We can generalise this idea of 'directional' interpolation by prescribing a vector field $V = (v_1, v_2)$ of directions of significant structures in the inpainted image u in whose direction we want to diffuse image grey values from ∂D inwards. To do so, we solve the following evolution equation:

$$u_t = V^{\perp} D^2 u V^{\perp} + \lambda \mathbb{1}_{\Omega \setminus D}(g - u), \tag{4.25}$$

where $\lambda > 0$ and $V^{\perp} = (-v_2, v_1)$. Equation (4.25) diffuses known image information from $\Omega \setminus D$ into the unknown parts D of the image in the direction that is given in $V^{\perp}$. Of course, in this case knowledge of V – at least up to a certain accuracy – is needed. Although this cannot always be expected in applications, whenever we have this information, the design of V enables us to customise (4.25) for a given image (or indeed a specific application) (see [GS12] for its application to fingerprint enhancement). In his book [Wei98], Joachim Weickert proposes various anisotropic second-order diffusion equations for image smoothing. One of them is coherence-enhancing diffusion, which is able to close small gaps in linelike structures. Figure 4.12 shows fingerprint enhancement with coherence-enhancing and directed diffusion. Tschumperlé [Tsc06] also picked up the idea of diffusion in a coherent direction and designed a class of non-linear PDEs that diffuse image contents along a vector field of coherent directions. Among other imaging applications, he showed that these PDEs perform very well for image inpainting. We will re-encounter the coherent inpainting idea of Tschumperlé as a main ingredient of the inpainting method by coherent transport in Section 6.2.

Another example of a second-order diffusion inpainting method is seen in [GS03], where the authors proposed the complex Ginzburg-Landau equation for image inpainting. We also refer to the books by Guillermo Sapiro [Sap06] and Frédéric Cao [Cao03] for a discussion of geometrical PDEs for image analysis and processing.

Curvature-Driven Diffusion To remedy the shortcomings of TV inpainting discussed in Section 4.3, a number of curvature-based inpainting models have been suggested in the literature. The basic idea is to integrate more geometrical information (in terms of the curvature of level lines) into the inpainting procedure to resolve the issue of the good continuation of level lines (cf. Chapter 3). Most of the discussion on this will be the topic of Chapter 5. However, there is one inpainting method within this class which – although it is of third differential order – fits better to the discussion of this chapter, and that is inpainting by curvature-driven diffusion (CDD) [CS01b]. Here the non-linear diffusion process of TV inpainting (4.18) is amended by integrating curvature information into the interpolation process. As such, CCD inpainting is able to connect level lines across the inpainting domain, reflecting Kanizsa's amodal completion law in Chapter 3. This is done as follows: let d: $\mathbb{R}^+ \to \mathbb{R}^+$ be a continuously increasing function with

$$d(s) = \begin{cases} 0, & s = 0, \\ +\infty, & s = +\infty, \\ \text{continuously increasing}, & 0 < s < \infty. \end{cases}$$

Further, we define $\kappa = \operatorname{div}(\nabla u/|\nabla u|)$ as the curvature of u (more explanation on this will be given in Section 5.1). Then the CDD inpainting model computes the inpainted image u of the given image g as a stationary solution of the following evolution equation:

$$u_t = \operatorname{div}\left(\frac{d(|\kappa|)}{|Du|} Du\right), \quad \text{in } D, \tag{4.26}$$

$$u = g, \quad \text{in } \Omega \setminus D. \tag{4.27}$$

Compared with the non-linear diffusion term of TV inpainting (4.18), where the strength of the diffusion was determined by $1/|Du|$, here the diffusion is additionally guided by the size of the curvature κ. More precisely, the flow (4.26) is a third-order non-linear diffusion equation with diffusivity coefficient $g(|\kappa|/|Du|)$. This kind of non-linear diffusion falls into the class of so-called curvature-driven diffusions [cf. [GH$^+$86, Gra87 and OS88] for analytical investigations of (4.26)]. Heuristically, the CDD flow resolves the connectedness failure of TV inpainting by increasing the diffusion for larger curvatures. In particular, this means that the TV inpainting result in Figure 4.8, where the curvature is $+\infty$ in the four endpoints of the edges, is not supported by the CDD flow (4.26) as a steady state.

Choice of the Data Fidelity Term Finally, let us mention that the data fidelity term in the variational approaches for image inpainting, for example, for TV inpainting in Section 4.3, does not have to be chosen as the squared L^2 norm of the difference $\mathbb{1}_{\Omega\setminus D}(u-g)$. Another popular choice is to take the L^1 norm, that is, $\|\mathbb{1}_{\Omega\setminus D}(u-g)\|_{L^1}$ (see, e.g., [THC11]). The advantage of an L^1 fidelity for inpainting is that it is more robust to outliers, that is, a possibly not perfect identification of the inpainting domain D. This is also the reason why, in the framework of image de-noising, the L^1 fidelity is a good model for impulse noise [Nik02, Nik04, DAG09], which is noise concentrated on random pixels in the image domain where the image information has been lost. The disadvantage of using the L^1 term is that it is computationally more involved than the L^2 model because of its non-differentiability.

5

Higher-Order PDE Inpainting

The scope of this chapter is the presentation of inpainting methods which use fourth-order (and higher!) partial differential equations (PDEs) to fill in missing image contents in gaps in the image domain. In the following section, we first motivate the use of higher-order flows for image inpainting.

5.1. Second- Versus Higher-Order Approaches

In this section we want to emphasise the difference between second-order diffusions as discussed in Chapter 4 and higher-order – in particular, fourth-order – diffusions in inpainting. As we have seen already, second-order inpainting methods (in which the order of the method is determined by the derivatives of highest order in the PDE), such as total variation (TV) inpainting, have drawbacks when it comes to the connection of edges over large distances and the smooth propagation of level lines into the damaged domain – qualities that we agreed an image interpolator which follows the good continuation principle from Chapter 3 should have. The disability, in general, of second-order methods to connect structures across the inpainting domain was demonstrated for harmonic inpainting in Figure 4.1 and for TV inpainting in Figure 4.8. An example of the lack of smoothness of interpolated level lines is given in Figure 4.7 for TV inpainting. In the case of TV inpainting, this behaviour of the interpolator is explained using the co-area formula, Theorem 4.3.6. To remind ourselves, TV inpainting seeks an interpolator whose level lines have minimal length, thus connecting level lines from the boundary of the inpainting domain via the shortest distance (linear interpolation). In [MM98, Mas98 and Mas02], Masnou and Morel propose an extension of the length penalisation in TV inpainting by an additional curvature term that should be small for interpolating level lines. In particular, for positive parameters $\alpha, \beta > 0$, they considered the minimisation of

$$\min_{\Gamma_\lambda} \int_{-\infty}^{\infty} \int_{\Gamma_\lambda} (\alpha + \beta(\kappa(s))^2)\, ds\, d\lambda, \tag{5.1}$$

where Γ_λ is the λ level line of the image function u, and κ is its curvature. We will define these mathematical terms more precisely in Section 5.2. If u is smooth and non-constant, problem, (5.1) can be equivalently written in functionalised form as

$$\min_u \int_\Omega \left(\alpha + \beta \left(\operatorname{div}\left(\frac{\nabla u}{|\nabla u|}\right)\right)^2\right) |\nabla u|\, dx. \tag{5.2}$$

We will discuss this in more detail in Section 5.2. For now, let us stress, firstly, that in (5.1) not only the length of the level lines but also their curvature is penalised (the penalisation of each depends on the ratio b/a) and, secondly, that in comparison with TV inpainting, the differential order of the method has increased. Indeed, when differentiating (5.2), we obtain as the corresponding Euler-Lagrange equation for a minimiser a non-linear fourth-order diffusion equation. The application of (5.1) to image inpainting then results in an interpolation method that features a smooth continuation of level lines over the inpainting domain as well as over large distances (see Figures 5.1 and 5.2). Note, however, that one obvious drawback – in particular, numerically but also in terms of limitations in its analysis – of Euler elastica inpainting is that the functional in (5.2) is non-convex and the corresponding PDE is non-linear and fourth order.

The performance of higher-order inpainting methods, such as Euler elastica inpainting, also can be interpreted via the second boundary condition, necessary for the well-posedness of the corresponding Euler-Lagrange equation of fourth order. For example, Bertozzi et al. showed in [BEglG07] that the Cahn-Hilliard inpainting model in fact favours the continuation of the image gradient into the inpainting domain. The Cahn-Hilliard inpainting model is based on computation of the inpainted image as a stationary solution of

$$u_t = -\epsilon \Delta^2 u + \Delta F'(u) + \lambda \mathbb{1}_{\Omega \setminus D}(g - u),$$

for $1 \gg \epsilon > 0$ and a double-well potential $F(u) = (u^2 - 1)^2$. Here Δ^2 denotes the bi-Laplacian, and $\lambda > 0$ is a typically large constant that forces the inpainted image u to stay close to g outside the inpainting domain D. For this model, the authors in [BEglG07] proved that in the limit $\lambda \to \infty$ a stationary solution of

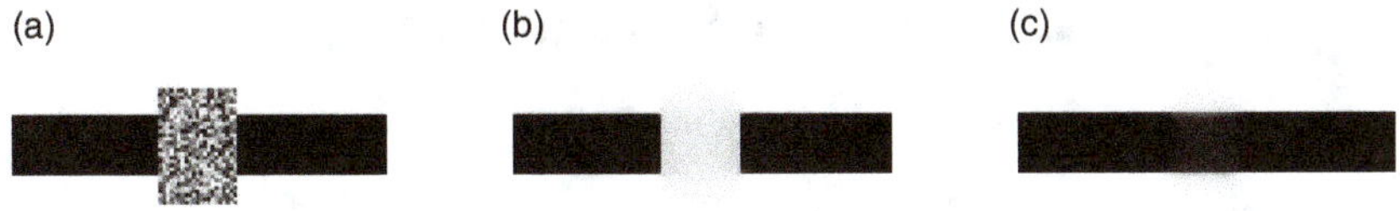

Figure 5.1. An example of Euler elastica inpainting compared with TV inpainting. In the case of large aspect ratios, TV inpainting fails to comply with the connectedness principle. (a) Broken stripe. (b) TV inpainting. (c) Euler elastica inpainting.

Figure 5.2. An example of Euler elastica inpainting (center) compared with TV inpainting (right). Despite the presence of high curvature, TV inpainting truncates the circle inside the inpainting domain (linear interpolation of level lines). Depending on the weights a and b, Euler's elastica inpainting returns a smoothly restored object, taking the curvature of the circle into account (curvature preservation).

the Cahn-Hilliard inpainting equation fulfils

$$u = g, \qquad \text{on } \partial D$$
$$\nabla u = \nabla g, \quad \text{on } \partial D$$

for a given image g regular enough ($g \in C^2$). This means that not only the grey values of the image are specified on the boundary of the inpainting domain but also the gradient of the image function, namely, the direction of the level lines, is given. The Cahn-Hilliard inpainting model and its generalisation for grey value images – called TV-H^{-1} *inpainting* – will be discussed in Section 5.3.

In an attempt to derive PDE inpainting methods that fulfil the good continuation principle, a number of fourth-order (and higher) diffusions have been suggested for image inpainting. In what follows we will describe some of them in more detail.

5.2. Curvature-Based Inpainting

An interpolant u in D is computed as a minimiser of

$$\mathcal{J}(u) = \int_D |\nabla u|(\alpha + \beta|\kappa|^p)\, dx, \tag{5.3}$$

where $p \geq 1$, α, β are two positive constants, κ is the curvature of u [defined soon in (5.4)] and $\mathcal{J}$ is minimised over an appropriate class of image functions with appropriate boundary conditions. The common idea of inpainting approaches using (5.3) is to constrain the interpolant u to have smooth level lines with minimal lengths and curvature. Minimal curvature interpolation has a long tradition in visual completion and computer vision, as was discussed in part in Chapter 3. Shaáshua and Ullmann [US88] sought a network of salient image structures by grouping (i.e., connecting) level lines based on the minimality of $\int_D |\kappa|^2$. Guy and Medioni picked up this idea in [GM93] to infer perceptually important structures in images by the computation of an extension field. In both articles, minimal curvature appears as a criterion for 'good continuation'.

We start this section in a rather unsexy way, with some notational conventions and mathematical preliminaries that we will use in what follows.

5.2.1. Preliminaries

Level Sets We quickly recapture our definitions in Section B.9 of Appendix B for level sets of an integrable function u. We define the (sub) level sets Γ_λ of u as

$$\Gamma_\lambda(u) = \{x \in \Omega : \ u(x) \leq \lambda\}.$$

Then the level-set representation of u is $\Gamma(u) = \{\Gamma_\lambda(u) : \ \lambda \in \mathbb{R}\}$. We also denote by γ_λ the boundaries of the level sets.

Curvature In this section we will consider interpolation methods that find the interpolant with minimal curvature. To formalise this, we first have to say what we mean by 'curvature', both in the sense of curves (associated with level sets) and in the sense of curvature of a function (i.e., mean curvature of the surface defined by the image function u). Based on the definitions in Section B.10 of Appendix B, we consider the curvature of a simple planar twice continuously differentiable curve γ parametrised on the interval $[0, 1]$ given by (B.4), that is, $|\kappa(s)| = |\gamma''(s)|$, where s is the arclength. We will see later in Section 5.2 that the curvature of a curve γ can be defined in a weaker sense and that the regularity assumption on γ can be relaxed.

In what follows we are in particular concerned with measuring the curvature of the boundary of level sets of an image function u. A general measure for a smooth curve γ of this type is given by

$$\int_\gamma \phi(\kappa)\, ds,$$

where $\phi(\kappa)$ is a suitable function of curvature. With $\phi(\kappa) = |\kappa|$, we recover the total absolute curvature of γ. The case $\phi(\kappa) = \alpha + \beta\kappa^2$ gives the so-called Euler elastica measure of γ that is the main topic of this section.

There is no co-area formula à la Theorem 4.3.6 for the curvature of level lines. There only is a conjecture of De Giorgi [MD97]. However, for smooth functions, the functional representation of the curvature can be derived via the mean curvature vector of the hypersurface defined by the boundary of level sets [AM03]. In particular, for a smooth function u (at least C^2, say) whose gradient does not vanish in a neighbourhood of a point $x \in \gamma_\lambda$, the curvature of $\gamma_\lambda(u) = \partial\Gamma_\lambda(u)$ at that point is defined as [MS95]

$$\kappa = \kappa(\gamma_\lambda(x)) = \operatorname{div}\left(\frac{\nabla u(x)}{|\nabla u(x)|}\right). \tag{5.4}$$

5.2.2. The Euler Elastica Inpainting Approach

In the particular case $p = 2$, the energy (5.3) reads

$$\mathcal{J}(u) = \int_D |\nabla u|(\alpha + \beta|\kappa|^2)\, dx \tag{5.5}$$

and is called *Euler's elastica energy* [Eul44, Mum94, NM90, NMS93]. The corresponding inpainting approach discussed in [MM98, Mas98, Mas02 and CKS02] is called *Euler elastica inpainting*. The elastica energy was first studied by Euler in [Eul44] to model the shape of a thin elastic rod. Much later in 1965, Birkhoff and De Boor [BDB65] introduced it as a non-linear spline for function interpolation. Since the pioneering works of Mumford and Nitzberg [Mum94, NM90, NMS93], Masnou and Morel [MM98] and Chan, Kang and Shen [CKS02], Euler's elastica energy has gained increasing importance in image processing, in particular, in image segmentation and inpainting. Before we go deeper into the study of this energy for inpainting, let us start with a simple definition.

Definition 5.2.1 A parametrised curve $\gamma\colon [0,1] \to \mathbb{R}^2$ is a Euler elastica curve if it is a minimising curve of

$$\mathcal{J}(\gamma) = \int_0^{\text{length}(\gamma)} (\alpha + \beta\kappa(s)^2)\, ds \tag{5.6}$$

over all either closed curves γ or curves γ with specified boundary conditions in $s = 0$ and $s = \text{length}(\gamma)$. Here s denotes the arc length, $\kappa(s)$ is the scalar curvature of γ and α, β are two positive parameters.

Euler elastica curves and their existence have been studied by [Fun62 and Lov13]. The use of (5.5) for image interpolation is motivated by works of Nitzberg, Mumford and Shiota [NM90, NMS93], where they proposed the use of $\mathcal{J}$ in (5.5) for the segmentation of partially occluded objects. The occluded part of an object is defined as the one whose edge set (i.e., the boundary of the object) has minimal length and curvature. Given two endpoints (T-junctions) p_1 and p_2 of a boundary edge with normal vectors $\vec{n}_1, \vec{n}_2$ on ∂D, respectively, the continuation γ of this edge into D is computed as a minimiser of

$$\int_\gamma (\alpha + \beta|\kappa|^2)\, ds, \tag{5.7}$$

subject to the boundary conditions at the beginning and endpoints: $\gamma(0) = p_1$ with normal vector $\vec{n_1}$ and $\gamma(\text{length}(\gamma)) = p_2$ with normal vector $\vec{n_2}$. Here $\int_\gamma$ denotes the integration along the curve γ, that is, $\int_\gamma = \int_0^{\text{length}(\gamma)}$. Moreover, κ is the curvature of γ parameterised by arc length s as in (B.4). Motivated by this, Masnou and Morel [MM98, Mas98] extended (5.7) for image inpainting based on the smooth continuation of level lines. They called their approach *level-line-based disocclusion* because they lifted the variational criterion (5.7) from the edge set of an image to all of its level sets. In their approach, the inpainted image u is reconstructed from a continuation of all level lines γ, arriving at ∂D that is optimal in terms of (5.7). More precisely, these optimal level lines are solutions of

$$\min_{\gamma \in \mathcal{A}_\lambda} \int_{\gamma_\lambda} (\alpha + \beta|\kappa|^p)\, ds, \tag{5.8}$$

where $\mathcal{A}_\lambda$ is the admissible set for the level λ defined by

$$\mathcal{A}_\lambda = \{\gamma : \gamma(0) = \gamma_\lambda(p_1),\ \gamma(1) = \gamma_\lambda(p_2), \text{and normals } \vec{n_1} \text{ and } \vec{n_2} \text{ in } p_1 \text{ and } p_2\},$$

and γ_λ is the boundary of the level set Γ_λ in (B.1). Compare Figure 5.3 for an illustration.

The work in [MM98] was followed by a series of other contributions and extensions (e.g., [CKS02, Mas02, BBC$^+$01, BCV03]). In particular, in [CKS02], the functionalised formulation (5.5) was studied. In the following we shall give an account to this approach, its main ideas and algorithmic ingredients. In connection to our Bayesian discussion in Section 2.1, let us start our current considerations with a Bayesian interpretation of Euler elastica

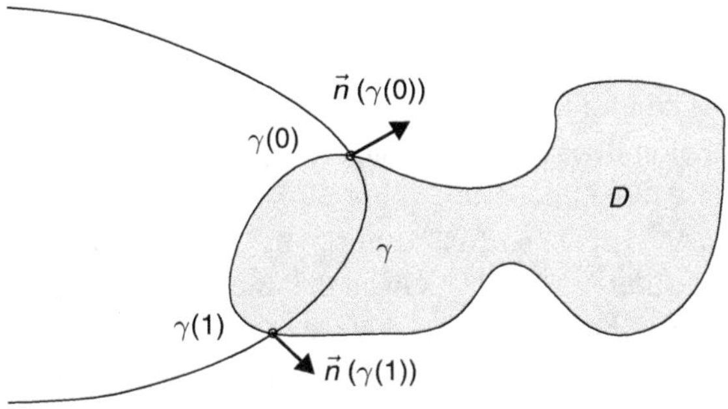

Figure 5.3. Level-line-based disocclusion as in (5.8). The good continuation principle is enforced by penalising the angle between the normals to the given part of the level line and its optimal continuation γ at the endpoints $\gamma(0)$ and $\gamma(1)$.

curves. This interpretation was first given in the seminal paper by Mumford [Mum94] and after that re-explained in more detail in [CKS02].

Bayesian Interpretation of Euler Elastica In [Mum94 and CKS02], the Bayesian framework of the Euler elastica energy (5.6) was discussed. Interestingly enough, there is an analogy between the walk of a (modestly) drunken person and a discrete Euler elastica curve! More precisely, it can be shown that minimising (5.6) for polygonal curves γ is equivalent to maximising the likelihood of a polygonal random walk with certain smoothness constraints. To do so, for a fixed number $N \in \mathbb{N}$ we define N random step sizes h_k, $k = 1, \ldots, N$. The step sizes are independently identically distributed random numbers with probability density $p(h) = \lambda e^{-\lambda h}$ and mean $1/\lambda$ for a fixed $\lambda \in \mathbb{R}$. This part of the model will give us the length term in the likelihood of γ. Further, we introduce the orientation θ_k of the kth step as the angle between the walking direction and the x-axis. Then the change of direction for two subsequent steps is given by $\Delta\theta_k = \theta_k - \theta_{k-1}$, $k = 1, \ldots, N$ with initial value $\theta_0 = 0$:

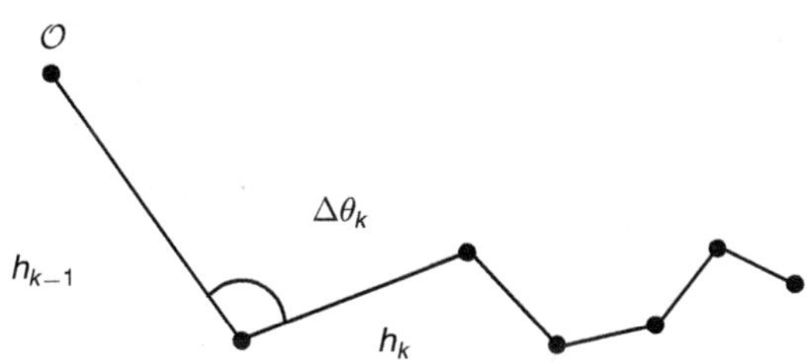

The distribution of $\Delta\theta_k$ is determined by the assumption that in each step the larger the step size h_k, the more uncertain the change of orientation $\Delta\theta_k$ will become. This condition on the polygonal walk, that is, demanding that the change of orientation depends on the step size, will result into an additional curvature term in the likelihood for γ. For a fixed positive σ, this is modelled by $\Delta\theta_k \in N(0, h_k\sigma^2)$; that is, $\Delta\theta_k$ is normally distributed with probability density $p(\theta) = 1/\sqrt{2h\sigma^2}e^{-\theta^2/(2h\sigma^2)}$. Given that the distributions of step sizes h and ratios $\Delta\theta/\sqrt{h}$ are independent of each other, their N values uniquely define the N-step polygonal walk γ. The preceding modelling assumptions give the following likelihood for a polygonal walk γ:

$$
\begin{aligned}
p(\gamma) &= p((h_1,\ldots,h_N))\cdot p((\Delta\theta_1/\sqrt{h_1},\ldots,\Delta\theta_N/\sqrt{h_N})) \\
&= \lambda^N e^{-\lambda\sum_k h_k}\, dh_1\cdots dh_N \cdot \frac{1}{(\sqrt{2\pi}\,\sigma)^N} e^{-1/2\sigma^2\sum_k(\Delta\theta_k)^2/h_k} \\
&\quad \times d\frac{\Delta\theta_1}{\sqrt{h_1}}\ldots d\frac{\Delta\theta_N}{\sqrt{h_N}},
\end{aligned}
$$

which up to a constant factor equals

$$
e^{-\lambda\sum_k \mathrm{length}_k(\gamma) - (1/2\sigma^2)\sum_k \kappa_k(\gamma)^2},
$$

where $\mathrm{length}_k(\gamma) = h_k\, dh$ is the length of the line of the kth step, and $\kappa_k(\gamma)$ is the curvature that two polygonal lines of γ make between the $k-1$th and kth steps. Indeed, this is a discrete version of the Euler elastica energy (5.6).

5.2.3. Well-Posedness of Minimal Curvature Interpolation via Level Lines

For the following considerations, let $D \subset \mathbb{R}^2$ be an open, bounded and simply connected domain which defines the occluded part of the image with Jordan (simple and closed) boundary ∂D. Moreover, we assume that the given image $g \in BV(\mathbb{R}^2 \setminus \bar{D})$; that is, for simplicity, we consider images on the entire plane $\mathbb{R}^2$. We are seeking an inpainted image $u \in BV(\mathbb{R}^2)$ with $u = g$ in $\mathbb{R}^2 \setminus D$ and u fulfilling the variational criterion (5.3) in D. The minimisation of the functional $J(u)$ in (5.3) is well defined only for a certain class of smooth functions u. In order to deal with a more general class of functions such as functions of bounded variation (BV), the problem has to be relaxed. For the case $p > 1$, the authors in [BDMP93] considered a relaxed version of $\mathcal{J}$, embedding the functional into its lower semi-continuous envelope within a subspace of $BV(D)$. Their result was extended in [AM03] to higher dimensions $d \geq 2$, consulting the theory of varifolds. The relaxation procedure is a standard tool in the calculus of variations [Dac89,FL07], and the theory of varifolds has a long-standing tradition in geometrical measure theory [AFP00, FF69]. For an illustration of the concept of relaxation within this setting, see Example 15 in Appendix A. However, here we shall discuss the well-posedness of the model via its level-line representation in (5.8). In particular, we shall review the results obtained in [Mas98] for the well-posedness of the level-line approach (5.8) for all $p \geq 1$. There the inpainted image was created by interpolating the level lines of the image function (rather than interpolating the image function itself). This is a very important mathematical concept of thinking about the inpainting of an image, not least because the first, most pioneering inpainting models have been developed within this framework. Moreover, the formulation (5.8) in terms of level lines brings the inpainting approach closer to Kanizsa's amodal completion theory [Kan79] (cf. Section 3.2). Eventually, the inpainted image can be recovered from its level sets. In fact, an image function $u \in BV(\mathbb{R}^2)$ can be equivalently represented as the union of its level sets as in (B.1). More precisely, let $\Gamma_\lambda(u) = \{x : u(x) \leq \lambda\}$ be the level set of u with grey level $\lambda \in \mathbb{R}$; then

$$u(x) = \inf\{\lambda : x \in \Gamma_\lambda\}, \qquad \text{for almost every } x \in \mathbb{R}^2.$$

In order to make precise mathematical sense of the functional in (5.8), let us introduce the class of curves over which we are minimising. For the following considerations, we assume that ∂D is a smooth, simple and closed curve such that the one-dimensional restriction $u|_{\partial D}$ of u to ∂D has bounded variation – that is, $\int_{\partial D} |u|_{\partial D}| d\mathcal{H}^1 < \infty$ and $\int_0^{\text{length}(\partial D)} |(u|_{\partial D})'| < \infty$ – and the level lines of u are transverse. Such a domain D is called an *admissible occlusion* [Mas98, Lemma 3.2.2]. Note that this assumption is for the ease of notation only. In fact, Masnou [Mas98] proved that for any domain D as earlier and $u \in BV(\mathbb{R}^2 \setminus \bar{D})$ there exists a simple rectifiable curve Γ arbitrarily close to D such that Γ has these desired properties and could take the role of ∂D in the following considerations. We consider all possible level sets in D and their corresponding level lines $\gamma_\lambda = \partial \Gamma_\lambda$, parameterised such that their endpoints $\gamma_\lambda(0), \gamma_\lambda(1)$ are on the boundary of D and constrained by the requirement that they have finite length and curvature, that is,

$$\int_0^{\text{length}(\gamma_\lambda)} (1 + |\gamma_\lambda''(s)|^p)\, ds < \infty,$$

where γ_λ'' is the second distributional derivative of the curve γ_λ, and s is the arc length. For $p > 1$, this means that we only look for optimal curves $\gamma = \gamma(s) \in W^{2,p}(0, \text{length}(\gamma))$, whereas for $p = 1$, we assume that $\gamma \in W^{1,1}(0, \text{length}(\gamma))$ with $\gamma' \in BV(0, \text{length}(\gamma))$. In this case, $\int_0^{\text{length}(\gamma)} |\gamma''(s)|\, ds$

denotes the total variation of the measure γ''. Then we define the admissible set of disoccluding level lines as

$$\begin{aligned}
\mathcal{A}_p =& \{\gamma \in W^{2,p}(0,\text{length}(\gamma)) : \gamma(0), \gamma(1) \in \partial\Gamma_\lambda, \text{ for a } \lambda \in \mathbb{R}\}, \quad p > 1 \\
\mathcal{A}_1 =& \{\gamma \in W^{1,1}(0,\text{length}(\gamma)), \gamma' \in BV(0,\text{length}(\gamma)) : \gamma(0), \gamma(1) \in \partial\Gamma_\lambda, \\
& \text{for a } \lambda \in \mathbb{R}\}, \quad p = 1.
\end{aligned}$$

With this admissible set of potential disoccluding curves, we can make the variational criterion (5.8) more precise. To do so, we have to treat the cases $p = 1$ and $p > 1$ separately. This is so because for $p > 1$ we have from $\gamma \in W^{2,p}(0,\text{length}(\gamma))$ that γ is continuously differentiable on $[0,1]$, and hence, the boundary conditions in (5.8) can be defined in terms of γ'. In the case $p = 1$, a curve $\gamma \in \mathcal{A}_1$ does not necessarily have to be continuously differentiable, and boundary terms have to be set up differently.

We start with $p > 1$ and define the following energy over $\mathcal{A}_p$:

$$\mathcal{J}(\gamma_\lambda) = \int_0^{\text{length}(\gamma)} (\alpha + \beta|\gamma_\lambda''(s)|^p)\, ds + |\angle(\vec{n}(0), \gamma_\lambda'(0_+))| + |\angle(\vec{n}(1), \gamma_\lambda'(1_-))|, \tag{5.9}$$

where $\alpha, \beta > 0$, $\vec{n}(0)$ is the normal vector to the given part $\gamma_\lambda(0)$ of the level line $\partial\Gamma_\lambda$ (where the boundary here must be understood in a weaker sense; compare the essential boundary in [AFP00]), γ_λ' is the first distributional derivative of γ_λ (pointing in the direction normal to γ_λ) and $\angle(\cdot,\cdot)$ denotes the angle between the two vectors in $\mathbb{R}^2$ modulo 2π. The latter two terms replace the exact boundary conditions on the normals in the admissible set for (5.8) with approximate ones. They make sure that the level lines are smooth up to the boundary ∂D. In particular, these terms enforce the direction of the level lines to be approximately orthogonal to the boundary of the inpainting domain D and as such guarantee that the good continuation property (cf. Section 3.2) is satisfied.

For $p = 1$, the boundary terms have to be defined differently. To do so, we define $\forall\lambda \in \mathbb{R}$ the line of level λ arriving at $x \in \partial\Gamma_\lambda$ and of length $\alpha \ll 1$ as $S(x,\lambda)$. For this line, we denote $\tau(0)$ the angle that S encloses with ∂D at x. Then, for each curve $\gamma \in \mathcal{A}_1$, we associate the extended curve $\tilde{\gamma} : [0, \text{length}(\gamma) + 2\alpha] \to \mathbb{R}^2$ such that

$$\tilde{\gamma} = \begin{cases} S(\gamma(0), \lambda), & \text{on } [0,\alpha], \\ \gamma, & \text{on } [\alpha, \text{length}(\gamma) + \alpha], \\ S(\gamma(1), \lambda), & \text{on } [\text{length}(\gamma) + \alpha, \text{length}(\gamma) + 2\alpha]. \end{cases}$$

If length(γ) $= 0$, only the first and last cases have effect. With these definitions, we introduce the energy for $p = 1$

$$\mathcal{J}(\gamma_\lambda) = \int_0^{\text{length}(\gamma)+2\alpha} (\alpha + \beta|\tilde{\gamma}_\lambda''(s)|^p)\, ds$$

Then we can now formalise the inpainting problem as follows: instead of (5.3), an optimal disocclusion is computed as a minimiser of the total energy over all level lines (5.8), that is, the minimisation of

$$\mathcal{J}(u) = \int_{\mathbb{R}} \mathcal{J}(\gamma_\lambda)\, d\lambda.$$

The existence of a minimiser $u \in BV(D)$ is proven in Masnou's thesis [Mas98] for any $p \geq 1$. More precisely, Masnou proves the following theorem:

Theorem 5.2.2 *Let $D \subset \mathbb{R}^2$ be an admissible occlusion, $g \in BV(\mathbb{R}^2 \setminus D)$ such that $|g| < C$ is bounded by a constant $C > 0$. Then*

1. *There exists an optimal occlusion $u \in BV(\mathbb{R}^2)$ of g for $p = 1$ such that $u = g$ in $\mathbb{R}^2 \setminus \bar{D}$.*
2. *Further, if we assume that the restriction of g to ∂D attains only a finite number of level lines on ∂D, then there also exists an optimal occlusion $u \in BV(\mathbb{R}^2)$ for $p > 1$ such that $u = g$ in $\mathbb{R}^2 \setminus \bar{D}$.*

Remark 5.2.3

- The proof of Theorem 5.2.2 can be found in [Mas98]. Let us just remark that for $p > 1$, the restriction to images g which attain only finitely many values λ on ∂D is a technical assumption only which is needed to derive a uniform bound on the energy.
- Also note that instead of adding constraints on the continuity of the solution at the boundary as in (5.9), another way of guaranteeing the good continuation principle would be to enlarge the domain of the level sets from D to, say, $D \subset \tilde{D}$, restrict them to the given level lines in $\tilde{D} \setminus D$ and replace $\mathcal{J}$ in (5.9) with

$$\mathcal{J}(\gamma_\lambda) = \int_0^1 (\alpha + \beta|\gamma_\lambda''(s)|^p)\, ds. \tag{5.10}$$

The Functionalised Inpainting Model For the numerical solution of Euler elastica inpainting, the level-line formulation (5.8) can be rather problematic. Potential problems are that two level lines might intersect inside D and that the collection of all optimal level-line interpolants does not have to weave the entire inpainting domain. Therefore, let us return to the functionalised definition of the energy that was derived and discussed in [CKS02]. That is, let us return from (5.8) to (5.3) for $p = 2$. This can be done if we assume that an admissible inpainting u on D is smooth. Then we can represent the curvature of the level line γ_s: $u = s$ by

$$\kappa = \text{div}\, \vec{n} = \text{div}\left(\frac{\nabla u}{|\nabla u|}\right).$$

Moreover, for the length element dl along the normal direction $\vec{n}$, we have

$$\frac{ds}{dl} = |\nabla u|.$$

Therefore, we can write

$$\begin{aligned}\mathcal{J}(u) &= \mathcal{E}(\mathcal{F}) \\ &= \int_0^1 \int_{\gamma_s:\, u=s} (\alpha + \beta\kappa^2)\, dt\, ds \\ &= \int_D \left(\alpha + \beta\left(\text{div}\left(\frac{\nabla u}{|\nabla u|}\right)\right)^2\right) |\nabla u|\, dx.\end{aligned}$$

Remark 5.2.4 Note that this derivation from the level-line formulation to the functionalised model works in the smooth (and non-constant) case only. A result as general as the co-area formula for the total variation does not exist for the curvature (conjecture of De Giorgi).

From our considerations so far, the smoothness assumption on u – which we need for the curvature to be well defined – is not a very realistic assumption for an image function. We would rather desire to have an inpainting approach that is defined for image functions, even if they are discontinuous. This can be done as suggested in [CKS02] by introducing a 'weak' form of the curvature and formulating Euler's elastica inpainting for image functions in BV. Consider $g \in BV(\Omega)$ with $\int_{\partial D} |Dg| = 0$. The latter condition can be rewritten as

$$\int_{\partial D} |Dg| = \int_{\partial D} |g^+ - g^-|\, d\mathcal{H}^1 = 0,$$

which means that $g^+ = g^-$ a.e. (w.r.t. to $\mathcal{H}^1$) along ∂D. In other words, we assume that g does not jump across the boundary of D; that is, there is no essential overlap between ∂D and image edges. Then we consider the minimisation of

$$\mathcal{J}(u) = \int_D \left(\alpha + \beta\left(\text{div}\left(\frac{Du}{|Du|}\right)\right)^2\right) |Du|\, dx, \tag{5.11}$$

under the conditions

$$u|_{\Omega\setminus D} = g|_{\Omega\setminus D}, \quad \int_{\partial D} |Du| = 0, \quad |\kappa| < \infty \text{ a.e. along } \partial D,$$

where a.e. is meant in the sense of the Hausdorff measure. The second and third conditions enforce a certain regularity on u on the boundary ∂D, namely, that u does not have essential discontinuities across the boundary and that its curvature κ is finite.

Now, for a function $u \in BV(\Omega)$, the Euler elastica energy in (5.11) is still defined only formally because a general BV function lacks the necessary regularity for the curvature to be well defined. To be able to rigorously define (5.11) for functions of bounded variation, we shall – in the following – introduce the weak form of curvature. For that, let us denote for a function $u \in BV(D)$,

$$d_u \nu(S) = |Du|(S), \quad S \subset D$$

the total variation of u in a subset S of D, which is a Radon measure on D. Let, further, $\text{supp}(d_u\nu)$ be the support of the TV measure. Then, for any $p \in \text{supp}(d_u\nu)$, we have

$$d_u \nu(N_p) = |Du|(N_p) > 0$$

on any small neighbourhood $N_p \subset D$ of p. Now let

$$\rho_\sigma(x) = \frac{1}{\sigma^2}\rho\left(\frac{x}{\sigma}\right), \quad u_\sigma = \rho_\sigma * u, \tag{5.12}$$

where ρ is a fixed radially symmetric non-negative mollifier [i.e., ρ is smooth, and $\lim_{\sigma\to 0}\rho_\sigma(x) = \delta_x$, where the limit is understood in the space of distributions] with compact support and unit integral. Then, for $p \in \text{supp}(d_u\nu)$, we define the *weak absolute curvature*

$$\tilde{\kappa}(p) = \limsup_{\sigma\to 0}\left|\text{div}\left(\frac{Du_\sigma}{|Du_\sigma|}\right)(p)\right|,$$

where for those σ which give $|Du_\sigma(p)| = 0$, we define $\text{div}(Du_\sigma/|Du_\sigma|) = \infty$. For any point p outside the support $\text{supp}(d_u\nu)$, we define $\tilde{\kappa}(p) = 0$, since u is a.e. a constant function near a neighbourhood of p.

With this concept of weak curvature, the functionalised elastica energy in (5.11) now can be rigorously defined for BV functions u with $\tilde{\kappa} \in L^2(D, d_u\nu)$, that is, square integrable with respect to the TV measure $d_u\nu$. For more properties of weak curvature and the equivalence between classical and weak curvature over certain classes of functions, see [CKS02]. In what follows, we will continue by deriving the first variation of the Euler elastica energy (5.11) in the weak sense and, in turn, an interpretation of its inpainting dynamics in terms of transport and diffusion of grey value information in g. For the following derivation, we assume that the inpainted image u is smooth enough at least $W^{2,1}(\Omega)$ and that the curvature is well defined.

Theorem 5.2.5 *Let $\phi \in C_c^1(\mathbb{R},[0,\infty))$, and define*

$$R(u) = \int_\Omega \phi(\kappa) \cdot |\nabla u|\, dx$$

for $u \in W^{2,1}(\Omega)$. Then the first variation of R over the set of $C_c^\infty(\Omega)$ is given by

$$\nabla_u R = -\mathrm{div}\vec{V},$$

where

$$\vec{V} = \phi(\kappa) \cdot \vec{n} - \frac{\vec{t}}{|\nabla u|} \frac{\partial\left(\phi'(\kappa)|\nabla u|\right)}{\partial \vec{t}}.$$

Here $\vec{n} = \nabla u/|\nabla u|$, and $\vec{t} \perp \vec{n}$ is the unit tangent vector.

Proof Let $v \in C_c^\infty(\Omega)$, a compactly supported test function on Ω; then the first variation of R is defined for $\tau \in \mathbb{R}$ as

$$\int_\Omega \nabla_u R \cdot v\, dx = \left(\frac{d}{d\tau} R(u+\tau v)\right)\Big|_{\tau=0}.$$

We first compute

$$\frac{d}{d\tau} R(u+\tau v) = \int_\Omega \phi(\kappa_{u+\tau v}) \cdot \frac{\nabla(u+\tau v)}{|\nabla(u+\tau v)|} \nabla v + \phi'(\kappa_{u+\tau v}) \left(\frac{d}{d\tau}\kappa_{u+\tau v}\right) |\nabla(u+\tau v)|\, dx.$$

Then we determine the derivative of $\kappa_{u+\tau v} = \mathrm{div}\left(\nabla(u+\tau v)/|\nabla(u+\tau v)|\right)$ as

$$\begin{aligned}
&\frac{d}{d\tau}\left(\mathrm{div}\left(\frac{\nabla(u+\tau v)}{|\nabla(u+\tau v)|}\right)\right) \\
&= \mathrm{div}\left(\frac{\nabla v|\nabla(u+\tau v)| - \nabla(u+\tau v) \cdot \frac{(\nabla(u+\tau v))^{\mathrm{T}}}{|\nabla(u+\tau v)|}\nabla v}{|\nabla(u+\tau v)|^2}\right) \\
&= \mathrm{div}\left(\frac{1}{|\nabla(u+\tau v)|}\left[Id - \frac{\nabla(u+\tau v)}{|\nabla(u+\tau v)|} \otimes \frac{\nabla(u+\tau v)}{|\nabla(u+\tau v)|}\right]\nabla v\right),
\end{aligned}$$

where $\vec{a} \otimes \vec{a} = \vec{a}\vec{a}^{\mathrm{T}}$ is the orthogonal projection on the vector $\vec{a} \in \mathbb{R}^2$. Then

$$\left(\frac{d}{d\tau} R(u+\tau v)\right)\Big|_{\tau=0} = \int_\Omega \phi(\kappa)\vec{n}\nabla v + \phi'(\kappa)\mathrm{div}\left(\frac{1}{|\nabla u|}\left[Id - \vec{n}\otimes\vec{n}\right]\nabla v\right)|\nabla u|\, dx,$$

where we have used $\vec{n} = \nabla u/|\nabla u|$. The second term in the preceding addition can be simplified by integration by parts to

$$-\int_\Omega [Id - \vec{n}\otimes\vec{n}] \frac{1}{|\nabla u|} \nabla\left(\phi'(\kappa)|\nabla u|\right) \nabla v \, dx$$
$$= \int_\Omega \operatorname{div}\left([Id - \vec{n}\otimes\vec{n}] \frac{1}{|\nabla u|} \nabla\left(\phi'(\kappa)|\nabla u|\right) \right) v \, dx.$$

Integrating the first term by parts as well and using $\vec{n}\otimes\vec{n} + \vec{t}\otimes\vec{t} = Id$ and $\vec{t}\otimes\vec{t}\nabla f = \vec{t}\frac{\partial f}{\partial \vec{t}}$, we derive

$$\int_\Omega \nabla_u R \cdot v \, dx = -\int_\Omega \operatorname{div}\left(\phi(\kappa)\vec{n} - \frac{1}{|\nabla u|} \frac{\partial\left(\phi'(\kappa)|\nabla u|\right)}{\partial \vec{t}} \vec{t} \right) v \, dx$$
$$= -\int_\Omega \operatorname{div}\vec{V} \cdot v \, dx,$$

which completes the proof. ■

In particular, for the elastica inpainting model with

$$R(u) = \int_\Omega (a + b\kappa^2)|\nabla u| \, dx,$$

Theorem 5.2.5 gives the following:

Corollary 5.2.6 *For the case of Euler elastica inpainting, that is, $\phi(\kappa) = (a + b\kappa^2)$ for non-negative constants a, b, the preceding theorem gives the following expression for the first variation of the respective regularising energy R:*

$$\nabla_u R = -\operatorname{div}\vec{V}, \quad \vec{V} = (a + b\kappa^2)\vec{n} - \frac{2b}{|\nabla u|} \frac{\partial \kappa |\nabla u|}{\partial \vec{t}} \vec{t}.$$

In [CKS02], Chan, Kang and Shen split the preceding equation into transport and diffusion terms and then gave an explanation of the main interpolation mechanisms of Euler elastica inpainting in terms of transport and diffusion. We will discuss this in more detail – alongside the mechanisms present in other higher-order inpainting approaches – in Chapter 8.

Remark 5.2.7 Natural boundary conditions Taking the variation of R in Theorem 5.2.5 over $C^1(\mathbb{R}, (0, \infty))$, the resulting natural boundary conditions along $\partial\Omega$ are

$$\frac{\partial u}{\partial \nu} = 0, \quad \frac{\partial(\phi'(\kappa)|\nabla u|)}{\partial \nu} = 0,$$

where ν denotes the outward-pointing normal on $\partial\Omega$.

5.2.4. A Remark on Other Powers of the Curvature

As we have seen in our discussion of level-line-based disocclusion in Section 5.2, other powers of the curvature of u also can be employed in the curvature regularising energy. Hence, the question arises about the influence of different choices for powers p of the curvature; that is, how does the behaviour of minimisers of

$$\mathcal{J}_p(u) = \int_\Omega (a + b|\kappa|^p)\ |\nabla u|\ dx \tag{5.13}$$

change for different $p \geq 1$? For the choice $p = 1$ in (5.13), a possible solution is the one that fulfils $|\nabla u| \cdot \operatorname{curv} u$ and hence results in a straight-line connection, as the discussion of TV inpainting in Section 4.3 has shown. In particular, in this case, level lines do not have to be smooth but are allowed to suddenly turn. In contrast, the choice $p = 2$ results in elastica curves which are smooth. At this point, one might wonder what the effect might be of other choices of p. One can imagine that the larger the power of p, the heavier would be the penalisation of large curvatures, that is, sharp turns and less penalisation on small curvature parts. The next theorem from [CKS02] shows that in fact the energy (5.13) for $p \geq 3$ looses its natural applicability to images.

Theorem 5.2.8 *Let u be C^2 in a neighbourhood of a point $z \in \Omega$ where*

$$\nabla u(z) = 0, \quad H_u(z) \textit{ is non-singular,}$$

where $H_u(z)$ is the Hessian of u in z. Then $\mathcal{J}_p(u)$ in (5.13) *is infinite for all $p \geq 3$.*

Proof Without loss of generality, let $z = (0,0)$ and $u(z) = 0$. For a point x in a small neighbourhood of z, that is, $x \in B_\epsilon(z)$, $\epsilon > 0$, we can write

$$u(x) = \frac{1}{2} x H_u(z) x^{\mathrm{T}} + \mathcal{O}(\|x\|^3),$$

where in what follows we can drop the higher-order terms because when analysing $\mathcal{J}_p(u)$ we are only interested in at most second-order features (the curvature of u). Moreover, the matrix H_u is symmetric, and there exists an orthonormal matrix Q such that $H_u = QDQ^{\mathrm{T}}$, where $D = \operatorname{diag}(\lambda_1, \lambda_2)$ is the matrix whose diagonal entries are the eigenvalues of H_u. Since both $|\nabla u|$ and κ are invariant with respect to orthogonal transformations in our considerations, we can replace H_u with the diagonal matrix of its eigenvalues. Since H_u is non-singular, we can assume that $\lambda_1 \geq \lambda_2 > 0$. We start with the case $\lambda_1 = \lambda_2$ and without loss of generality set $\lambda_1 = 1$. Then

$$u(x) = \tfrac{1}{2} x x^{\mathrm{T}} = x_1^2 + x_2^2 = r^2.$$

Now consider the curvature energy in a small ball $B_{\epsilon/2}(z)$, and we get

$$\begin{aligned}\int_{B_{\epsilon/2}} \kappa^p |\nabla u|\, dx &= \int_0^{\epsilon/2} \int_{u=\lambda} \kappa^p \, ds\, d\lambda \\ &= \int_0^{\sqrt{\epsilon/2}} 2\pi r \frac{1}{r^p} 2r\, dr \\ &= 4\pi \int_0^{\sqrt{\epsilon/2}} \frac{1}{r^{p-2}}\, dr.\end{aligned}$$

Hence $\mathcal{J}_p$ is finite in $B_{\epsilon/2}$ if and only if $p < 3$. The general result follows from a similar argument for $\lambda_1 > \lambda_2$. ■

5.2.5. A Relaxation of Euler's Elastica

A relaxation of Masnou and Morel's approach of level-line-based disocclusion [MM98] presented in Section 5.2 has been proposed by Ballester et al. [BBC+01]. By weakly decoupling the interpolant u from the vector field of the directions of level lines of u, they developed a well-defined interpolation model of the following form.

Let $D \subset \Omega \subset \mathbb{R}^2$ be the inpainting domain and the image domain, respectively. Further, let D_ϵ denote an $\epsilon-$extension of D such that $D \subset D_\epsilon$ (see Figure 5.4). We assume D, D_ϵ to be open domains with Lipschitz boundary and that D_ϵ does not touch the boundary of Ω. The following inpainting algorithm will be acting on the extension D_ϵ of the inpainting domain, where interpolants u are constrained to share certain image features derived from the given image g in the band $B = D_\epsilon \setminus \bar{D}$. These image features are the grey values g in

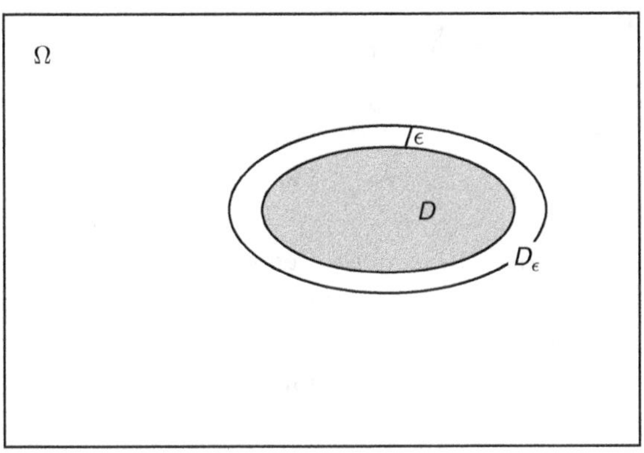

Figure 5.4. The domain setup for the joint interpolation algorithm.

B as well as the direction $\theta_g \in \mathbb{R}^2$ orthogonal to the level lines defined by $\theta_g \cdot \nabla g = |\nabla g|$ and $|\theta_g| \leq 1$. Then the interpolant u and the vector field θ that encodes the gradient information ∇u are computed under the preceding constraints as minimisers of the functional

$$\mathcal{J}(u,\theta) = \int_{D_\epsilon} |\text{div}\theta|^p (\alpha + \beta |\nabla u_\sigma|)\, dx + \lambda \int_{D_\epsilon} (|\nabla u| - \theta \cdot \nabla u)\, dx, \qquad (5.14)$$

where $u_\sigma = \rho_\sigma * u$ denotes the image function u convolved with a mollifier of width $\sigma \geq 0$ as in (5.12). Note that if we set $\sigma = 0$ and impose $\theta = \nabla u / |\nabla u|$, the second term in the functional would vanish, and we are back to the smooth level-line interpolation of Masnou and Morel [MM98]. By introducing the normalised gradient of u as an additional unknown in the interpolation problem, the curvature term $\text{div}(\nabla u / |\nabla u|)$ appears only in a regularised form $\text{div}(\theta)$. This results in a relaxed version of the original model, for which a beautiful existence analysis can be carried out (see the following paragraph). Moreover, connections can be seen to the numerical approach of operator splitting methods discussed later in this book.

Preliminaries Most of the preliminary results we need in this section can be found in a more detailed discussion in [Anz83]. As before, we denote by $BV(\Omega)$, Ω an open subset of $\mathbb{R}^d$ with Lipschitz boundary, the space of functions of bounded variation on Ω. Recall that for a function $u \in BV(\Omega)$, almost all of its level sets $\Gamma_\lambda = \{x \in \Omega : u(x) \geq \lambda\}$ are sets of finite perimeter [i.e., its indicator function $\mathbf{1}_{\Gamma_\lambda} \in BV(\Omega)$]. Moreover, we can define the essential boundary $\partial^* \Gamma_\lambda$ on which we can compute for $\mathcal{H}^{d-1}$ almost all points $x \in \partial^* \Gamma_\lambda$ of the normal vector $\theta(x)$. The latter formally satisfies $\theta \cdot \nabla u = |\nabla u|$ and $|\theta| \leq 1$ almost everywhere. For the upcoming well-posedness analysis of (5.14), we introduce the following spaces:

$$W^{1,p}(\text{div}, \Omega) = \{\theta \in L^p(\Omega)^d : \text{div}(\theta) \in L^p(\Omega)\}, \quad 1 \leq p < \infty,$$

and the larger space

$$BV(\text{div}, \Omega) = \{\theta \in L^1(\Omega)^d : \text{div}(\theta) \text{ is a Radon measure in } \Omega\}.$$

For a vector $\theta \in W^{1,p}(\text{div}, \Omega)$, its normal component with respect to ∂D_ϵ, that is, $\theta \cdot \eta|_{\partial D_\epsilon}$ is well defined. To introduce the right framework for dealing with the second term in $\mathcal{J}$ in (5.14), we have to introduce some additional concepts. Let $p, p' \geq 1$ with $1/p + 1/p' = 1$, and define the space of bounded vector fields with divergence in L^p as

$$V(\Omega)_p = \{z \in L^\infty(\Omega, \mathbb{R}^d) : \text{div} z \in L^p(\Omega)\}.$$

Then, for $z \in V(\Omega)_p$ and $w \in BV(\Omega) \cap L^{p'}(\Omega)$, we define a linear functional (z, Dw) by

$$\langle (z, Dw), \varphi \rangle = -\int_\Omega w\varphi \mathrm{div} z \, dx - \int_\Omega wz \cdot \nabla \varphi \, dx, \quad \text{for } \varphi \in C_c^\infty(\Omega).$$

Then (z, Dw) is a Radon measure in Ω, and for any Borel subset $E \subset \Omega$, we have

$$\left| \int_E (z, Dw) \right| \leq \int_E |(z, Dw)| \leq \|z\|_\infty \int_E |Dw|. \tag{5.15}$$

Moreover, we have the following integration-by-parts formula:

Theorem 5.2.9 *For $z \in V(\Omega)_p$, there exists a linear operator $\gamma : V(\Omega)_p \to L^\infty(\partial\Omega)$ such that $\|\gamma(z)\|_{L^\infty(\partial\Omega)} \leq \|z\|_{L^\infty(\Omega)}$ and $\gamma(z) = z(x) \cdot \vec{n}(x)$ for all $x \in \partial\Omega$ if $z \in C^1(\bar{\Omega}, \mathbb{R}^d)$, and $\vec{n}$ denotes the outward-pointing unit normal on $\partial\Omega$. Moreover, for $w \in BV(\Omega) \cap L^{p'}(\Omega)$, the following integration-by-parts formula holds:*

$$\int_\Omega w \, \mathrm{div} z \, dx + \int_\Omega (z, Dw) = \int_{\partial\Omega} \gamma(z) \, w \, d\mathcal{H}^{d-1}.$$

Well-Posedness of the Model With the machinery just attained, we present the well-posedness analysis of the relaxed Euler elastica functional derived in [BBC+01]. For $p > 1$, we consider the following refined version of the minimisation problem in (5.14):

$$\min_{\mathcal{A}} \left\{ \mathcal{J}(u, \theta) = \int_{D_\epsilon} |\mathrm{div}(\theta)|^p (\alpha + \beta |\nabla u_\sigma|) \, dx + \lambda \int_{D_\epsilon} (|Du| - (\theta, Du)) \right\}, \tag{5.16}$$

where $\alpha, \lambda > 0$, $\beta \geq 0$ and $u_\sigma = \rho_\sigma * u$, with ρ_σ as a mollifier of width $\epsilon > \sigma > 0$ as in (5.12) and with

$$\begin{aligned} \mathcal{A} = \big\{ (u, \theta) : \; & \theta \in W^{1,p}(\mathrm{div}, D_\epsilon), \; u \in BV(D_\epsilon), \; u = g \text{ in } B, \\ & \gamma(\theta)|_{\partial D_\epsilon} = \gamma(\theta_g)|_{\partial D_\epsilon}, \text{ and } |\theta| \leq 1, \; |u| \leq \|g\|_{L^\infty(B)} \text{ a.e. in } D_\epsilon \big\}. \end{aligned} \tag{5.17}$$

Here the boundary function γ is as defined in Theorem 5.2.9.

Theorem 5.2.10 *If $p > 1$, there is a minimum $(u, \theta) \in \mathcal{A}$ for problem* (5.16).

Proof Because $|\theta| \leq 1$, we have

$$|Du| - (\theta, Du) \geq 0,$$

and hence, $\mathcal{J}$ is bounded from below by 0. Therefore, we can consider a minimising sequence (θ_n, u_n), that is,

$$\text{For } m := \inf_{(\theta,u)\in\mathcal{A}} \mathcal{J}(u,\theta), \qquad \text{we have that} \qquad \mathcal{J}(u_n,\theta_n) \to m \text{ as } n \to \infty.$$

Since $|\theta_n| \leq 1$ is uniformly bounded, the sequence is weakly compact in $L^q(D_\epsilon)^2$ for $1 < q < \infty$, and therefore (up to a subsequence), θ_n weakly converges to a θ in $L^q(D_\epsilon)^2$ for all $1 < q < \infty$. Moreover, we have

$$\infty > \mathcal{J}(u_n,\theta_n) \geq \int_{D_\epsilon} |\text{div } \theta_n|^p \left(\alpha + \beta|\nabla(u_n)_\sigma|\right) dx \geq C \int_{D_\epsilon} |\text{div } \theta_n|^p \, dx,$$

and hence θ_n weakly converges in $W^{1,p}(\text{div}, D_\epsilon)$. Next, we consider

$$\begin{aligned}\int_{D_\epsilon} (\theta_n, Du_n) &= -\int_{D_\epsilon} \text{div } \theta_n u_n \, dx + \int_{\partial D_\epsilon} \gamma(\theta_n) u_n \, d\mathcal{H}^{d-1} \\ &= -\int_{D_\epsilon} \text{div } \theta_n \cdot u_n \, dx + \int_{\partial D_\epsilon} \gamma(\theta_g) g \, d\mathcal{H}^{d-1},\end{aligned}$$

where we have used the integration-by-parts formula from Theorem 5.2.9. Taking the modulus in the preceding equation, using (5.15) and applying Hölder again, we obtain

$$\left| \int_{D_\epsilon} (\theta_n, Du_n) \right| \leq \|\text{div } \theta_n\|_{L^p} \cdot \|u_n\|_{L^{p'}} + \int_{\partial D_\epsilon} |g|.$$

Moreover, the minimising sequence $(u_n, \theta_n) \in \mathcal{A}$ and hence u_n are uniformly bounded in $L^\infty(D_\epsilon)$. Therefore, from the preceding inequality we get that $\left|\int_{D_\epsilon} (\theta_n, Du_n)\right|$ is uniformly bounded and therefore also that $\int_{D_\epsilon} |Du_n|$ is uniformly bounded. Hence, u_n is uniformly bounded in $BV(D_\epsilon)$, which can be compactly embedded in $L^1(D_\epsilon)$. It follows that u_n converges (up to a subsequence) strongly in $L^1(D_\epsilon)$ and, in fact, (because of the L^∞ bound) converges strongly in $L^q(D_\epsilon)$ for all $1 \leq q < \infty$. This gives that $\nabla(u_n)_\sigma$ converges uniformly to $\nabla(u)_\sigma$ in D_ϵ. In particular, from these convergences we get the following lower semi-continuity properties:

$$\int_{D_\epsilon} |\text{div } \theta|^p \left(\alpha + \beta|\nabla u_\sigma|\right) dx \leq \liminf_n \int_{D_\epsilon} |\text{div } \theta_n|^p \left(\alpha + \beta|\nabla(u_n)_\sigma|\right) dx$$

$$\int_{D_\epsilon} |Du| \leq \liminf_n \int_{D_\epsilon} |Du_n|.$$

Moreover,

$$\left|\int_{D_\epsilon} (\operatorname{div}\theta_n u_n - \operatorname{div}\theta u)\,dx\right| \le \left|\int_{D_\epsilon} \operatorname{div}\theta_n (u_n - u)\,dx\right| + \left|\int_{D_\epsilon} \operatorname{div}(\theta_n - \theta) u\,dx\right|$$

$$\le \underbrace{\|\operatorname{div}\theta_n\|_{L^p}}_{\le C} \cdot \underbrace{\|u_n - u\|_{L^{p'}}}_{\xrightarrow{n\to\infty} 0} + \underbrace{\left|\int_{D_\epsilon} \operatorname{div}(\theta_n - \theta) u\,dx\right|}_{\xrightarrow{n\to\infty} 0}$$

and hence

$$\lim_n \int_{D_\epsilon} (\theta_n, Du_n) = \lim_n -\int_{D_\epsilon} \operatorname{div}\theta_n \cdot u_n\,dx + \int_{\partial D_\epsilon} \gamma(\theta_g) g\,d\mathcal{H}^{d-1}$$

$$= -\int_{D_\epsilon} \operatorname{div}\theta \cdot u\,dx + \int_{\partial D_\epsilon} \gamma(\theta_g) g\,d\mathcal{H}^{d-1} = \int_{D_\epsilon} (\theta, Du).$$

In sum, we get

$$\mathcal{J}(u,\theta) \le \liminf_n \mathcal{J}(u_n,\theta_n)$$

and hence (u,θ) is a minimiser of $\mathcal{J}$. Since the admissible set $\mathcal{A}$ is a convex and closed subset of $W^{1,p}(\operatorname{div}, D_\epsilon) \times BV(D_\epsilon)$, a minimiser is also admissible. ∎

5.2.6. Numerical discussion

The numerical solution of Euler elastica inpainting is known to be notoriously difficult. This is due to the non-convexity of (5.5) and its functionalised formulation (5.11), as well as its differential order – resulting in a Euler-Lagrange equation that constitutes a fourth-order non-linear PDE. In [MM98], the authors consider the level set formulation (5.8) and find the inpainted image u as the collection of all its inpainted level lines. There the inpainting problem (5.8) is solved for each level line separately via a quadratic programming approach. As learned already earlier, such an approach needs careful realisation that prevents inpainted level lines from crossing each other and leaving untreated holes in the inpainting domain. The functionalised formulation (5.11), however, does not have these issues. Nevertheless, minimising (5.11) requires the optimisation of a second-order non-convex functional, which is problematic in itself. In their seminal paper [CKS02], Chan, Kang and Shen computed the Euler elastica inpainted image as a solution of the corresponding Euler-Lagrange equation

$$\nabla_u R + \lambda \mathbb{1}_{\Omega\setminus D}(g-u) = 0, \quad \text{in } \Omega,$$

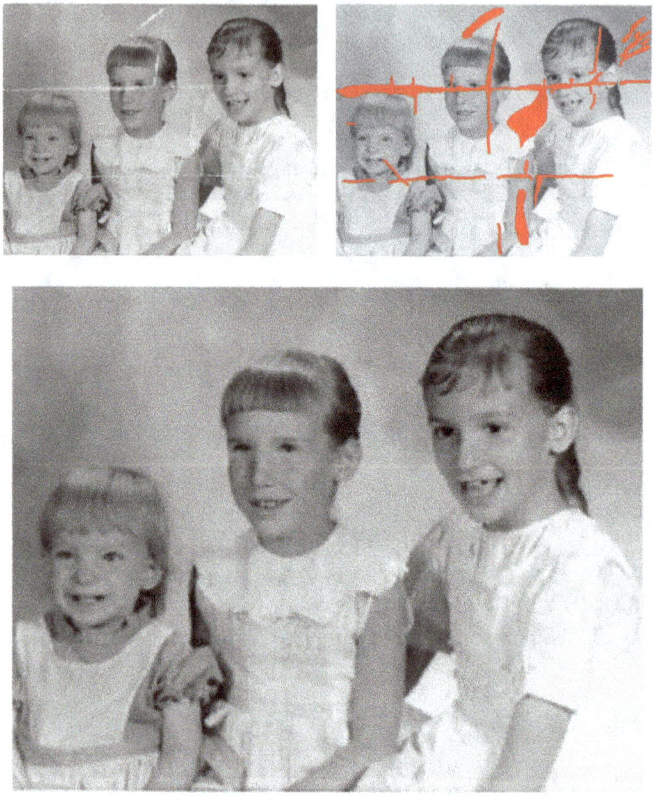

Figure 5.5. Euler elastica inpainting of an old photograph with $\alpha = 1$ and $\beta = 10$, computed using [THC11]. Top: Old photograph and photograph with the inpainting mask coloured in red. Bottom: Inpainted image. Photo courtesy of Marcelo Bertalmio.

where $\nabla_u R$ is the variational derivative of the Euler elastica functional given in Corollary 5.2.6, and $\lambda \mathbb{1}_{\Omega \setminus D}(g-u)$ is a fitting term that for large parameter λ keeps the inpainted image u close to the given one in $\Omega \setminus D$. This equation is solved numerically by employing a steepest-descent approach with an explicit Euler scheme which, due to the non-linearity and fourth differential order, can only progress extremely slowly. For computation of the subsequent numerical examples in Figures (5.5)–(5.8), we have used a recent approach by Hahn et al. [THC11] in which the authors solved the minimisation problem for the Euler elastica energy (5.11) with a strict constraint $u = g$ in $\Omega \setminus D$ by an elaborate augmented Lagrangian and an operator splitting approach. For the details on their implementation, we refer to [THC11]. Let us just note here that while this approach is able to give qualitatively excellent results in a computationally

Figure 5.6. Euler elastica inpainting with 60 per cent missing pixels using $\alpha = 1$ and $\beta = 10$, computed using [THC11]. Top: Original photograph and photograph with the inpainting mask coloured in red. Bottom: Inpainted image.

very efficient way, it has its caveats owing to the difficulty of finding a choice of parameters that work. Let us also mention that more recent approaches aim to compute minimisers of a convex approximation of the Euler elastica energy which produce very promising first results (see, e.g., [BPW13]).

5.3. Cahn-Hilliard and TV-H^{-1} Inpainting

The Cahn-Hilliard equation is a non-linear fourth-order diffusion equation originating in material science for modelling phase separation and phase coarsening in binary alloys. Another approach in the class of fourth-order inpainting algorithms is inpainting of binary images using a modified Cahn-Hilliard

Figure 5.7. Euler elastica inpainting with $\alpha = 0.1$ and $\beta = 1$, computed using [THC11]. Left: Broken line with the inpainting mask coloured in red. Right: Inpainted line.

Figure 5.8. Euler elastica inpainting compared to TV inpainting with 70 per cent missing pixels. Top: The original image (left) and the image with the inpainting mask coloured in red (right). Bottom: TV inpainting from Section 4.3 (left). (Photo courtesy of Kostas Papafitsoros.) Euler elastica inpainting using $\alpha = 1$ and $\beta = 20$, computed using [THC11] (right).

equation, as proposed by Bertozzi, Esedoglu and Gillette [BEG07]. The inpainted version u of $f \in L^2(\Omega)$ assumed with any (trivial) extension to the inpainting domain is constructed by following the evolution of

$$u_t = \Delta\left(-\epsilon\Delta u + \frac{1}{\epsilon}F'(u)\right) + \lambda \mathbb{1}_{\Omega\setminus D}(f-u), \quad \text{in } \Omega, \tag{5.18}$$

where $F(u)$ is a so-called double-well potential, for example, $F(u) = u^2(u-1)^2$, and $\lambda \gg 1$. The two wells of F correspond to values of u that are taken by most of the grey scale values. By choosing a potential with wells at the values 0 (black) and 1 (white), (5.18) therefore provides a simple model for the inpainting of binary images. The parameter $\epsilon > 0$ determines the steepness of the transition between 0 and 1.

The Cahn-Hilliard equation is a relatively simple fourth-order PDE used for this task rather than more complex models involving curvature terms such as Euler elastica inpainting, which we discussed in the preceding section. In fact, the numerical solution of (5.18) was shown to be of at least an order of magnitude faster than competing inpainting models (cf. [BEG07]). Still, the Cahn-Hilliard inpainting model has many of the desirable properties of curvature-based inpainting models such as the smooth continuation of level lines into the missing domain. In fact, the mainly numerical paper [BEG07] was followed by a very careful analysis of (5.18) in [BEglG07]. Therein the authors proved that in the limit $\lambda \to \infty$, a stationary solution of (5.18) solves

$$\begin{aligned} \Delta\left(\epsilon\Delta u - \frac{1}{\epsilon}F'(u)\right) &= 0, && \text{in } D,\\ u &= f, && \text{on } \partial D,\\ \nabla u &= \nabla f, && \text{on } \partial D, \end{aligned} \tag{5.19}$$

for f regular enough ($f \in C^2$). This, once more, supports the claim that fourth-order methods are superior to second-order methods with respect to a smooth continuation of the image contents into the missing domain. Whereas the fact that Cahn-Hilliard inpainting can be used for inpainting of binary images only might seem a bit restrictive, combined with other inpainting strategies, it can constitute a very powerful method for inpainting of the structural part of an image. We will pursue this idea further when discussing two applications in Sections 9.1 and 9.2.

An extension of Cahn-Hilliard inpainting to grey value images is discussed in [BHS09], where the authors call this approach TV-H^{-1} *inpainting*. It is motivated by a Γ-convergence result [Bra02] that says that the sharp-interface limit of the Cahn-Hilliard energy (i.e., as $\epsilon \to 0$) Γ converges to the total variation of a binary function (compare Theorem 5.3.6). The generalisation to grey value images is then realised by using higher-order sub-gradients of the TV functional within the inpainting flow, which leads to structure inpainting with smooth curvature of level sets. More precisely, the TV-H^{-1} inpainted image u of $f \in L^2(\Omega)$ is a stationary solution of

$$u_t = \Delta p + \lambda \mathbb{1}_{\Omega \setminus D}(f - u), \quad p \in \partial TV_\infty(u), \tag{5.20}$$

with

$$TV_\infty(u) = \begin{cases} |Du|(\Omega), & \text{if } |u(x)| \leq 1 \text{ a.e. in } \Omega, \\ +\infty, & \text{otherwise.} \end{cases} \tag{5.21}$$

Here $\partial TV_\infty(u)$ denotes the subdifferential of the functional $TV(u)$; that is, $\partial \mathcal{J}(u) = \{p \in X^* \mid \langle v - u, p \rangle \leq \mathcal{J}(v) - \mathcal{J}(u), \forall v \in X\}$, where here we can take $X = X^* = L^2(\Omega)$. The L^∞ bound in the definition of the TV functional (5.21) is quite natural as we are only considering digital images u whose grey value can be scaled to $[-1, 1]$. It is further motivated by the Γ-convergence result of Theorem 5.3.6, which is discussed in the Section 5.3.2.

5.3.1. Cahn-Hilliard Inpainting: Well-Posedness and Numerical Discussion

In [BEglG07], the authors proved the global existence of a unique weak solution of the evolution equation (5.18). More precisely, for λ sufficiently large, the solution u is proven to be an element in $C([0,T]; L^2(\Omega)) \cap L^2([0,T]; V)$, where $V = \left\{\phi \in H^2(\Omega) \mid \partial\phi/\partial\nu = 0 \text{ on } \partial\Omega\right\}$, and ν is the outward-pointing normal on $\partial\Omega$. Their proof is mainly based on a priori estimates on the solution and Grönwall's lemma. We do not go into more detail on their proof here but rather present the existence analysis for a stationary solution of (5.18), which indeed guarantees the existence of an inpainted image. In [BEglG07], the solution of the stationary equation

$$\Delta\left(-\epsilon\Delta u + \frac{1}{\epsilon}F'(u)\right) + \lambda \mathbb{1}_{\Omega \setminus D}(f - u) = 0, \quad \text{in } \Omega \tag{5.22}$$

remains unaddressed. The difficulty in dealing with the stationary equation is the lack of an energy functional for (5.18); that is, the modified Cahn-Hilliard equation (5.18) cannot be easily represented by a gradient flow of an energy functional over a certain Banach space. In fact, the most evident variational approach would be to minimise the functional

$$\int_\Omega \left(\frac{\epsilon}{2}|\nabla u|^2 + \frac{1}{\epsilon}F(u)\right) dx + \frac{1}{2}\left\|\lambda \mathbb{1}_{\Omega \setminus D}(u - f)\right\|_{-1}^2, \tag{5.23}$$

where we denote by $H^{-1}(\Omega)$ the dual space of $H_0^1(\Omega)$ with corresponding norm $\|\cdot\|_{-1}$. For a function $f \in H^{-1}(\Omega)$, the norm is defined as

$$\|f\|_{-1}^2 = \left\|\nabla\Delta^{-1}f\right\|_2^2 = \int_\Omega \left(\nabla\Delta^{-1}f\right)^2\, dx.$$

Thereafter, the operator Δ^{-1} denotes the inverse to the negative Dirichlet-Laplacian; that is, $u = \Delta^{-1}f$ is the unique solution to

$$\begin{aligned} -\Delta u &= f, && \text{in } \Omega, \\ u &= 0, && \text{on } \partial\Omega. \end{aligned}$$

Now the minimisation problem (5.23) exhibits the optimality condition

$$0 = -\epsilon\Delta u + \frac{1}{\epsilon}F'(u) + \lambda\mathbb{1}_{\Omega\setminus D}\Delta^{-1}\left(\lambda\mathbb{1}_{\Omega\setminus D}(u-f)\right),$$

which splits into

$$\begin{aligned} 0 &= -\epsilon\Delta u + \frac{1}{\epsilon}F'(u), && \text{in } D, \\ 0 &= -\epsilon\Delta u + \frac{1}{\epsilon}F'(u) + \lambda^2\Delta^{-1}(u-f), && \text{in } \Omega\setminus D. \end{aligned}$$

Hence the minimisation of (5.23) translates into a second-order diffusion inside the inpainting domain D, whereas a solution of (5.22) fulfils

$$\begin{aligned} 0 &= \Delta\left(-\epsilon\Delta u + \frac{1}{\epsilon}F'(u)\right), && \text{in } D, \\ 0 &= \Delta\left(-\epsilon\Delta u + \frac{1}{\epsilon}F'(u)\right) + \lambda(f-u), && \text{in } \Omega\setminus D. \end{aligned}$$

This problem arises because the fidelity term $\lambda\mathbb{1}_{\Omega\setminus D}(f-u)$ is not symmetric with respect to the H^{-1} inner product. However, making use of fixed-point arguments in [BHS09], the authors were indeed able to prove the existence of a stationary solution.

Theorem 5.3.1 *Equation* (5.22) *admits a weak solution in* $H^1(\Omega)$ *provided that* $\lambda \geq C/\epsilon^3$ *for a positive constant* C *depending on* $|\Omega|, |D|$ *and* F *only.*

Remark 5.3.2 In the numerical examples in [BHS09], we can see that the condition $\lambda \geq C/\epsilon^3$ in Theorem 5.3.1 is naturally fulfiled, since in order to obtain good visual results in inpainting approaches, λ has to be chosen rather large in general (cf. Figure 5.10). Note that the same condition also appears in [BEglG07] where it is needed to prove the global existence of solutions of (5.18).

Existence of a Stationary Solution Let us proceed by proving the existence of a weak solution of the stationary equation (5.22); that is, we will verify Theorem 5.3.1. Let $\Omega \subset \mathbb{R}^2$ be a bounded Lipschitz domain and $f \in L^2(\Omega)$ given. In order to be able to impose boundary conditions in the equation, we assume f to be constant in a small neighbourhood of $\partial\Omega$. This assumption is for technical purposes only and does not influence the inpainting process as long as the inpainting domain D does not touch the boundary of the image domain Ω. Instead of Neumann boundary data, as in the original Cahn-Hilliard inpainting approach (cf. [BEglG07]), we use Dirichlet boundary conditions for our analysis; that is, we consider

$$\begin{cases} u_t = \Delta\left(-\epsilon\Delta u + \frac{1}{\epsilon}F'(u)\right) + \lambda\mathbb{1}_{\Omega\setminus D}(f-u), & \text{in } \Omega, \\ u = f, -\epsilon\Delta u + \frac{1}{\epsilon}F'(u) = 0, & \text{on } \partial\Omega. \end{cases} \tag{5.24}$$

This change from a Neumann to a Dirichlet problem makes it easier to deal with the boundary conditions in our proofs but does not have a significant impact on the inpainting process as long as we assume that $\overline{D} \subset \Omega$. In the appendix of [BHS09], the authors nevertheless proposed a setting to extend the presented analysis for (5.18) to the originally proposed model with Neumann boundary data. In the current setting, we define a weak solution of (5.22) as a function $u \in H = \{u \in H^1(\Omega),\ u|_{\partial\Omega} = f|_{\partial\Omega}\}$ that fulfils

$$\langle \epsilon\nabla u, \nabla\phi\rangle_2 + \left\langle \frac{1}{\epsilon}F'(u), \phi\right\rangle_2 - \langle \lambda\mathbb{1}_{\Omega\setminus D}(f-u), \phi\rangle_{-1} = 0, \quad \forall \phi \in H_0^1(\Omega). \tag{5.25}$$

Remark 5.3.3 With $u \in H^1(\Omega)$ and the compact embedding $H^1(\Omega) \hookrightarrow\hookrightarrow L^q(\Omega)$ for every $1 \leq q < \infty$ and $\Omega \subset \mathbb{R}^2$, the weak formulation is well defined.

To see that (5.25) defines a weak formulation for (5.22) with Dirichlet boundary conditions, we integrate by parts in (5.25) and get

$$\begin{aligned} &\int_\Omega \left(-\epsilon\Delta u + \frac{1}{\epsilon}F'(u) - \Delta^{-1}\left(\lambda\mathbb{1}_{\Omega\setminus D}(f-u)\right)\right)\phi\, dx \\ &\quad - \int_{\partial\Omega} \Delta^{-1}\left(\lambda\mathbb{1}_{\Omega\setminus D}(f-u)\right)\nabla\Delta^{-1}\phi\cdot\nu\, d\mathcal{H}_1 = 0, \qquad \forall\phi \in H_0^1(\Omega). \end{aligned}$$

Since the preceding equality holds for all $\phi \in H_0^1(\Omega)$, it holds in particular for all ϕ in the subset $H_0^1(\Omega) \cap \{\nabla\Delta^{-1}\phi\cdot\nu = 0 \text{ on } \partial\Omega\}$. This yields

$$\begin{cases} \epsilon \Delta u - \frac{1}{\epsilon} F'(u) + \Delta^{-1}\left(\lambda \mathbb{1}_{\Omega \setminus D}(f-u)\right) = 0, & \text{in } \Omega, \\ \Delta^{-1}\left(\lambda \mathbb{1}_{\Omega \setminus D}(f-u)\right) = 0, & \text{on } \partial\Omega. \end{cases} \tag{5.26}$$

Assuming sufficient regularity on u, we can use the definition of Δ^{-1} to see that u solves

$$\begin{cases} -\epsilon \Delta \Delta u + \frac{1}{\epsilon} \Delta F'(u) + \lambda \mathbb{1}_{\Omega \setminus D}(f-u) = 0, & \text{in } \Omega, \\ \Delta^{-1}\left(\lambda \mathbb{1}_{\Omega \setminus D}(f-u)\right) = -\epsilon \Delta u + \frac{1}{\epsilon} F'(u) = 0, & \text{on } \partial\Omega, \end{cases}$$

Since additionally $u|_{\partial\Omega} = f|_{\partial\Omega}$, the function u solves (5.22) with Dirichlet boundary conditions.

For the proof of existence of a solution to (5.25), we take the following approach: we consider the fixed-point operator $\mathcal{A}$: $L^2(\Omega) \to L^2(\Omega)$, where $\mathcal{A}(v) = u$ fulfils for a given $v \in L^2(\Omega)$ the equation

$$\begin{cases} \frac{1}{\tau}\Delta^{-1}(u-v) = \epsilon \Delta u - \frac{1}{\epsilon} F'(u) + \Delta^{-1} \\ \qquad \times \left[\lambda \mathbb{1}_{\Omega \setminus D}(f-u) \quad + \lambda(1-\mathbb{1}_{\Omega \setminus D})(v-u)\right], & \text{in } \Omega, \\ u = f, \Delta^{-1}\left(\frac{1}{\tau}(u-v) - \lambda \mathbb{1}_{\Omega \setminus D}(f-u) - \lambda(1-\mathbb{1}_{\Omega \setminus D})(v-u)\right) = 0, & \text{on } \partial\Omega, \end{cases} \tag{5.27}$$

where $\tau > 0$ is a parameter. The boundary conditions of $\mathcal{A}$ are given by the second equation in (5.27). Note that actually the solution u will be in $H^1(\Omega)$, and hence the boundary condition is well defined in the trace sense, the operator $\mathcal{A}$ into $L^2(\Omega)$ then is obtained with further embedding. We define a weak solution of (5.27) as before by a function $u \in H = \left\{u \in H^1(\Omega),\, u|_{\partial\Omega} = f|_{\partial\Omega}\right\}$ that fulfils

$$\left\langle \frac{1}{\tau}(u-v), \phi \right\rangle_{-1} + \langle \epsilon \nabla u, \nabla \phi \rangle_2 + \left\langle \frac{1}{\epsilon} F'(u), \phi \right\rangle_2 - \langle \lambda \mathbb{1}_{\Omega \setminus D}(f-u)$$
$$+ \lambda(1-\mathbb{1}_{\Omega \setminus D})(v-u), \phi \rangle_{-1} = 0, \quad \forall \phi \in H_0^1(\Omega). \tag{5.28}$$

A fixed point of the operator $\mathcal{A}$, provided that it exists, then solves the stationary equation with Dirichlet boundary conditions as in (5.26).

Note that in (5.27) the indicator function $\lambda \mathbb{1}_{\Omega\setminus D}$ in the fitting term $\lambda \mathbb{1}_{\Omega\setminus D}(f-u)+\lambda(1-\mathbb{1}_{\Omega\setminus D})(v-u)=\lambda(v-u)+\lambda \mathbb{1}_{\Omega\setminus D}(f-v)$ only appears in combination with given functions f,v and is not combined with the solution u of the equation. For (5.27), that is, (5.28), we can therefore state a variational formulation. This is, for a given $v \in L^2(\Omega)$, that (5.27) is the Euler-Lagrange equation of the minimisation problem

$$u^* = \underset{u\in H^1(\Omega), u|_{\partial\Omega}=f|_{\partial\Omega}}{\operatorname{argmin}} \mathcal{J}^\epsilon(u,v) \tag{5.29}$$

with

$$\begin{aligned} \mathcal{J}^\epsilon(u,v) = & \int_\Omega \left(\frac{\epsilon}{2}|\nabla u|^2 + \frac{1}{\epsilon}F(u)\right) dx + \frac{1}{2\tau}\|u-v\|_{-1}^2 \\ & + \frac{\lambda}{2}\left\|u - \frac{\lambda \mathbb{1}_{\Omega\setminus D}}{\lambda}f - \left(1-\mathbb{1}_{\Omega\setminus D}\right)v\right\|_{-1}^2. \end{aligned} \tag{5.30}$$

We are going to use the variational formulation (5.30) to prove that (5.27) admits a weak solution in $H^1(\Omega)$. This solution is unique under additional conditions.

Proposition 5.3.4 *Equation* (5.27) *admits a weak solution in* $H^1(\Omega)$ *in the sense of* (5.28). *For* $\tau \le C\epsilon^3$, *where* C *is a positive constant depending on* $|\Omega|, |D|$ *and* F *only, the weak solution of* (5.27) *is unique.*

Further, we prove that operator $\mathcal{A}$ admits a fixed point under certain conditions.

Proposition 5.3.5 *Set* $\mathcal{A}$: $L^2(\Omega) \to L^2(\Omega)$, $\mathcal{A}(v) = u$, *where* $u \in H^1(\Omega)$ *is the unique weak solution of* (5.27). *Then* $\mathcal{A}$ *admits a fixed point* $\hat{u} \in H^1(\Omega)$ *if* $\tau \le C\epsilon^3$ *and* $\lambda \ge \frac{C}{\epsilon^3}$ *for a positive constant* C *depending on* $|\Omega|, |D|$ *and* F *only.*

Hence, the existence of a stationary solution of (5.18) follows under the condition $\lambda \ge C/\epsilon^3$. We begin with considering the fixed-point equation (5.27), that is, the minimisation problem (5.29). In the following, we prove the existence of a unique weak solution of (5.27) by showing the existence of a unique minimiser for (5.30).

Proof of Proposition 5.3.4 We want to show that $\mathcal{J}^\epsilon(u,v)$ has a minimiser in $H = \{u \in H^1(\Omega),\, u|_{\partial\Omega} = f|_{\partial\Omega}\}$. For this we consider a minimising sequence

$u_n \in H$ of $\mathcal{J}^\epsilon(u,v)$. To see that u_n is uniformly bounded in $H^1(\Omega)$, we show that $\mathcal{J}^\epsilon(u,v)$ is coercive in $H^1(\Omega)$. With $F(u) \geq C_1(u-C)^2 - C_2$ for a constant C and positive constants $C_1, C_2 > 0$ and the triangular inequality in the $H^{-1}(\Omega)$ space, we obtain

$$\mathcal{J}^\epsilon(u,v) \geq \frac{\epsilon}{2}\|\nabla u\|_2^2 + \frac{C_1}{\epsilon}\|u\|_2^2 - \frac{C_2}{\epsilon} + \frac{1}{2\tau}\left(\frac{1}{2}\|u\|_{-1}^2 - \|v\|_{-1}^2\right)$$

$$+\frac{\lambda}{2}\left(\frac{1}{2}\|u\|_{-1}^2 - \left\|\mathbb{1}_{\Omega\setminus D}f + (1-\mathbb{1}_{\Omega\setminus D})v\right\|_{-1}^2\right)$$

$$\geq \frac{\epsilon}{2}\|\nabla u\|_2^2 + \frac{C_1}{\epsilon}\|u\|_2^2 + \left(\frac{\lambda}{4} + \frac{1}{4\tau}\right)\|u\|_{-1}^2 - C_3(v,f,\lambda,\epsilon,\Omega,D).$$

Therefore, a minimising sequence u_n is bounded in $H^1(\Omega)$, and it follows that $u_n \rightharpoonup u^*$ up to a subsequence in $H^1(\Omega)$. To finish the proof of existence for (5.27), we have to show that $\mathcal{J}^\epsilon(u,v)$ is weakly lower semi-continuous in $H^1(\Omega)$. For this we divide the sequence $\mathcal{J}^\epsilon(u_n,v)$ of (5.30) in two parts. We denote the first term by

$$a_n = \underbrace{\int_\Omega \left(\frac{\epsilon}{2}|\nabla u_n|^2 + \frac{1}{\epsilon}F(u_n)\right)\,dx}_{CH(u_n)}$$

and the second term by

$$b_n = \underbrace{\frac{1}{2\tau}\|u_n - v\|_{-1}^2}_{D(u_n,v)} + \underbrace{\frac{\lambda}{2}\left\|u_n - \mathbb{1}_{\Omega\setminus D}f - (1-\mathbb{1}_{\Omega\setminus D})v\right\|_{-1}^2}_{FIT(u_n,v)}.$$

Since $H^1 \hookrightarrow\hookrightarrow L^2$, it follows $u_n \to u^*$ up to a subsequence in $L^2(\Omega)$. Further, we know that if $\lim_n b_n$ exists, then

$$\liminf(a_n + b_n) = \liminf a_n + \lim b_n. \tag{5.31}$$

We begin with consideration of the last term in (5.30). We denote $\tilde{f} := \mathbb{1}_{\Omega\setminus D}f + (1-\mathbb{1}_{\Omega\setminus D})v$. We want to show that

$$\left\|u_n - \tilde{f}\right\|_{-1}^2 \longrightarrow \left\|u^* - \tilde{f}\right\|_{-1}^2$$

or, equivalently, that

$$\langle \Delta^{-1}(u_n - \tilde{f}), u_n - \tilde{f}\rangle_2 \longrightarrow \langle \Delta^{-1}(u^* - \tilde{f}), u^* - \tilde{f}\rangle_2.$$

For this we consider the absolute difference of the two terms

$$
\begin{aligned}
&|\langle \Delta^{-1}(u_n - \tilde{f}), u_n - \tilde{f}\rangle_2 - \langle \Delta^{-1}(u^* - \tilde{f}), u^* - \tilde{f}\rangle_2| \\
&= |\langle \Delta^{-1}(u_n - u^*), u_n - \tilde{f}\rangle_2 - \langle \Delta^{-1}(u^* - \tilde{f}), u_n - u^*\rangle_2| \\
&\le |\langle u_n - u^*, \Delta^{-1}(u_n - \tilde{f})\rangle_2| + |\langle \Delta^{-1}(u^* - \tilde{f}), u^* - u_n\rangle_2| \\
&\le \underbrace{\left\| u_n - u^* \right\|_2}_{\to 0} \cdot \left\| \Delta^{-1}(u_n - \tilde{f}) \right\|_2 + \underbrace{\left\| u_n - u^* \right\|_2}_{\to 0} \cdot \left\| \Delta^{-1}(u^* - \tilde{f}) \right\|_2 .
\end{aligned}
$$

Since the operator $\Delta^{-1} : H^{-1}(\Omega) \to H_0^1(\Omega)$ is linear and continuous, it follows that

$$
\left\| \Delta^{-1} w \right\|_2 \le \left\| \Delta^{-1} \right\| \cdot \|w\|_2, \quad \text{for all } w \in H^{-1}(\Omega).
$$

Thus

$$
\begin{aligned}
&|\langle \Delta^{-1}(u_n - \tilde{f}), u_n - \tilde{f}\rangle_2 - \langle \Delta^{-1}(u^* - \tilde{f}), u^* - \tilde{f}\rangle_2| \\
&\le \underbrace{\left\| u_n - u^* \right\|_2}_{\to 0} \underbrace{\left\| \Delta^{-1} \right\|}_{\text{const}} \underbrace{\left\| u_n - \tilde{f} \right\|_2}_{\text{bounded}} + \underbrace{\left\| u_n - u^* \right\|_2}_{\to 0} \underbrace{\left\| \Delta^{-1} \right\|}_{\text{const}} \underbrace{\left\| u^* - \tilde{f} \right\|_2}_{\text{const}} \\
&\longrightarrow 0 \quad \text{as} \quad n \to \infty,
\end{aligned}
$$

and we conclude that $FIT(u_n, v)$ converges to $FIT(u^*, v)$. With the same argument, it follows that $D(u_n, v)$ converges and, consequently, that the sequence b_n converges in $L^2(\Omega)$. Further, $CH(\cdot)$ is weakly lower semi-continuous, which follows from the lower semi-continuity of the Dirichlet integral and from the continuity of F by applying Fatou's lemma. Hence, we obtain

$$
\mathcal{J}^\epsilon(u^*, v) \le \liminf \mathcal{J}^\epsilon(u_n, v).
$$

Therefore, $\mathcal{J}^\epsilon$ has a minimiser in H^1, that is,

$$
\exists u^* \text{ with } u^* = \underset{u \in H^1(\Omega)}{\operatorname{argmin}} \mathcal{J}^\epsilon(u, v).
$$

We next assert that u^* fulfils the boundary condition $u^*|_{\partial\Omega} = f|_{\partial\Omega}$. To see this, note that for an admissible function $w \in H$, $u_n - w \in H_0^1(\Omega)$. Now $H_0^1(\Omega)$ is a closed linear subspace of $H^1(\Omega)$ and so, by Mazur's theorem (cf. [Eva98, appendix D.4], for example), is weakly closed. Hence, $u^* - w \in H_0^1(\Omega)$, and consequently, the trace of u^* on $\partial\Omega$ is equal to f.

For simplicity, let $u = u^*$ in the following. To see that the minimiser u is a weak solution of (5.27), we compute the corresponding Euler-Lagrange equation to the minimisation problem. For this we choose any test function

$\phi \in H_0^1(\Omega)$ and compute the first variation of $\mathcal{J}^\epsilon$, that is,

$$\frac{d}{d\tau}\mathcal{J}^\epsilon(u+\delta\phi, v)\bigg|_{\tau=0},$$

which has to be zero for a minimiser u. Thus, we have

$$\begin{aligned}&\epsilon\,\langle \nabla u, \nabla\phi\rangle_2 + \frac{1}{\epsilon}\langle F'(u),\phi\rangle_2\\&\quad+\left\langle \frac{1}{\tau}(u-v)+\lambda\left[u-\mathbb{1}_{\Omega\setminus D}f-\left(1-\mathbb{1}_{\Omega\setminus D}\right)v\right],\phi\right\rangle_{-1}=0.\end{aligned}$$

Integrating by parts in both terms, we get

$$\begin{aligned}&\left\langle -\epsilon\Delta u+\frac{1}{\epsilon}F'(u)-\Delta^{-1}\left(\frac{1}{\tau}(u-v)+\lambda\left[u-\mathbb{1}_{\Omega\setminus D}f-\left(1-\mathbb{1}_{\Omega\setminus D}\right)v\right]\right),\phi\right\rangle_2\\&\quad+\int_{\partial\Omega}\nabla u\cdot\nu\phi\,ds+\int_{\partial\Omega}\Delta^{-1}\left(\frac{1}{\tau}(u-v)+\lambda\left[u-\mathbb{1}_{\Omega\setminus D}f-\left(1-\mathbb{1}_{\Omega\setminus D}\right)v\right]\right)\\&\quad\times\nabla\Delta^{-1}\phi\cdot\nu\,ds=0.\end{aligned}$$

Since ϕ is an element in $H_0^1(\Omega)$, the first boundary integral vanishes. Further, a minimiser u fulfils the boundary condition $u=f$ on the boundary $\partial\Omega$. Hence, we obtain that u fulfils the weak formulation (5.28) of (5.27).

For the uniqueness of the minimiser, we need to prove that $\mathcal{J}^\epsilon$ is strictly convex. To do so, we prove that for any $u_1,u_2\in H^1(\Omega)$,

$$\mathcal{J}^\epsilon(u_1,v)+\mathcal{J}^\epsilon(u_2,v)-2\mathcal{J}^\epsilon\left(\frac{u_1+u_2}{2},v\right)>0, \tag{5.32}$$

based on an assumption that $F(\cdot)$ satisfies $F(u_1)+F(u_2)-2F(\frac{u_1+u_2}{2})>-C(u_1-u_2)^2$ for a constant $C>0$. An example is when $F(u)=\frac{1}{8}(u^2-1)^2$, $C=\frac{1}{8}$. Denoting $u=u_1-u_2$, we have

$$\begin{aligned}&\mathcal{J}^\epsilon(u_1,v)+\mathcal{J}^\epsilon(u_2,v)-2\mathcal{J}^\epsilon\left(\frac{u_1+u_2}{2},v\right)\\&\quad>\frac{\epsilon}{4}\|u\|_1^2+\left(\frac{1}{4\tau}+\frac{\lambda}{4}\right)\|u\|_{-1}^2-\frac{C}{\epsilon}\|u\|_2^2.\end{aligned}$$

By using the inequality

$$\|u\|_2^2\le\|u\|_1\,\|u\|_{-1} \tag{5.33}$$

and the Cauchy-Schwarz inequality, for (5.32) to be fulfiled, we need

$$2\sqrt{\frac{\epsilon}{4}\left(\frac{1}{4\tau}+\frac{\lambda}{4}\right)}\ge\frac{C}{\epsilon},$$

that is,

$$\epsilon^3\left(\lambda+\frac{1}{\tau}\right)\geq C^2.$$

Therefore, $\mathcal{J}^\epsilon(u,v)$ is strictly convex in u, and our minimisation problem has a unique minimiser if τ is chosen smaller than $C\epsilon^3$ for a constant C depending on $|\Omega|,|D|$ and F only. Because of the convexity of $\mathcal{J}^\epsilon$ in ∇u and u, every weak solution of the Euler-Lagrange equation (5.27) is in fact a minimiser of $\mathcal{J}^\epsilon$. This proves the uniqueness of a weak solution of (5.27) provided that $\tau < C\epsilon^3$. ∎

Next, we want to prove Proposition 5.3.5, that is, the existence of a fixed point of (5.27) and with this the existence of a stationary solution of (5.18). To do so, we are going to apply Schauder's fixed-point theorem.

Proof of Proposition 5.3.5 We consider a solution $\mathcal{A}(v)=u$ of (5.27) with $v\in L^2(\Omega)$ given. In the following, we will prove the existence of a fixed point by using Schauder's fixed-point theorem. We start with proving that

$$\|\mathcal{A}(v)\|_2^2=\|u\|_2^2\leq\beta\,\|v\|_2^2+\alpha \tag{5.34}$$

for constants $\beta<1$ and $\alpha>0$. Having this, we have shown that $\mathcal{A}$ is a map from the closed ball $K=B(0,M)=\left\{u\in L^2(\Omega):\|u\|_2\leq M\right\}$ into itself for an appropriate constant $M>0$. We conclude the proof with showing the compactness of K and the continuity of the fixed-point operator $\mathcal{A}$. From Schauder's theorem, the existence of a fixed point follows.

Let us, for the time being, additionally assume that ∇u and Δu are bounded in $L^2(\Omega)$. Hence, we can multiply (5.27) by $-\Delta u$ and integrate over Ω to obtain

$$-\int_\Omega\Delta u\Delta^{-1}\left[\frac{1}{\tau}(u-v)-\lambda\mathbb{1}_{\Omega\setminus D}(f-u)-\lambda(1-\mathbb{1}_{\Omega\setminus D})(v-u)\right]dx$$
$$=-\epsilon\,\langle\Delta u,\Delta u\rangle_2+\frac{1}{\epsilon}\int_\Omega F'(u)\Delta u\,dx.$$

After integration by parts, we find with the shorthand notation

$$w:=\frac{1}{\tau}(u-v)-\lambda\mathbb{1}_{\Omega\setminus D}(f-u)-\lambda(1-\mathbb{1}_{\Omega\setminus D})(v-u)$$

that

$$\int_\Omega uw\,dx-\int_{\partial\Omega}\left[\nabla u\cdot\nu(\Delta^{-1}w+\frac{1}{\epsilon}F'(u))-u\nabla(\Delta^{-1}w)\cdot\nu\right]d\mathcal{H}_1$$
$$=-\epsilon\,\|\Delta u\|_2^2-\frac{1}{\epsilon}\int_\Omega F''(u)\,|\nabla u|^2\,dx.$$

Now we insert the boundary conditions $\Delta^{-1}w = 0$, $u = f =: f_1$ and $F'(u) = F'(f) = f_2$ on $\partial\Omega$ with constants f_1 and f_2 on the left-hand side, that is,

$$\int_\Omega uw\,dx - \int_{\partial\Omega}\left[\frac{f_2}{\epsilon}\nabla u\cdot\nu - f_1\nabla(\Delta^{-1}w)\cdot\nu\right]d\mathcal{H}_1$$
$$= -\epsilon\,\|\Delta u\|_2^2 - \frac{1}{\epsilon}\int_\Omega F''(u)\,|\nabla u|^2\,dx.$$

An application of Gauss's theorem to the boundary integral implies that

$$\int_{\partial\Omega}\left[\frac{f_2}{\epsilon}\nabla u\cdot\nu - f_1\nabla(\Delta^{-1}w)\cdot\nu\right]d\mathcal{H}_1 = \frac{f_2}{\epsilon}\int_\Omega \Delta u\,dx + f_1\int_\Omega w\,dx,$$

and we get

$$\int_\Omega uw\,dx = -\epsilon\,\|\Delta u\|_2^2 - \frac{1}{\epsilon}\int_\Omega F''(u)\,|\nabla u|^2\,dx + \frac{f_2}{\epsilon}\int_\Omega \Delta u\,dx + f_1\int_\Omega w\,dx.$$

By further applying Young's inequality to the last two terms, we get

$$\int_\Omega uw\,dx \le \left(\frac{f_2\delta}{2\epsilon} - \epsilon\right)\|\Delta u\|_2^2 - \frac{1}{\epsilon}\int_\Omega F''(u)\,|\nabla u|^2\,dx$$
$$+\frac{f_1\delta}{2}\,\|w\|_2^2 + C(f_1,f_2,|\Omega|,\epsilon,\delta)$$

for a constant $\delta > 0$. Using the identity $\lambda\mathbb{1}_{\Omega\setminus D}(f-u) + \lambda(1-\mathbb{1}_{\Omega\setminus D})(v-u) = \lambda\mathbb{1}_{\Omega\setminus D}(f-v) + \lambda(v-u)$ in the definition of w yields

$$\int_\Omega u\cdot\frac{1}{\tau}(u-v)\,dx \le \left(\frac{f_2\delta}{2\epsilon} - \epsilon\right)\|\Delta u\|_2^2 - \frac{1}{\epsilon}\int_\Omega F''(u)\,|\nabla u|^2\,dx + \frac{f_1\delta}{2}\,\|w\|_2^2$$
$$+\lambda\left(\int_{\Omega\setminus D} u(f-u)\,dx + \int_D u(v-u)\,dx\right)$$
$$+C(f_1,f_2,|\Omega|,\epsilon,\delta).$$

By applying the standard inequality $(a+b)^2 \le 2(a^2+b^2)$ to the L^2 norm of $w = ((1/\tau)+\lambda)u - ((1/\tau)+\lambda-\lambda\mathbb{1}_{\Omega\setminus D})v - \lambda\mathbb{1}_{\Omega\setminus D}f$ and by using $(1-\mathbb{1}_{\Omega\setminus D}) \le 1$

in the resulting L^2 norm of v, we get

$$\int_\Omega u \cdot \frac{1}{\tau}(u-v)\,dx \leq \left(\frac{f_2\delta}{2\epsilon} - \epsilon\right) \|\Delta u\|_2^2 - \frac{1}{\epsilon}\int_\Omega F''(u)\,|\nabla u|^2\,dx$$
$$+ f_1\delta\left(\frac{1}{\tau}+\lambda\right)^2 \|u\|_2^2 + 2f_1\delta\left(\frac{1}{\tau}+\lambda\right)^2 \|v\|_2^2$$
$$+\lambda\left(\int_{\Omega\setminus D} u(f-u)\,dx + \int_D u(v-u)\,dx\right)$$
$$+C(f,f_1,f_2,|\Omega|,\epsilon,\delta,\lambda).$$

With $F''(u) \geq C_1 u^2 - C_2$ for some constants $C_1, C_2 > 0$ and for all $u \in \mathbb{R}$, and by further applying the Cauchy-Schwarz inequality to the last two integrals, we obtain

$$\int_\Omega u \cdot \frac{1}{\tau}(u-v)\,dx \leq \left(\frac{f_2\delta}{2\epsilon} - \epsilon\right) \|\Delta u\|_2^2 - \frac{C_1}{\epsilon}\|u\ |\nabla u|\ \|_2^2 + \frac{C_2}{\epsilon}\|\nabla u\|_2^2$$
$$+ f_1\delta\left(\frac{1}{\tau}+\lambda\right)^2 \|u\|_2^2 + 2f_1\delta\left(\frac{1}{\tau}+\lambda\right)^2 \|v\|_2^2$$
$$+\lambda\left[-\left(1-\frac{\delta_2}{2}\right)\int_{\Omega\setminus D} u^2\,dx + \left(\frac{\delta_1}{2}-1\right)\int_D u^2\,dx + \frac{1}{2\delta_1}\int_D v^2\,dx\right]$$
$$+C(f,f_1,f_2,|\Omega|,|D|,\epsilon,\lambda,\delta,\delta_2)$$

for constants $\delta_1, \delta_2 > 0$. Setting $\delta_2 = 1$ and $\delta_1 = 2$, we see that

$$\int_\Omega u \cdot \frac{1}{\tau}(u-v)\,dx \leq \left(\frac{f_2\delta}{2\epsilon} - \epsilon\right) \|\Delta u\|_2^2 - \frac{C_1}{\epsilon}\|u\ |\nabla u|\ \|_2^2 + \frac{C_2}{\epsilon}\|\nabla u\|_2^2$$
$$+ f_1\delta\left(\frac{1}{\tau}+\lambda\right)^2 \|u\|_2^2 + 2f_1\delta\left(\frac{1}{\tau}+\lambda\right)^2 \|v\|_2^2$$
$$+\lambda\left[-\frac{1}{2}\int_{\Omega\setminus D} u^2\,dx + \frac{1}{4}\int_D v^2\,dx\right] + C(f,f_1,f_2,|\Omega|,|D|,\epsilon,\delta,\lambda).$$

We follow the argument used in the proof of existence for (5.18) in [BEglG07] by observing the following property: a standard interpolation inequality for ∇u reads

$$\|\nabla u\|_2^2 \leq \delta_3 \|\Delta u\|_2^2 + \frac{C_3}{\delta_3}\|u\|_2^2. \tag{5.35}$$

The domain of integration in the second integral of this inequality can be taken to be smaller than Ω by taking a larger constant C_3. Further, we use of the L^1 version of the Poincaré inequality applied to the function u^2 (cf. Theorem B.5.6 in Appendix B). Then, assuming that $u \neq 0$ in $\Omega \setminus D$, we choose the constant

C_4 (which depends on the size of D compared to Ω) large enough such that

$$\int_\Omega u^2\,dx - C_4\int_{\Omega\setminus D} u^2\,dx \le \int_\Omega \left|u^2-(u^2)_\Omega\right|\,dx \le C_4\int_\Omega \left|\nabla u^2\right|\,dx$$

or, in other words,

$$\|u\|_2^2 \le C_4\left\|\nabla u^2\right\|_{L^1(\Omega)} + C_4\int_{\Omega\setminus D} u^2\,dx. \tag{5.36}$$

By Hölder's inequality, we also have

$$\left\|\nabla u^2\right\|_{L^1(\Omega)} \le \frac{\alpha}{2}\,\|u\ |\nabla u|\ \|_2^2 + \frac{C_5}{2\alpha}, \tag{5.37}$$

for a constant $\alpha > 0$. Putting the last three inequalities (5.35)–(5.37) together, we obtain

$$\|\nabla u\|_2^2 \le \delta_3\,\|\Delta u\|_2^2 + \frac{C_3C_4\alpha}{2\delta_3}\,\|u|\nabla u|\|_2^2 + \frac{C_3C_4}{\delta_3}\int_{\Omega\setminus D} u^2\,dx + \frac{C_3C_4C_5}{2\alpha\delta_3}.$$

We now use the last inequality to bound the gradient term in our estimates from earlier to get

$$\begin{aligned}\int_\Omega u\cdot\frac{1}{\tau}(u-v)\,dx \le{}& \left(\frac{f_2\delta+2C_2\delta_3}{2\epsilon}-\epsilon\right)\|\Delta u\|_2^2\\ &+\left(\frac{C_2C_3C_4\alpha}{2\delta_3\epsilon}-\frac{C_1}{\epsilon}\right)\|u\ |\nabla u|\ \|_2^2\\ &+\left(f_1\delta\left(\frac{1}{\tau}+\lambda\right)^2+\frac{C_2C_3C_4}{\delta_3\epsilon}-\frac{C_4\lambda}{2}\right)\|u\|_2^2\\ &+\left(\frac{\lambda}{4}+2f_2\delta\left(\frac{1}{\tau}+\lambda\right)^2\right)\|v\|_2^2\\ &+C(f,f_1,f_2,|\Omega|,|D|,\epsilon,\delta,\lambda).\end{aligned} \tag{5.38}$$

With $\delta_3 < (2\epsilon^2 - f_2\delta)/2C_2$ and α,δ small enough, the first two terms can be estimated as zero. Applying the Cauchy-Schwarz inequality on the left-hand side and rearranging the terms on both sides of the inequality, we conclude that

$$\begin{aligned}&\left(\frac{1}{2\tau}+\frac{C_4\lambda}{2}-f_1\delta\left(\frac{1}{\tau}-\lambda\right)^2-\frac{C_2C_3C_4}{\delta_3\epsilon}\right)\|u\|_2^2\\ &\quad\le\left(\left(\frac{\lambda}{4}+\frac{1}{2\tau}\right)+2f_2\delta\left(\frac{1}{\tau}+\lambda\right)^2\right)\|v\|_2^2+C(f,f_1,f_2,|\Omega|,|D|,\epsilon,\delta,\lambda).\end{aligned}$$

Choosing δ small enough, C_4 large enough and $\lambda \geq CC_4(1/\epsilon^3)$, the solutions u and v fulfil

$$\|u\|_2^2 \leq \beta \|v\|_2^2 + C, \tag{5.39}$$

with $\beta < 1$ and a constant C independent of v. Hence, u is bounded in $L^2(\Omega)$.

To see that our regularity assumptions on u from the beginning of the proof are automatically fulfiled, we consider (5.38) with appropriate constants δ_3, δ and α as specified in the paragraph following (5.38). But now we only estimate the second term on the right-hand side as zero and keep the first term. By applying the Cauchy-Schwarz inequality and rearranging the terms as before, we obtain

$$\left(\frac{1}{2\tau} + \frac{C_4\lambda}{2} - f_1\delta\left(\frac{1}{\tau} - \lambda\right)^2 - \frac{C_2C_3C_4}{\delta_3\epsilon}\right)\|u\|_2^2 + \left(\epsilon - \frac{f_2\delta + 2C_2\delta_3}{2\epsilon}\right)\|\Delta u\|_2^2$$
$$\leq \left(\left(\frac{\lambda}{4} + \frac{1}{2\tau}\right) + 2f_2\delta\left(\frac{1}{\tau} + \lambda\right)^2\right)\|v\|_2^2 + C(f, f_1, f_2, |\Omega|, |D|, \epsilon, \delta, \lambda),$$

with the coefficient $\epsilon - (f_2\delta + 2C_2\delta_3)/2\epsilon > 0$ due to our choice of δ_3. Therefore, not only the L^2 norm of u is uniformly bounded but also the L^2 norm of Δu. By the standard interpolation inequality (5.35), the boundedness of u in $H^1(\Omega)$ follows. From the last result, we additionally get that the operator $\mathcal{A}$ is a compact map because $\mathcal{A}$: $L^2(\Omega) \to H^1(\Omega) \hookrightarrow\hookrightarrow L^2(\Omega)$. Therefore, $\mathcal{A}$ maps into a compact and convex subset of $L^2(\Omega)$.

It remains to show that the operator $\mathcal{A}$ is continuous. Indeed, if $v_k \to v$ in $L^2(\Omega)$, then $\mathcal{A}(v_k) = u_k$ is bounded in $H^1(\Omega)$ for all $k = 0, 1, 2, \ldots$. Thus, we can consider a weakly convergent subsequence $u_{k_j} \rightharpoonup u$ in $H^1(\Omega)$. Because $H^1(\Omega) \hookrightarrow\hookrightarrow L^q(\Omega)$, $1 \leq q < \infty$, the sequence u_{k_j} also strongly converges up to a subsequence to u in $L^q(\Omega)$. Hence, a weak solution $\mathcal{A}(v_k) = u_k$ of (5.27) weakly converges to a weak solution u of

$$\frac{1}{\tau}(-\Delta^{-1})(u - v) = \epsilon\Delta u - \frac{1}{\epsilon}F'(u) - \Delta^{-1}\left[\lambda \mathbb{1}_{\Omega\setminus D}(f - u) + \lambda(1 - \mathbb{1}_{\Omega\setminus D})(v - u)\right],$$

where u is the weak limit of $\mathcal{A}(v_k)$ as $k \to \infty$. Because the solution of (5.27) is unique provided that $\tau \leq C\epsilon^3$ (cf. Proposition 5.3.4), $u = \mathcal{A}(v)$, and therefore $\mathcal{A}$ is continuous. Applying Schauder's theorem (cf. Theorem B.5.5 in Appendix B), we have shown that the fixed-point operator $\mathcal{A}$ admits a fixed-point $\hat{u}$ in $L^2(\Omega)$ which fulfils

$$\langle \epsilon \nabla \hat{u}, \nabla \phi \rangle_2 + \left\langle \frac{1}{\epsilon} F'(\hat{u}), \phi \right\rangle_2 - \langle \lambda \mathbb{1}_{\Omega \setminus D}(f - \hat{u}), \phi \rangle_{-1}$$
$$+ \int_{\partial\Omega} \Delta^{-1} \left(\lambda \mathbb{1}_{\Omega \setminus D}(f - \hat{u}) \right) \nabla \Delta^{-1} \phi \cdot \nu \, d\mathcal{H}_1 = 0$$

for all $\phi \in H_0^1(\Omega)$. Because the solution of (5.27) is also an element of H, the fixed point $\hat{u}$ belongs to H. ■

Following the arguments from the beginning of this section, we conclude the existence of a stationary solution for (5.18).

By modifying the setting and the preceding proof in an appropriate way, we can also prove the existence of a stationary solution for (5.18) under Neumann boundary conditions, that is,

$$\nabla u \cdot \nu = \nabla \Delta u \cdot \nu = 0, \quad \text{on } \partial\Omega.$$

Numerical Discussion Having guaranteed the existence of an inpainting computed as the stationary solution of the modified Cahn-Hilliard equation (5.18), we are now ready to present some numerical results of its application for the inpainting of binary structures. The numerical scheme used here is called *convexity splitting* and has been thoroughly analysed in [SB11]. Let U_k be the approximate solution of (5.18) at time $t = k\tau$, where τ denotes the size of the discrete time steps. The numerical solution U_{k+1} of (5.18) is computed via a time-marching scheme of the form

$$\frac{U_{k+1} - U_k}{\tau} + \epsilon \Delta \Delta U_{k+1} - C_1 \Delta U_{k+1} + C_2 U_{k+1}$$
$$= \frac{1}{\epsilon} \Delta F'(U_k) - C_1 \Delta U_k + \lambda \mathbb{1}_{\Omega \setminus D}(f - U_k) + C_2 U_k,$$

with constants $C_1 > 1/\epsilon$ and $C_2 > \lambda$. The condition on the constants arises as a stability condition from the numerical analysis of the scheme (see [SB11]).

In Figures 5.9 and 5.10, Cahn-Hilliard inpainting was applied to two different binary images. In all the examples, we follow the procedure of [BEG07]; that is, the inpainted image is computed in a two-step process. In the first step, Cahn-Hilliard inpainting is solved with a rather large value of ϵ, for example, $\epsilon = 0.1$, until the numerical scheme is close to steady state. In this step, the level lines are continued into the missing domain. In a second step, the result of the first step is put as an initial condition into the scheme for a small ϵ, for example, $\epsilon = 0.01$, in order to sharpen the contours of the image contents. The reason for this two-step procedure is twofold. First of all,

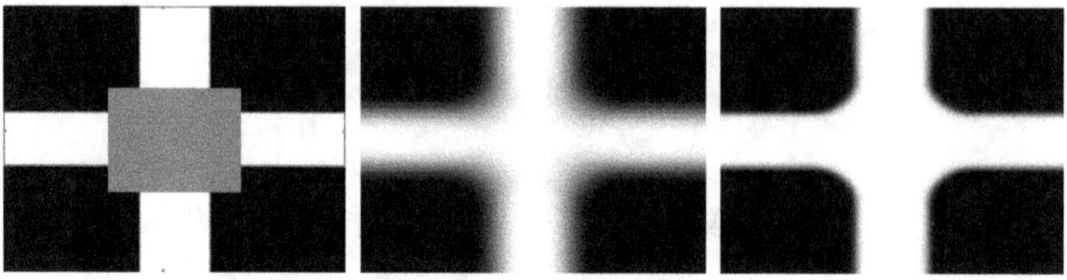

Figure 5.9. Destroyed binary image and the solution of Cahn-Hilliard inpainting with switching the ϵ value: $u(1,200)$ with $\epsilon = 0.1$, $u(2,400)$ with $\epsilon = 0.01$.

Figure 5.10. Destroyed binary image and the solution of Cahn-Hilliard inpainting with $\lambda = 10^9$ and switching ϵ value: $u(800)$ with $\epsilon = 0.8$, $u(1,600)$ with $\epsilon = 0.01$.

in [BEglG07], the authors gave numerical evidence that the steady state of the modified Cahn-Hilliard equation (5.18) is not unique; that is, it depends on the initial condition for the equation. As a consequence, computing the inpainted image by the application of Cahn-Hilliard inpainting with a small ϵ only might not prolong the level lines into the missing domain as desired. (See also [BEglG07] for a bifurcation diagram based on the numerical computations of the authors). The second reason for solving Cahn-Hilliard inpainting in two steps is that it is computationally less expensive. Solving the preceding time-marching scheme for, say, $\epsilon = 0.1$ is faster than solving it for $\epsilon = 0.01$. This is so because of a damping introduced by ϵ into the scheme (cf. [SB11] for details).

5.3.2. TV-H^{-1} *Inpainting*

The inpainting approach in (5.20) is called TV-H^{-1} *inpainting*. The total variation $|Du|(\Omega)$ and the space $BV(\Omega)$ are defined in Definition 4.3.1.

We start our analysis for TV-H^{-1} inpainting (5.20) with a Γ-convergence result for the Cahn-Hilliard energy that served as a motivation for the proposal of TV-H^{-1} inpainting. See Section B.8 in Appendix B for a brief recap on Γ-convergence. In [BHS09], the authors prove that the sequence of functionals for an appropriate time-discrete Cahn-Hilliard inpainting approach

Γ-converges to a functional regularised with the total variation for binary arguments $u = \mathbb{1}_E$, where E is some Borel measurable subset of Ω. This is stated in more detail in the following theorem:

Theorem 5.3.6 *Let $f, v \in L^2(\Omega)$ be two given functions and $\tau > 0$ a positive parameter. Let further $\|\cdot\|_{-1}$ be the norm in $H^{-1}(\Omega)$ and $\epsilon > 0$ a parameter. Then the sequence of functionals*

$$\mathcal{J}^\epsilon(u,v) = \int_\Omega \left(\frac{\epsilon}{2} |\nabla u|^2 + \frac{1}{\epsilon} F(u) \right) dx + \frac{1}{2\tau} \|u - v\|_{-1}^2 + \frac{\lambda}{2} \left\| u - \mathbb{1}_{\Omega \setminus D} f - \left(1 - \mathbb{1}_{\Omega \setminus D}\right) v \right\|_{-1}^2$$

Γ-converges for $\epsilon \to 0$ in the topology of $L^1(\Omega)$ to

$$\mathcal{J}(u,v) = TV_b(u) + \frac{1}{2\tau} \|u - v\|_{-1}^2 + \frac{\lambda}{2} \left\| u - \mathbb{1}_{\Omega \setminus D} f - \left(1 - \mathbb{1}_{\Omega \setminus D}\right) v \right\|_{-1}^2,$$

where

$$TV_b(u) = \begin{cases} C_0 \, |Du| \, (\Omega) & \text{if } u = \mathbb{1}_E \text{ for some Borel measurable subset } E \subset \Omega \\ +\infty & \text{otherwise,} \end{cases}$$

$\mathbb{1}_E$ denoting the indicator function of E and $C_0 = 2 \int_0^1 \sqrt{F(s)}\, ds$.

Remark 5.3.7 Setting $v = u_k$ and a minimiser u of the functional $\mathcal{J}^\epsilon(u,v)$ to be $u = u_{k+1}$, the minimisation of $\mathcal{J}^\epsilon$ can be seen as an iterative approach with step size τ to solve (5.18).

Remark 5.3.8 The Γ-convergence of the Ginzburg-Landau energy $\int_\Omega (\epsilon/2 |\nabla u|^2 + 1/\epsilon F(u))\, dx$ to the total variation of an indicator function for a subset of Ω goes back to Modica and Mortola [MM77a, MM77b]. Their result also can be understood in terms of a sharp interface limit. In other words, the limit of the energy of a phase field model (the Ginzburg-Landau energy), where black (0) and white (1) regions are separated by a transition layer of order ϵ, was shown to be a model that describes the motion of the sharp interface between these regions without a transition layer, that is, $\epsilon = 0$. Similar results have been shown for the Cahn-Hilliard equation directly. Here the corresponding sharp interface limit is known as the *Mullins-Sekerka problem* or the *Hele-Shaw problem* (cf. [MS63, Peg89, ABC94 and Sto96]).

Now, by extending the validity of the TV functional $TV_b(u)$ from functions $u = 0$ or 1 to functions $|u| \leq 1$, we receive the functional TV_∞ in (5.21) and an inpainting approach for grey value images rather than binary images which is given in (5.20).

Remark 5.3.9 Similar to the use of the functional $\mathcal{J}^\epsilon$ from Theorem 5.3.6 as an iterative approach to solve (5.18) (cf. Remark 5.3.7), the functional $\mathcal{J}$ from Theorem 5.3.6 might serve as an iterative process to solve (5.20).

A similar form of the TV-H^{-1} inpainting approach already appeared in the context of decomposition and restoration of grey value images (see, e.g., [VO03, OSV03 and LV08]). Different from the situation of TV-H^{-1} de-noising/decomposition, the TV-H^{-1} inpainting approach does not exhibit an obvious variational form, and hence, its analytical and numerical treatment is different. Further, in Bertalmio at al. [BVSO03], an application of the model from [VO03] to image inpainting has been proposed. In contrast to the inpainting approach (5.20), the authors in [BVSO03] only use a different form of the TV-H^{-1} approach for a decomposition of the image into cartoon and texture prior to the inpainting process, which is accomplished with the method presented in [BSCB00].

Using the same methodology as in the proof of Theorem 5.3.1, we obtain the following existence theorem (cf. [BHS09]):

Theorem 5.3.10 *Let $f \in L^2(\Omega)$. The stationary equation*

$$\Delta p + \lambda \mathbb{1}_{\Omega \setminus D}(f - u) = 0, \quad p \in \partial TV_\infty(u) \tag{5.40}$$

admits a solution $u \in BV(\Omega)$.

Moreover, in [BHS09], a characterisation of elements in the subdifferential $\partial TV_\infty(u)$ is given. Here $TV_\infty(u)$ is defined as in (5.21); that is, $TV_\infty(u) = |Du|(\Omega) + \chi_1(u)$, where

$$\chi_1(u) = \begin{cases} 0, & \text{if } |u| \leq 1 \text{ a.e. in } \Omega, \\ +\infty, & \text{otherwise.} \end{cases}$$

In particular, the following theorem can be proven:

Theorem 5.3.11 *Let $\tilde{p} \in \partial \chi_1(u)$. An element $p \in \partial TV_\infty(u)$ with $|u(x)| \leq 1$ in Ω fulfils the following set of equations:*

$$\begin{aligned}
p &= -\nabla \cdot \left(\frac{\nabla u}{|\nabla u|} \right), & & \textit{a.e. in } \{x : |u(x)| < 1\}, \\
p &= -\nabla \cdot \left(\frac{\nabla u}{|\nabla u|} \right) + \tilde{p}, \quad \tilde{p} \leq 0, & & \textit{a.e. in } \{x : u(x) = -1\}, \\
p &= -\nabla \cdot \left(\frac{\nabla u}{|\nabla u|} \right) + \tilde{p}, \quad \tilde{p} \geq 0, & & \textit{a.e. on } \{x : u = 1\}.
\end{aligned}$$

Additionally, estimates for the inpainting error and stability information are desirable for an inpainting procedure. Ideally, the deviation in the inpainted image for different input data should be bounded by the deviation in the data. It has been demonstrated by many authors (cf. [BO04, BRH07, Pös08, BB11]) that Bregman distances related to the regularisation functional are natural error measures for variational regularisation methods when the regularisation is convex. In particular, Pöschl [Pös08] derived estimates for variational regularisation methods with powers of metrics, which apply to the TV functional we consider here. In what follows, we present the analysis demonstrated in [BHS09], which uses the Bregman distance for error estimates in the TV-H^{-1} inpainting case. Let f_{true} be the original image and $\hat{u}$ a stationary solution of (5.20). In their considerations, they used the symmetric Bregman distance defined as

$$D_{TV_\infty}^{\text{symm}}(\hat{u},f_{\text{true}}) = \langle \hat{u} - f_{\text{true}}, \hat{p} - \xi \rangle_2, \quad \hat{p} \in \partial TV_\infty(\hat{u}), \quad \xi \in \partial TV_\infty(f_{\text{true}}),$$

and proved the following result:

Theorem 5.3.12 *Let f_{true} fulfil the so-called source condition, namely, that*

$$\text{There exists } \xi \in \partial TV_\infty(f_{\text{true}}) \text{ such that } \xi = \Delta^{-1} q$$
$$\text{for a source element } q \in H^{-1}(\Omega),$$

and $\hat{u} \in BV(\Omega)$ is a stationary solution of (5.20) given by $\hat{u} = u^s + u^d$, where u^s is a smooth function and u^d is a piecewise constant function. Then the inpainting error reads

$$D_{TV_\infty}^{\text{symm}}(\hat{u},f_{\text{true}}) + \frac{\lambda}{2}\left\|\hat{u} - f_{\text{true}}\right\|_{-1}^2 \leq \frac{1}{\lambda}\|\xi\|_1^2 + C\lambda\,|D|^{(r-2)/r}\,\text{err}_{\text{inpaint}},$$

with $2 < r < \infty$, constant $C > 0$ and

$$\text{err}_{\text{inpaint}} := K_1 + K_2\left(|D|\,C(M(u^s),\beta) + 2\left|R(u^d)\right|\right),$$

where K_1 and K_2 are appropriate positive constants, and C is a constant depending on the smoothness bound $M(u^s)$ for u^s and β that is determined from the shape of D. The error region $R(u^d)$ is determined by the level lines of u^d.

Let us go in more detail through these analytical results before we finish this section with numerical results and briefly explain the numerical implementation of TV-H^{-1} inpainting.

Γ-Convergence of the Cahn-Hilliard Energy In the following we want to motivate our new inpainting approach (5.20) by considering the Γ-limit for

$\epsilon \to 0$ of an appropriate time-discrete Cahn-Hilliard inpainting approach, that is, the Γ-limit of the functionals from our fixed point approach in (5.30). More precisely, we want to prove Theorem 5.3.6.

Proof of Theorem 5.3.6 Modica and Mortola have shown in [MM77a and MM77b] that the sequence of Ginzburg-Landau functionals

$$CH(u) = \int_\Omega \left(\frac{\epsilon}{2} |\nabla u|^2 + \frac{1}{\epsilon} F(u) \right) dx$$

Γ-converges in the topology $L^1(\Omega)$ to

$$TV_\infty(u) = \begin{cases} C_0 |Du|(\Omega) & \text{if } u = \mathbb{1}_E \text{ for some Borel measurable subset } E \subset \Omega \\ +\infty & \text{otherwise} \end{cases}$$

as $\epsilon \to 0$, where $C_0 = 2\int_0^1 \sqrt{F(s)}\, ds$. Now, for a given function $v \in L^2(\Omega)$, the functional $\mathcal{J}^\epsilon$ from our fixed-point approach (5.27), that is,

$$\mathcal{J}^\epsilon(u,v) = \underbrace{\int_\Omega \left(\frac{\epsilon}{2} |\nabla u|^2 + \frac{1}{\epsilon} F(u) \right) dx}_{:=CH(u)} + \underbrace{\frac{1}{2\tau} \|u - v\|_{-1}^2}_{:=D(u,v)}$$
$$+ \underbrace{\frac{\lambda}{2} \left\| u - \mathbb{1}_{\Omega\setminus D} f - (1 - \mathbb{1}_{\Omega\setminus D}) v \right\|_{-1}^2}_{:=FIT(u,v)},$$

is the sum of the regularising term $CH(u)$, the damping term $D(u,v)$ and the fitting term $FIT(u,v)$. We recall the fact from Theorem B.8.2 in Appendix B that for a continuous functional G and a sequence of functionals (F_n) that Γ-converges to F, the sum $(F_n + G)$ Γ-converges to $F + G$. Since the H^{-1} norm is continuous in $H^{-1}(\Omega)$ and hence in particular in $L^1(\Omega)$, the two terms in $\mathcal{J}^\epsilon$ that are independent of ϵ, that is, $D(u,v)$ and $FIT(u,v)$, are continuous in $L^1(\Omega)$. ∎

Together with the Γ-convergence result of Modica and Mortola for the Ginzburg-Landau energy, we have proven that the Cahn-Hilliard inpainting functional $\mathcal{J}^\epsilon$ can be seen as a regularised approximation in the sense of Γ-convergence of the TV functional

$$\mathcal{J}(u,v) = |Du|(\Omega) + D(u,v) + FIT(u,v),$$

for functions $u \in BV(\Omega)$ with $u(x) = \mathbb{1}_E(x)$ for a Borel measurable subset $E \subset \Omega$. In fact, we have gone from a smooth transition layer between 0 and 1 in the Cahn-Hilliard inpainting approach (depending on the size of ϵ) to a sharp interface limit in which the image function now jumps from 0 to 1.

This sharp interface limit motivates the extension of $\mathcal{J}(u,v)$ to grey value functions such that $|u| \leq 1$ on Ω and hence leads us from the Cahn-Hilliard inpainting approach for binary images to a generalisation for grey value images, namely, our so-called TV-H^{-1} inpainting method (5.20) and (5.21).

Existence of a Stationary Solution Our strategy for proving the existence of a stationary solution for TV-H^{-1} inpainting (5.20) is similar to our existence proof for a stationary solution of the modified Cahn-Hilliard equation (5.18) in Section 5.3.1. As in our analysis for (5.18) in Section 5.3.1, we consider (5.20) with Dirichlet boundary conditions, namely,

$$\begin{aligned} u_t =& \Delta p + \lambda \mathbb{1}_{\Omega\setminus D}(f-u), && \text{in } \Omega, \\ u =& f, && \text{on } \partial\Omega, \end{aligned}$$

for $p \in \partial TV_\infty(u)$.

Now let $f \in L^2(\Omega)$, $|f| \leq 1$ be the given grey value image. For $v \in L^r(\Omega)$, $1 < r < 2$, we consider the minimisation problem

$$u^* = \arg \min_{u \in BV(\Omega)} \mathcal{J}(u,v),$$

with functionals

$$\mathcal{J}(u,v) := TV_\infty(u) + \underbrace{\frac{1}{2\tau}||u-v||^2_{-1}}_{D(u,v)} + \underbrace{\frac{\lambda}{2}||u - \mathbb{1}_{\Omega\setminus D}f - (1-\mathbb{1}_{\Omega\setminus D})v||^2_{-1}}_{FIT(u,v)}, \tag{5.41}$$

with $TV_\infty(u)$ defined as in (5.21). Note that $L^r(\Omega)$ can be continuously embedded in $H^{-1}(\Omega)$. Hence the functionals in (5.41) are well defined.

Firstly, we will show that for a given $v \in L^r(\Omega)$, the functional $\mathcal{J}(\cdot, v)$ attains a unique minimiser $u^* \in BV(\Omega)$ with $|u^*(x)| \leq 1$ a.e. in Ω.

Proposition 5.3.13 *Let $f \in L^2(\Omega)$ be given with $|f(x)| \leq 1$ a.e. in Ω and $v \in L^r(\Omega)$. Then the functional $\mathcal{J}(\cdot, v)$ has a unique minimiser $u^* \in BV(\Omega)$ with $|u^*(x)| \leq 1$ a.e. in Ω.*

Proof Let $(u_n)_{n\in\mathbb{N}}$ be a minimising sequence for $\mathcal{J}(u,v)$, that is,

$$\mathcal{J}(u_n, v) \overset{n\to\infty}{\to} \inf_{u \in BV(\Omega)} \mathcal{J}(u,v).$$

Then $u_n \in BV(\Omega)$, and $|u_n(x)| \leq 1$ in a.e. Ω [because otherwise $TV_\infty(u_n)$ would not be finite]. Therefore,

$$|Du_n|(\Omega) \leq M, \quad \text{for an } M \geq 0 \text{ and for all } n \geq 1,$$

and, because of the uniform boundedness of $|u(x)|$ for every point $x \in \Omega$,

$$\|u_n\|_{L^p(\Omega)} \leq \tilde{M}, \text{ for an } \tilde{M} \geq 0, \quad \forall n \geq 1, \quad \text{and} \quad 1 \leq p \leq \infty.$$

Thus, u_n is uniformly bounded in $L^p(\Omega)$ and in particular in $L^1(\Omega)$. Together with the boundedness of $|Du_n|(\Omega)$, the sequence u_n is also bounded in $BV(\Omega)$, and there exists a subsequence, still denoted u_n, and a $u \in BV(\Omega)$ such that $u_n \rightharpoonup u$ weakly in $L^p(\Omega)$, $1 \leq p < \infty$, and weakly* in $BV(\Omega)$. Because $L^2(\Omega) \subset L^2(\mathbb{R}^2) \subset H^{-1}(\Omega)$ (by zero extensions of functions on Ω to $\mathbb{R}^2$), $u_n \rightharpoonup u$ also weakly in $H^{-1}(\Omega)$. Because $|Du|(\Omega)$ is lower semi-continuous in $BV(\Omega)$, and by the lower semi-continuity of the H^{-1} norm, we get

$$\begin{aligned}
\mathcal{J}(u,v) &= TV_\infty(u) + D(u,v) + FIT(u,v) \\
&\leq \liminf_{n\to\infty}(TV_\infty(u_n) + D(u_n,v) + FIT(u_n,v)) \\
&= \liminf_{n\to\infty}\mathcal{J}(u_n,v).
\end{aligned}$$

So u is a minimiser of $\mathcal{J}(u,v)$ over $BV(\Omega)$.

To prove the uniqueness of the minimiser, we show that $\mathcal{J}$ is strictly convex (as in the proof of Theorem 5.3.4). Namely, we prove that for all $u_1, u_2 \in BV(\Omega)$, $u_1 \neq u_2$,

$$\mathcal{J}(u_1,v) + \mathcal{J}(u_2,v) - 2\mathcal{J}\left(\frac{u_1+u_2}{2}, v\right) > 0.$$

We have

$$\begin{aligned}
&\mathcal{J}(u_1,v) + \mathcal{J}(u_2,v) - 2\mathcal{J}\left(\frac{u_1+u_2}{2}, v\right) \\
&= \left(\frac{1}{2\tau} + \frac{\lambda}{2}\right)\left(\|u_1\|_{-1}^2 + \|u_2\|_{-1}^2 - 2\left\|\frac{u_1+u_2}{2}\right\|_{-1}^2\right) \\
&\quad + TV_\infty(u_1) + TV_\infty(u_2) - 2TV_\infty\left(\frac{u_1+u_2}{2}\right) \\
&\geq \left(\frac{1}{4\tau} + \frac{\lambda}{4}\right)\|u_1 - u_2\|_{-1}^2 > 0.
\end{aligned}$$

This finishes the proof. ∎

Next, we shall prove the existence of a stationary solution for (5.20). For this end, we consider the corresponding Euler-Lagrange equation to (5.41), that is,

$$\begin{aligned}
&\Delta^{-1}\left(\frac{u-v}{\tau}\right) + p - \Delta^{-1}\left(\lambda \mathbb{1}_{\Omega\setminus D}(f-u) + \lambda(1 - \mathbb{1}_{\Omega\setminus D})(v-u)\right) = 0, \\
&p \in \partial TV_\infty(u),
\end{aligned}$$

with weak formulation

$$\left\langle \frac{1}{\tau}(u-v), \phi \right\rangle_{-1} + \langle p, \phi \rangle_2$$
$$- \left\langle \lambda \mathbb{1}_{\Omega \setminus D}(f-u) + \lambda(1-\mathbb{1}_{\Omega \setminus D})(v-u), \phi \right\rangle_{-1} = 0 \ \forall \phi \in H_0^1(\Omega).$$

A fixed point of the preceding equation, that is, a solution $u = v$, is then a stationary solution for (5.20). Thus, to prove the existence of a stationary solution of (5.20), that is, to prove Theorem 5.3.10, as before, we are going to use a fixed-point argument. Let $\mathcal{A}$: $L^r(\Omega) \to L^r(\Omega)$, $1 < r < 2$, be the operator which maps a given $v \in L^r(\Omega)$ to $\mathcal{A}(v) = u$ under the condition that $\mathcal{A}(v) = u$ is the minimiser of the functional $\mathcal{J}(\cdot, v)$ defined in (5.41). The choice of the fixed-point operator $\mathcal{A}$ over $L^r(\Omega)$ was made in order to obtain the necessary compactness and continuity properties for the application of Schauder's theorem.

Since here the treatment of the boundary conditions is similar to that in Section 5.3.1, we will leave this part of the analysis in the upcoming proof to the reader and just carry out the proof without explicitly taking care of the boundary.

Proof of Theorem 5.3.10 Let $\mathcal{A}$: $L^r(\Omega) \to L^r(\Omega)$, $1 < r < 2$, be the operator that maps a given $v \in L^r(\Omega)$ to $\mathcal{A}(v) = u$, where u is the unique minimiser of the functional $\mathcal{J}(\cdot, v)$ defined in (5.41). Existence and uniqueness of u follow from Proposition 5.3.13. Since u minimises $\mathcal{J}(\cdot, v)$ we have $u \in L^\infty(\Omega)$ and hence $u \in L^r(\Omega)$. Additionally, we have $\mathcal{J}(u, v) \leq \mathcal{J}(0, v)$, that is,

$$\frac{1}{2\tau}||u-v||_{-1}^2 + \frac{\lambda}{2}||u - \mathbb{1}_{\Omega \setminus D} f - (1-\mathbb{1}_{\Omega \setminus D})v||_{-1}^2 + TV_\infty(u)$$
$$\leq \frac{1}{2\tau}||v||_{-1}^2 + \frac{\lambda}{2}||\mathbb{1}_{\Omega \setminus D} f + (1-\mathbb{1}_{\Omega \setminus D})v||_{-1}^2 \leq \frac{|\Omega|}{2\tau} + \lambda(|\Omega| + |D|).$$

Here the last inequality was obtained because $L^r(\Omega) \hookrightarrow H^{-1}(\Omega)$ and hence $||v||_{-1} \leq C$ and $||\lambda \mathbb{1}_{\Omega \setminus D} v||_{-1} \leq C$ for a $C > 0$. [In fact, since $H^1(\Omega) \hookrightarrow L^{r'}(\Omega)$ for all $1 \leq r' < \infty$, it follows from duality that $L^r(\Omega) \hookrightarrow H^{-1}(\Omega)$ for $1 < r < \infty$.] By the last estimate we find that u is bounded in $BV(\Omega)$. Since $BV(\Omega) \hookrightarrow\hookrightarrow L^r(\Omega)$ compactly for $1 \leq r < 2$ and $\Omega \subset \mathbb{R}^2$, the operator $\mathcal{A}$ maps $L^r(\Omega) \to BV(\Omega) \hookrightarrow\hookrightarrow L^r(\Omega)$, that is, $\mathcal{A}$: $L^r(\Omega) \to K$, where K is a compact subset of $L^r(\Omega)$. Thus, for $v \in B(0,1)$ [where $B(0,1)$ denotes the ball in $L^\infty(\Omega)$ with centre 0 and radius 1], the operator $\mathcal{A}$: $B(0,1) \to B(0,1) \cap K = \tilde{K}$, where $\tilde{K}$ is a compact and convex subset of $L^r(\Omega)$.

Next, we have to show that $\mathcal{A}$ is continuous in $L^r(\Omega)$. Let $(v_n)_{n\geq 0}$ be a sequence which converges to v in $L^r(\Omega)$. Then $u_n = A(v_n)$ solves

$$\Delta p_n = \frac{u_n - v_n}{\tau} - \left(\lambda 1\!\!1_{\Omega\setminus D}(f - u_n) + \lambda(1 - 1\!\!1_{\Omega\setminus D})(v_n - u_n)\right),$$

where $p_k \in \partial TV_\infty(u_n)$. Thus, u_n is uniformly bounded in $BV(\Omega)\cap L^\infty(\Omega)$ [and hence in $L^r(\Omega)$], and since the right-hand side of the preceding equation is uniformly bounded in $L^r(\Omega)$, Δp_n is also bounded in $L^r(\Omega)$. Thus, there exists a subsequence p_{n_l} such that $\Delta p_{n_l} \rightharpoonup \Delta p$ in $L^r(\Omega)$ and a subsequence u_{n_l} that converges weakly* to a u in $BV(\Omega)\cap L^\infty(\Omega)$. Since $BV(\Omega) \hookrightarrow\hookrightarrow L^r(\Omega)$, we have $u_{n_l} \to u$ strongly in $L^r(\Omega)$ up to a further subsequence. Therefore, the limit u solves

$$\Delta p = \frac{u - v}{\tau} - \left(\lambda 1\!\!1_{\Omega\setminus D}(f - u) + \lambda(1 - 1\!\!1_{\Omega\setminus D})(v - u)\right). \tag{5.42}$$

If we additionally apply Poincaré inequality to Δp_n, we conclude that

$$\left\| \nabla p_n - (\nabla p_n)_\Omega \right\|_{L^r(\Omega)} \leq C \left\| \nabla \cdot \left(\nabla p_n - (\nabla p_n)_\Omega\right) \right\|_{L^r(\Omega)}.$$

In addition, since $p_n \in \partial TV_\infty(u_n)$, it follows that $(p_n)_\Omega = 0$ and $\|p_n\|_{BV^*(\Omega)} \leq 1$. Thus, $(\nabla p_n)_\Omega < \infty$, and p_n is uniformly bounded in $W^{1,r}(\Omega)$. There exists, therefore, a subsequence p_{n_l} such that $p_{n_l} \rightharpoonup p$ in $W^{1,r}(\Omega)$. In addition, $L^{r'}(\Omega) \hookrightarrow BV^*(\Omega)$ for $2 < r' < \infty$ (this follows again from a duality argument for the compactness of BV in L^r), and $W^{1,r}(\Omega) \hookrightarrow\hookrightarrow L^q(\Omega)$ for $1 \leq q < 2r/(2-r)$ (Rellich-Kondrachov compactness theorem; see [Alt85, theorem 8.7, p. 243]). By choosing $2 < q < 2r/(2-r)$, we therefore have $W^{1,r}(\Omega) \hookrightarrow\hookrightarrow BV^*(\Omega)$. Thus, $p_{n_l} \to p$ strongly in $BV^*(\Omega)$. Hence, the element p in (5.42) is an element in $\partial TV_\infty(u)$.

Because the minimiser of (5.41) is unique, $u = \mathcal{A}(v)$, and therefore, $\mathcal{A}$ is continuous in $L^r(\Omega)$. The existence of a stationary solution follows from Schauder's fixed-point theorem. ■

Characterisation of Solutions Finally, we want to compute elements $p \in \partial TV_\infty(u)$. In particular, we shall prove Theorem 5.3.11. As in [BRH07], the model for the regularising functional is the sum of a standard regulariser plus the indicator function of the L^∞ constraint. In particular, we have $TV_\infty(u) = |Du|(\Omega) + \chi_1(u)$, where $|Du|(\Omega)$ is the total variation of Du and

$$\chi_1(u) = \begin{cases} 0, & \text{if } |u| \leq 1 \text{ a.e. in } \Omega, \\ +\infty, & \text{otherwise.} \end{cases} \tag{5.43}$$

We want to compute the subgradients of TV_∞ by assuming $\partial TV_\infty(u) = \partial |Du|(\Omega) + \partial \chi_1(u)$. This means that we can separately compute the subgradients of χ_1. To guarantee that the splitting is allowed, we have to consider a regularised functional of the total variation, such as $\int_\Omega \sqrt{|\nabla u|^2 + \delta}\, dx$. This is sufficient because both $|D\cdot|(\Omega)$ and χ_1 are convex, and $|D\cdot|(\Omega)$ is continuous (compare [ET76, proposition 5.6., pp. 26]).

The subgradient $\partial |Du|(\Omega)$ is already well described, as, for instance, in [AK06 and Ves01]. We will just briefly recall its characterisation. Thus, we do not insist on the details of the rigorous derivation of these conditions; we limit ourselves to mentioning the main facts.

It is well known [Ves01, proposition 4.1] that $p \in \partial|Du|(\Omega)$ implies for $Du \neq 0$ that

$$\begin{cases} p = -\nabla \cdot \left(\dfrac{Du}{|\nabla D|} \right), & \text{in } \Omega, \\ \dfrac{Du}{|Du|} \cdot \nu = 0, & \text{on } \partial\Omega. \end{cases}$$

The preceding conditions do not fully characterise $p \in \partial|Du|(\Omega)$; additional conditions would be required [AK06, Ves01]. Since we are considering a regularised version of $|Du|(\Omega)$, the subdifferential becomes a gradient which reads

$$\begin{cases} p = -\nabla \cdot \left(\dfrac{\nabla u}{\sqrt{|\nabla u|^2 + \delta}} \right), & \text{in } \Omega, \\ \dfrac{\nabla u}{\sqrt{|\nabla u|^2 + \delta}} \cdot \nu = 0, & \text{on } \partial\Omega. \end{cases}$$

The subgradient of χ_1 is computed as in the following lemma:

Lemma 5.3.14 *Let $\chi_1 \colon L^r(\Omega) \to \mathbb{R} \cup \{\infty\}$ be defined by* (5.43), *and let $1 \leq r \leq \infty$. Then $p \in L^{r^*}(\Omega)$ for $r^* = r/(r-1)$, is a subgradient $p \in \partial\chi_1(u)$ for $u \in L^r(\Omega)$ with $\chi_1(u) = 0$ if and only if*

$$p = 0, \quad \textit{a.e. in } \{x : \ |u(x)| < 1\},$$

$$p \leq 0, \quad \textit{a.e. in } \{x : \ u(x) = -1\},$$

$$p \geq 0, \quad \textit{a.e. in } \{x : \ u(x) = 1\}.$$

Proof Let $p \in \partial\chi_1(u)$. Then we can choose $v = u + \epsilon w$ for w as any bounded function supported in $\{|u| < 1 - \alpha\}$ for arbitrary $0 < \alpha < 1$. If ϵ is sufficiently small, we have $|v| \leq 1$. Hence,

$$0 \geq \langle v - u, p \rangle_2 = \epsilon \int_{\{|u| < 1 - \alpha\}} wp\, dx.$$

Since we can choose ϵ either positive or negative, we obtain

$$\int_{\{|u|<1-\alpha\}} wp\,dx = 0.$$

Because $0<\alpha<1$ and w are arbitrary, we conclude that $p=0$ a.e. in $\{|u|<1\}$. If we choose $v=u+w$ with w an arbitrary bounded function with

$$\begin{cases} 0 \le w \le 1, & \text{in } \{-1 \le u \le 0\}, \\ w = 0, & \text{in } \{0 < u \le 1\}, \end{cases}$$

then v is still between -1 and 1, and

$$0 \ge \langle v-u,p\rangle_2 = \int_{\{u=-1\}} wp\,dx + \int_{\{u=1\}} wp\,dx = \int_{\{u=-1\}} wp\,dx.$$

Because w is arbitrary and positive on $\{u=-1\}$, it follows that $p\le 0$ a.e. on $\{u=-1\}$. If we choose now $v=u+w$ with w as an arbitrary bounded function with

$$\begin{cases} w = 0, & \text{in } \{-1 \le u \le 0\}, \\ -1 \le w \le 0, & \text{in } \{0 < u \le 1\}, \end{cases}$$

then v is still between -1 and 1, and

$$0 \ge \langle v-u,p\rangle_2 = \int_{\{u=-1\}} wp\,dx + \int_{\{u=1\}} wp\,dx = \int_{\{u=1\}} wp\,dx.$$

Analogously to before, since w is arbitrary and negative on $\{u=1\}$, it follows that $p\ge 0$ a.e. on $\{u=1\}$. However, assume that

$$\begin{aligned} p &= 0, \quad \text{a.e. in } \{|u|<1\}, \\ p &\le 0, \quad \text{a.e. in } \{u=-1\}, \\ p &\ge 0, \quad \text{a.e. on } \{u=1\}. \end{aligned}$$

We need to verify the sub-gradient property

$$\langle v-u,p\rangle_2 \le \chi_1(v) - \chi_1(u) = \chi_1(v), \qquad \text{for all } v \in L^r(\Omega)$$

only for $\chi_1(v)=0$ because it is trivial for $\chi_1(v)=\infty$. So let $v\in L^r(\Omega)$ be a function between -1 and 1 almost everywhere on Ω. Then, with p as earlier, we obtain

$$\begin{aligned} \langle v-u,p\rangle_2 &= \int_{\{u=-1\}} p(v-u)\,dx + \int_{\{u=1\}} p(v-u)\,dx \\ &= \int_{\{u=-1\}} p(v+1)\,dx + \int_{\{u=1\}} p(v-1)\,dx. \end{aligned}$$

Since $-1 \leq v \leq 1$, the first and second terms are always ≤ 0 because $p \leq 0$ for $\{u = -1\}$ and $p \geq 0$ for $\{u = 1\}$, respectively. Therefore, $\langle v - u, p \rangle_2 \leq 0$, and we are done. ■

Error Estimation and Stability Analysis with the Bregman Distance In the following analysis we want to present estimates for the error we actually make both in inpainting an image with our TV-H^{-1} approach (5.20) [see (5.51)] and in the stability of solutions for this problem [see (5.52)] in terms of the Bregman distance. This section is motivated by the error analysis for variational models in image restoration with Bregman iterations in [BRH07] and the error estimates for inpainting models developed in [CK06]. In [BRH07], the authors considered, among other things, the general optimality condition

$$p + \lambda A^*(Au - f_{\text{dam}}) = 0, \tag{5.44}$$

where $p \in \partial R(u)$ for a regularising term R, A is a bounded linear operator and A^* its adjoint. Now the error that is to be estimated depends on the form of smoothing of the image contained in (5.44). Considering this equation, one realises that smoothing consists of two steps. The first is created by the operator A and depends on the image restoration task at hand and actually smooths the subgradient p. The second smoothing step is the one that is directly implied by the regularising term, that is, its subgradient p, and depends on the relationship between the primal variable u and the dual variable p. A condition that represents this dual smoothing property of functions, that is, subgradients, is the so-called source condition. Letting f_{true} be the original image, the source condition for f_{true} reads

$$\text{There exists } \xi \in \partial R(f_{\text{true}}) \text{ such that } \xi = A^* q \text{ for a source element } q \in D(A^*), \tag{5.45}$$

where $D(A^*)$ is the domain of the operator A^*. It can be shown (cf. [BO04]) that this is equivalent to requiring from f_{true} to be a minimiser of

$$R(u) + \frac{\lambda}{2} \|Au - f_{\text{dam}}\|_2^2$$

for arbitrary $f_{\text{dam}} \in D(A^*)$ and $\lambda \in \mathbb{R}$. Now the source condition has a direct consequence for the Bregman distance, which gives rise to its use for subsequent error analysis. To be more precise, the Bregman distance is defined as

$$D_R^p(v, u) = R(v) - R(u) - \langle v - u, p \rangle_2, \quad p \in \partial R(u).$$

Then, if f_{true} fulfils the source condition with a particular subgradient ξ, we obtain

$$\begin{aligned} D_R^{\xi}(u,f_{\text{true}}) &= R(u) - R(f_{\text{true}}) - \langle u - f_{\text{true}}, \xi\rangle_2 \\ &= R(u) - R(f_{\text{true}}) - \langle q, Au - Af_{\text{true}}\rangle_2, \end{aligned}$$

and thus the Bregman distance can be related to both the error in the regularisation functional $[R(u) - R(f_{\text{true}})]$ and the output error $(Au - Af_{\text{true}})$. For the sake of symmetry properties in what follows, we shall consider the symmetric Bregman distance, which is defined as

$$\begin{aligned} D_R^{\text{symm}}(u_1,u_2) &= D_R^{p_1}(u_2,u_1) + D_R^{p_2}(u_1,u_2) \\ &= \langle u_1 - u_2, p_1 - p_2\rangle_2, \quad p_i \in \partial R(u_i). \end{aligned}$$

In addition to this error analysis, we shall get a control on the inpainting error $|u - f_{\text{true}}|$ inside the inpainting domain D by means of estimates from [CK06]. Therein the authors analysed the inpainting process by understanding how the regulariser continues level lines into the missing domain. The inpainting error then was determined by means of the definition of an error region, smoothness bounds on the level lines and quantities taking into the account the shape of the inpainting domain. In the following, we are going to implement both strategies, that is, [BRH07 and CK06], in order to prove Theorem 5.3.12.

Proof of Theorem 5.3.12 Let $f_{\text{dam}} \in L^2(\Omega)$ be the given damaged image with inpainting domain $D \subset \Omega$ and f_{true} the original image. We consider the stationary equation to (5.20), that is,

$$-\Delta p + \lambda \mathbb{1}_{\Omega\setminus D}(u - f_{\text{dam}}) = 0, \quad p \in \partial TV_\infty(u), \tag{5.46}$$

where $TV_\infty(u)$ is defined as in (5.21). In what follows, we want to characterise the error we make by solving (5.46) for u; that is, how large do we expect the distance between the restored image u and the original image f_{true} fulfiling a source condition (5.45) to be?

Now let Δ^{-1} be the inverse operator to $-\Delta$ with zero Dirichlet boundary conditions as before. In our case, the operator A in (5.44) is the embedding operator from $H_0^1(\Omega)$ into $H^{-1}(\Omega)$ and stands in front of the whole term $A(u - f_{\text{dam}})$ [cf. (5.46)]. The adjoint operator is $A^* = \Delta^{-1}$, which maps $H^{-1}(\Omega)$ into $H_0^1(\Omega)$. We assume that the given image f_{dam} coincides with f_{true} outside the inpainting domain, that is,

$$\begin{aligned} f_{\text{dam}} &= f_{\text{true}}, \quad \text{in } \Omega\setminus D, \\ f_{\text{dam}} &= 0, \quad \text{in } D. \end{aligned} \tag{5.47}$$

Further, we assume that f_{true} satisfies the source condition (5.45), that is,

$$\text{There exists } \xi \in \partial TV_\infty(f_{\text{true}}) \text{ such that } \xi = \Delta^{-1} q \text{ for a source element } q \in H^{-1}(\Omega). \tag{5.48}$$

For the following analysis, we first rewrite (5.46). For $\hat{u}$, a solution of (5.46), we get

$$\hat{p} + \lambda \Delta^{-1}(\hat{u} - f_{\text{true}}) = \Delta^{-1}\left[\lambda(1 - \mathbb{1}_{\Omega \setminus D})(\hat{u} - f_{\text{true}})\right], \quad \hat{p} \in \partial TV_\infty(\hat{u}).$$

Here we replaced f_{dam} by f_{true} using assumption (5.47). By adding a $\xi \in \partial TV_\infty(f_{\text{true}})$ from (5.48) to the perceding equation, we obtain

$$\hat{p} - \xi + \lambda \Delta^{-1}(\hat{u} - f_{\text{true}}) = -\xi + \lambda \Delta^{-1}\left[\left(1 - \mathbb{1}_{\Omega \setminus D}\right)(\hat{u} - f_{\text{true}})\right].$$

Taking the duality product with $\hat{u} - f_{\text{true}}$ [which is just the inner product in $L^2(\Omega)$ in our case], we get

$$\begin{aligned} D_{TV_\infty}^{\text{symm}}(\hat{u}, f_{\text{true}}) + \lambda \left\| \hat{u} - f_{\text{true}} \right\|_{-1}^2 &= \left\langle \nabla \xi, \nabla \Delta^{-1}(\hat{u} - f_{\text{true}}) \right\rangle_2 \\ &\quad + \lambda \left\langle \left(1 - \mathbb{1}_{\Omega \setminus D}\right)(\hat{u} - f_{\text{true}}), \hat{u} - f_{\text{true}} \right\rangle_{-1}, \end{aligned}$$

where

$$D_{TV_\infty}^{\text{symm}}(\hat{u}, f_{\text{true}}) = \left\langle \hat{u} - f_{\text{true}}, \hat{p} - \xi \right\rangle_2, \quad \hat{p} \in \partial TV_\infty(\hat{u}),\ \xi \in \partial TV_\infty(f_{\text{true}}).$$

An application of Young's inequality yields

$$D_{TV_\infty}^{\text{symm}}(\hat{u}, f_{\text{true}}) + \frac{\lambda}{2} \left\| \hat{u} - f_{\text{true}} \right\|_{-1}^2 \leq \frac{1}{\lambda} \|\xi\|_1^2 + \lambda \left\| \left(1 - \mathbb{1}_{\Omega \setminus D}\right)(\hat{u} - f_{\text{true}}) \right\|_{-1}^2 \tag{5.49}$$

For the last term, we obtain

$$\begin{aligned} \left\| \left(1 - \mathbb{1}_{\Omega \setminus D}\right) v \right\|_{-1} &= \sup_{\phi, \|\phi\|_{-1}=1} \left\langle \phi, \left(1 - \mathbb{1}_{\Omega \setminus D}\right) v \right\rangle_{-1} \\ &= \sup_{\phi, \|\phi\|_{-1}=1} -\left\langle \Delta^{-1}\phi, \left(1 - \mathbb{1}_{\Omega \setminus D}\right) v \right\rangle_2 \\ &= \sup_{\phi, \|\phi\|_{-1}=1} -\left\langle \left(1 - \mathbb{1}_{\Omega \setminus D}\right) \Delta^{-1}\phi, v \right\rangle_2 \\ &\underset{\text{Hölder}}{\leq} \|v\|_2 \cdot \sup_{\phi, \|\phi\|_{-1}=1} \left\| \left(1 - \mathbb{1}_{\Omega \setminus D}\right) \Delta^{-1}\phi \right\|_2. \end{aligned}$$

With $\Delta^{-1}: H^{-1} \to H^1 \hookrightarrow L^r$, $2 < r < \infty$, we get

$$\int_\Omega \left(\left(1 - \mathbb{1}_{\Omega\setminus D}\right) \Delta^{-1}\phi\right)^2 dx = \int_D \left(\Delta^{-1}\phi\right)^2 dx \underset{\text{Hölder}}{\leq} |D|^{1/q'} \cdot \left(\int_\Omega \left(\Delta^{-1}\phi\right)^{2q} dx\right)^{1/q}$$

$$\underset{\text{choose } q=r/2}{=} |D|^{(r-2)/r} \cdot \left\|\Delta^{-1}\phi\right\|_r^2 \leq_{H^1 \hookrightarrow L^r} C|D|^{(r-2)/r} \|\phi\|_{-1}^2 = C|D|^{(r-2)/r},$$

that is,

$$\left\|\left(1 - \mathbb{1}_{\Omega\setminus D}\right) v\right\|_{-1}^2 \leq C|D|^{(r-2)/r} \|v\|_2^2. \tag{5.50}$$

Applying (5.50) to (5.49), we see that

$$D_{TV_\infty}^{\text{symm}}(\hat{u}, f_{\text{true}}) + \frac{\lambda}{2} \left\|\hat{u} - f_{\text{true}}\right\|_{-1}^2 \leq \frac{1}{\lambda} \|\xi\|_1^2 + C\lambda\, |D|^{(r-2)/r} \left\|\hat{u} - f_{\text{true}}\right\|_2^2.$$

To estimate the last term, we use some error estimates for TV inpainting computed in [CK06]. Firstly, we have

$$\left\|\hat{u} - f_{\text{true}}\right\|_2^2 = \int_{\Omega\setminus D} (\hat{u} - f_{\text{true}})^2\, dx + \int_D (\hat{u} - f_{\text{true}})^2\, dx.$$

Since $\hat{u} - f_{\text{true}}$ is uniformly bounded in Ω [this follows from the L^∞ bound in the definition of $TV_\infty(u)$], we estimate the first term by a positive constant K_1 and the second term by the L^1 norm over D. We obtain

$$\left\|\hat{u} - f_{\text{true}}\right\|_2^2 \leq K_1 + K_2 \int_D \left|\hat{u} - f_{\text{true}}\right|\, dx.$$

Now let $\hat{u} \in BV(\Omega)$ be given by $\hat{u} = u^s + u^d$, where u^s is a smooth function and u^d is a piecewise constant function. Following the error analysis in [CK06, theorem 8] for functions $\hat{u} \in BV(\Omega)$, we have

$$\left\|\hat{u} - f_{\text{true}}\right\|_2^2 \leq K_1 + K_2 \text{err}(D) \leq K_1 + K_2 \left(|D|\, C\left(M(u^s), \beta\right) + 2\left|R(u^d)\right|\right),$$

where $M(u^s)$ is the smoothness bound for u^s, β is determined from the shape of D and the error region $R(u^d)$ is defined from the level lines of u^d. Note that in general the error region from higher-order inpainting models including the TV semi-norm is smaller than that from TV-L^2 inpainting (cf. [CK06, section 3.2.]).

Finally, we end up with

$$D_{TV_\infty}^{\text{symm}}(\hat{u}, f_{\text{true}}) + \frac{\lambda}{2} \left\|\hat{u} - f_{\text{true}}\right\|_{-1}^2 \leq \frac{1}{\lambda} \|\xi\|_1^2 + C\lambda\, |D|^{(r-2)/r} \text{err}_{\text{inpaint}}, \tag{5.51}$$

with

$$\text{err}_{\text{inpaint}} := K_1 + K_2 \left(|D|\, C\left(M(u^s), \beta\right) + 2\left|R(u^d)\right|\right).$$

The first term in (5.51) depends on the regulariser TV_∞, and the second term depends on the size of the inpainting domain D. ■

Remark 5.3.15 From inequality (5.51) we derive an optimal scaling for λ, that is, a scaling which minimises the inpainting error. It reads

$$\lambda^2 \, |D|^{(r-2)/r} \sim 1$$

and therefore,

$$\lambda \sim |D|^{-(r-2)/2r} .$$

In two space dimensions r can be chosen arbitrarily big, which gives $\lambda \sim 1/\sqrt{|D|}$ as the optimal order for λ.

Stability estimates for (5.46) also can be derived with an analogous technique. For u_i being the solution of (5.46) with $f_{\text{dam}} = f_i$ (again assuming that $f_i = f_{\text{true}}$ in $\Omega \setminus D$), the estimate

$$D_{TV_\infty}^{\text{symm}}(u_1, u_2) + \frac{\lambda}{2} \, \|u_1 - u_2\|_{-1}^2 \leq \frac{\lambda}{2} \int_D (f_1 - f_2)^2 \, dx \tag{5.52}$$

holds.

Numerical Results In this section, numerical results for the TV-H^{-1} inpainting approach (5.20) are presented. As for Cahn-Hilliard inpainting, we use convexity splitting for a numerical solution of (5.20). See also [SB11] for a numerical analysis of the resulting scheme. More precisely, we consider (5.20), where $p \in \partial TV_\infty(u)$ is replaced by its relaxed version $\nabla \cdot (\nabla u / |\nabla u|_\delta)$, where $|\nabla u|_\delta = \sqrt{|\nabla u|^2 + \delta}$. Namely, we consider

$$u_t = -\Delta \left(\nabla \cdot \left(\frac{\nabla u}{|\nabla u|_\delta} \right) \right) + \lambda \mathbb{1}_{\Omega \setminus D}(f - u). \tag{5.53}$$

Now let U_k be the approximate solution of (5.53) at time $t = k\tau$, where τ denotes the size of the discrete time steps. The numerical solution U_{k+1} of (5.53) is computed via a time-marching scheme of the form

$$\begin{aligned} &\frac{U_{k+1} - U_k}{\tau} + C_1 \Delta \Delta U_{k+1} + C_2 U_{k+1} \\ &= C_1 \Delta \Delta U_k - \Delta \left(\operatorname{div} \left(\frac{\nabla U_k}{|\nabla U_k|_\delta} \right) \right) + C_2 U_k + \lambda \mathbb{1}_{\Omega \setminus D}(f - U_k), \end{aligned}$$

with constants $C_1 > 1/\epsilon$ and $C_2 > \lambda$. As before in the case of Cahn-Hilliard inpainting in the preceding section, the condition on the constants arises as a stability condition from the numerical analysis of the scheme. The paper [SB11] is dedicated to a detailed discussion of this, including a derivation of and rigorous estimates for the preceding numerical scheme.

For the discretisation in space we used finite differences and spectral methods, that is, the discrete cosine transform, to symplify the inversion of the Laplacian Δ for the computation of U_{k+1}.

Figures 5.11 through 5.14 provide examples of the application of TV-H^{-1} inpainting to grey value images. In all examples, the total variation $|\nabla u|$ is approximated by its regularised version $\sqrt{|\nabla u|^2+\delta}$, with $\delta = 0.01$, and the time step size τ is chosen to be equal to 1. In Figure 5.12, a comparison of the TV-H^{-1} inpainting result with the result obtained by the second-order TV-L^2 inpainting model for a crop of the image in Figure 5.11 is

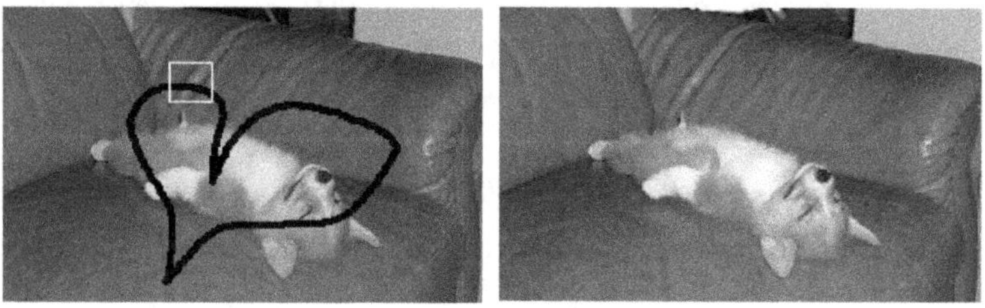

Figure 5.11. TV-H^{-1} inpainting example. Left: Picture of a dog damaged with a black marker. Right: TV-H^{-1} inpainting: $u(1,000)$ with $\lambda = 10^3$.

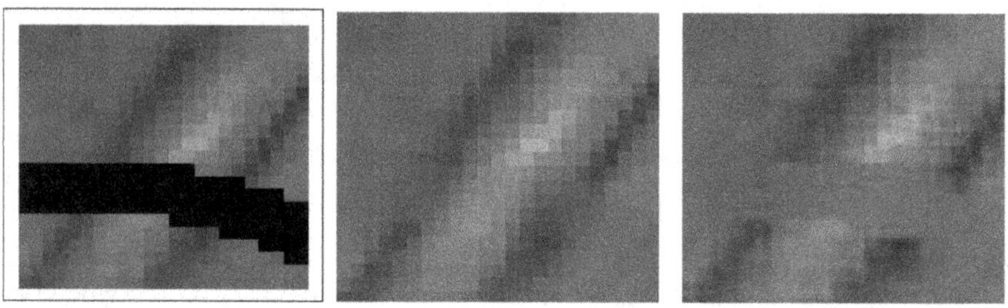

Figure 5.12. Comparison of TV-H^{-1} inpainting and TV inpainting (4.16) for the detail of Figure 5.11 framed in white. From left to right: Detail of damaged part, detail of TV-H^{-1} inpainted image $u(1,000)$ and detail of TV inpainted image $u(5,000)$.

Figure 5.13. Good continuation and connectedness: TV-H^{-1} inpainting compared to TV inpainting. From left to right: Broken stripe, TV-H^{-1} inpainted image $u(5,000)$ and TV inpainted image.

Figure 5.14. Good continuation and smoothness: TV-H^{-1} inpainting compared to TV inpainting. From left to right: Broken stripe, TV-H^{-1} inpainted image $u(5,000)$ and TV inpainted image.

presented. The superiority of the fourth-order TV-H^{-1} inpainting model to the second-order model with respect to the desired continuation of edges into the missing domain is clearly visible. Other examples which support this claim are presented in Figures 5.13 and 5.14, where the line is connected by the TV-H^{-1} inpainting model but clearly split by the TV model (4.16) discussed in Section 4.3. It would be interesting to strengthen this numerical observation with a rigorous result, as was done in [BEglG07] for Cahn-Hilliard inpainting [cf. (5.19)]. The author considers this to be another important contribution of future research.

5.4. Low Curvature Image Simplifiers

Another higher-order inpainting model proposed in [SB11] is inpainting with low curvature image simplifiers (LCISs). This approach is motivated by two famous second-order non-linear PDEs in image processing found in the work of Rudin, Osher and Fatemi [ROF92] and Perona and Malik [PM90]. These methods are based on a non-linear version of the heat equation

$$u_t = \nabla \cdot (g(|\nabla u|)\nabla u),$$

in which g is small in regions of sharp gradients. The LCIS approach represents a fourth-order relative of these non-linear second-order approaches. This method was proposed in [TT99] and later used by Bertozzi and Greer in [BG04] for the de-noising of piecewise linear signals. Related fourth-order equations combining diffusion and convection have been studied by the latter authors in [GB04a and GB04b]. In [SB11], the authors consider LCISs for image inpainting. With $f \in L^2(\Omega)$, our inpainted image u evolves in time as

$$u_t = -\nabla \cdot (g(\Delta u)\nabla \Delta u) + \lambda \mathbb{1}_{\Omega \setminus D}(f - u),$$

with thresholding function $g(s) = 1/(1+s^2)$ and $\lambda \gg 1$. Note that with $g(\Delta u)\nabla\Delta u = \nabla(\arctan(\Delta u))$, the preceding equation can be rewritten as

$$u_t = -\Delta(\arctan(\Delta u)) + \lambda \mathbb{1}_{\Omega\setminus D}(f-u). \tag{5.54}$$

Using the second derivative of the arctangent as a regularising term, spikes in the initial signal are preserved, whereas step functions are diffused (cf. [BG04]). This means that solutions to this equation are piecewise linear functions and not piecewise constant functions, like the ones produced by the TV model [ROF92]. Introducing a regularising parameter $0 < \delta \leq 1$ in $G'(y) = \arctan(y/\delta)$, we are able to control the smoothing effect. Namely, the smaller is δ, the less diffusive the equation will be.

In [BG04], the authors prove regularity of solutions of (5.54) for the case $D = \emptyset$. That is, they show that for smooth initial data $u(\cdot, t=0) = u_0$ and smooth data f, a unique smooth solution exists globally in time in one space dimension and locally in time in two dimensions. For numerical purposes, it is sufficient to have well-posedness and certain regularity properties of (5.54) on a finite time interval. Although the difference between the equation in [BG04] and our inpainting equation (5.54) is that the fidelity term is discontinuous; the results in [BG04] also suggest its validity in this case. Nevertheless, a rigorous analysis of the LCIS inpainting equation is still open to future research.

Numerical Results In this section, numerical results for the LCIS inpainting approach (5.54) are presented. The numerical scheme used is convexity splitting, the same as for Cahn-Hilliard and TV-H^{-1} inpainting in the preceding section (cf. also [SB11]).

For the comparison with TV-H^{-1} inpainting, we apply (5.54) to the same image as in Section 5.3.2. This example is presented in Figure 5.15. In Figure 5.16, the LCIS inpainting result is compared with TV-H^{-1} and

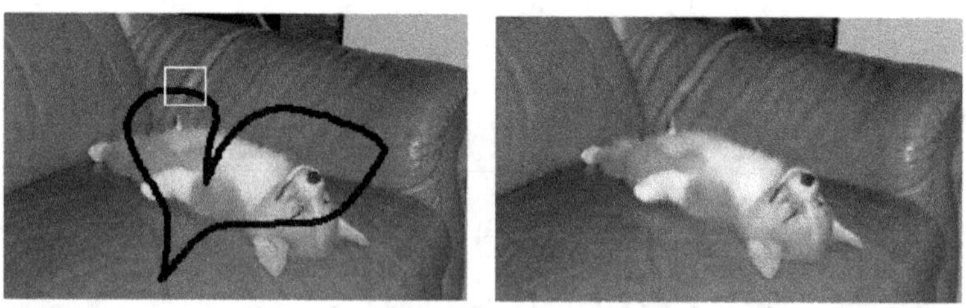

Figure 5.15. LCIS inpainting $u(500)$ with $\delta = 0.1$ and $\lambda_0 = 10^2$.

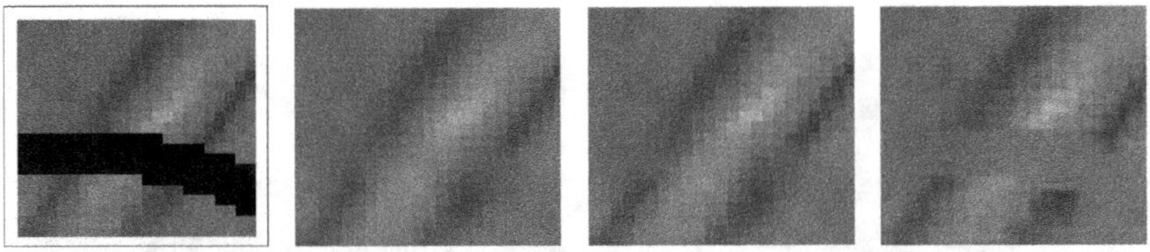

Figure 5.16. Comparison of the inpainting results with LCIS inpainting and TV-H^{-1} inpainting for a detail from the example of the dog picture in Figure 5.15. From left to right: Detail of damaged picture; $u(1,000)$ with LCIS inpainting, $u(1,000)$ with TV-H^{-1} inpainting and $u(5,000)$ with TV-L^2 inpainting.

TV-L^2 inpainting for a small part in the given image. Again, the result of this comparison indicates the continuation of the gradient of the image function into the inpainting domain. A rigorous proof of this observation is a matter for future research.

5.5. Second-Order Total Variation Inpainting

In this section we discuss a variational inpainting approach set in the space of functions of bounded Hessian that has been proposed in [PS14 and PSS13]. For a given damaged image $g \in L^2(\Omega)$ and two positive parameters α and β, an inpainted image u is computed as a minimiser of

$$\mathcal{J}(u) = \alpha \mathrm{TV}(u) + \beta \mathrm{TV}^2(u) + \tfrac{1}{2}\|\mathbb{1}_{\Omega \setminus D}(u-g)\|_2^2. \tag{5.55}$$

This and related models have been discussed for image smoothing in the context of de-noising and de-blurring previously by various authors [CL97, HS06, LT06, SS08, DWB09, LBU10, BP10]. For image inpainting, it was thoroughly studied in [PSS13]. It constitutes a straightforward higher-order extension of TV inpainting discussed in Section 4.3, to which we add a non-smooth second-order regulariser. It is the sum of the total variation $TV(u)$ plus the total variation of the gradient $TV^2(u) = TV(\nabla u)$. The appropriate Banach space for the minimisation (5.55) is the space of functions of bounded Hessian $BH(\Omega)$ (see [Dem84]). Hence, the variational problem is

$$u_r = \operatorname*{arg\,min}_{u \in \mathrm{BH}(\Omega)} \mathcal{J}(u). \tag{5.56}$$

Here, for sufficiently regular functions u, $TV^2(u)$ is the L^1 norm of the Hessian; that is, $TV^2(u) = \int_\Omega |\nabla^2 u|\, dx$.

The motivation for introducing this second-order term into the regulariser (5.55) is an attempt to propose an inpainting approach with the ability to

Figure 5.17. Removing text from an image using TV inpainting. Note that the absolute error is larger in the areas where the original image has texture. This is so because TV inpainting promotes piecewise constant reconstructions inside the inpainting domain. (a) Original image. (b) Image with text. (c) Restored image using TV inpainting with $\alpha = 0.01$. The inpainting domain was chosen to be all the white pixels. (d) Absolute difference between the original and the restored image – rescaled for better visibility. The PSNR inside the inpainting domain is $(29.47, 29.51, 31.54)$ for the RGB channels, respectively. Photo courtesy of Kostas Papafitsoros.

connect image structures across large gaps in the image while being convex and efficiently solvable. Indeed, the computational effort needed for the numerical solution of (5.56) is of the same order of magnitude as for solving the standard TV inpainting model (4.16). We will show that inpainting with (5.56) is able to re-connect broken structures even across large gaps when setting $\alpha = 0$ – however, producing a blurry rather than a sharp connection. This is due to the continuity of solutions $u \in BH(\Omega)$ of (5.56). We will give more details on the space of bounded Hessian functions in the next section. If the inpainting domain is thin enough, as it is, for example, in the case of text removal, TV inpainting performs satisfactorily (see Figure 5.17). In the task of interpolating large gaps along the inpainting domain, TV and harmonic inpainting – as we have already seen in Chapter 4 – perform poorly in comparison with TV^2 inpainting (see Figure 5.18). In the example in Figure 5.18, the task is to

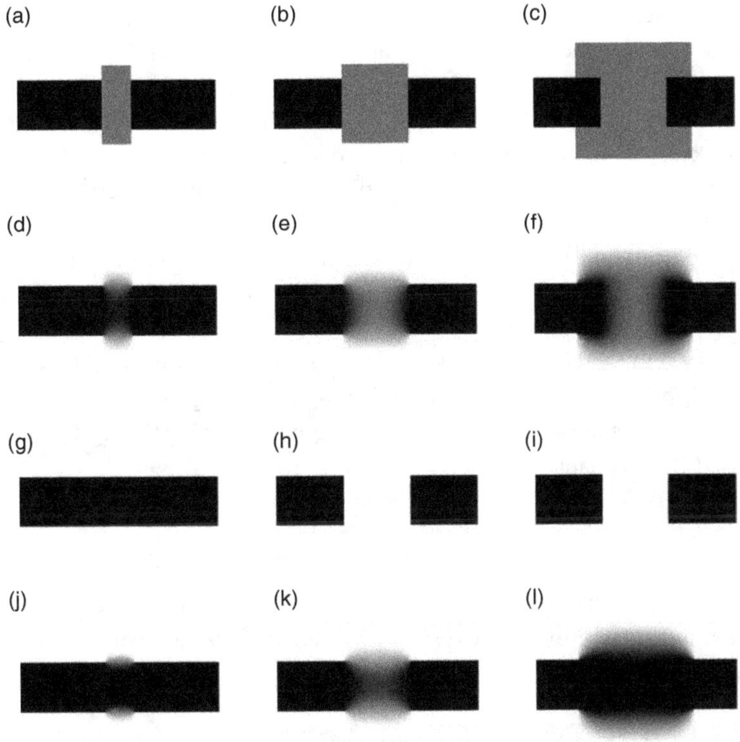

Figure 5.18. (a–c) Three different inpainting domains (grey area): (a) Small gap (b) Large gap 1. (c) Large gap 2. (d–f): Harmonic inpainting: (d) Small gap: result of harmonic inpainting with $\alpha = 0.01$. (e) Large gap 1: result of harmonic inpainting with $\alpha = 0.01$. (f) Large gap 2: result of harmonic inpainting with $\alpha = 0.01$. (g–i) TV inpainting: (g) Small gap: result of TV inpainting with $\alpha = 0.01$. (h) Large gap 1: result of TV inpainting with $\alpha = 0.01$. (i) Large gap 2: result of TV inpainting with $\alpha = 0.01$. (j–l) TV^2 inpainting: (j) Small gap: result of pure TV^2 inpainting with $\beta = 0.01$. (k) Large gap 1: result of pure TV^2 inpainting with $\beta = 0.01$. (l) Large gap 2: result of pure TV^2 inpainting with $\beta = 0.01$.

inpaint gaps of different widths in a black stripe. Harmonic inpainting achieves no connectedness, producing a rather smooth result. TV inpainting is able to connect and preserve the edge of the stripe only if the gap is small enough, whereas TV^2 inpainting is able to connect the large gap with the price of some blur. As we will see in the following discussion, this connectedness also depends on the size and geometry of the inpainting domain.

The role of the parameters α and β in combined TV-TV^2 inpainting is twofold. On the one hand, we want to keep both values of α and β small such that more weight is put on the fidelity term so that it remains close to zero and hence we essentially have $u_r = g$ in $\Omega \setminus D$. On the other hand, the values of α and β are also balancing the two regularising terms TV and TV^2. Although, as we will see in Section 5.5.2, connectedness along large gaps is essentially achieved with pure TV^2 inpainting ($\alpha = 0$), we introduce this combination in order to keep the method flexible, regarded as a superset of pure TV and pure TV^2 inpainting. In this way, we can study the effect that each term has on the inpainting image.

The well-posedness of the minimisation (5.55), that is, the existence of minimisers, and the stability results have been proven in [PS14]. Uniqueness of solutions is absent most of the time because of the non-strict convexity of the fidelity term and the regulariser. This non-uniqueness is not a fault of the variational problem but is correctly reflecting the ambiguity of the inpainting approach. This is again in line with other higher-order inpainting approaches discussed in this chapter.

5.5.1. Well-Posedness

For the well-posedness analysis, we follow [PS14]. Let us first present some facts about the space of bounded Hessian functions $BH(\Omega)$ over which we are minimising.

The space $BH(\Omega)$ The space $BH(\Omega)$ [often denoted by $BV^2(\Omega)$] is the space of *functions of bounded Hessian* . It was introduced by Demengel in [Dem84] and consists of all functions $u \in W^{1,1}(\Omega)$ whose distributional Hessian can be represented by an $\mathbb{R}^2 \times \mathbb{R}^2$-valued finite Radon measure. In other words,

$$BH(\Omega) = \{u \in W^{1,1}(\Omega) : \nabla u \in [BV(\Omega)]^2\}.$$

We set $D^2u := D(\nabla u)$. Again, it is immediate that $W^{2,1}(\Omega) \subseteq BH(\Omega)$. $BH(\Omega)$ is a Banach space equipped with the norm $\|u\|_{BH(\Omega)} = \|u\|_{BV(\Omega)} + |D^2u|(\Omega)$. If Ω has a Lipschitz boundary and is connected, then it can be shown that there exist positive constants C_1 and C_2 such that

$$\int_\Omega |\nabla u(x)|\, dx \leq C_1 |D^2u|(\Omega) + C_2 \int_\Omega |u|\, dx, \quad \forall u \in BH(\Omega).$$

Moreover, the embedding from $BH(\Omega)$ into $W^{1,1}(\Omega)$ is compact (see [Dem84]). We define the following notions of convergence in $BH(\Omega)$:

Definition 5.5.1 Weak* Convergence in BH(Ω) Let $(u_k)_{k\in\mathbb{N}}$, u belong to $BH(\Omega)$. We will say that u_k converges to u weakly* in $BH(\Omega)$ if

$$u_k \to u, \quad \text{in } L^1(\Omega),$$

and

$$\nabla u_k \rightharpoonup \nabla u \quad \text{weakly}^* \text{ in} \quad [BV(\Omega)]^2, \quad \text{as } k \to \infty,$$

or in other words

$$\|u_k - u\|_{L^1(\Omega)} \to 0,$$
$$\|\nabla u_k - \nabla u\|_{[L^1(\Omega)]^2} \to 0,$$
$$\int_\Omega \phi \, dD^2 u_k \to \int_\Omega \phi \, dD^2 u, \quad \forall \phi \in C_0(\Omega).$$

It is not hard to check that a basis for this topology consists of the following sets:

$$U(v,F,\epsilon) = \Bigg\{ u \in BH(\Omega) : \|v-u\|_{L^1(\Omega)} + \|\nabla v - \nabla u\|_{[L^1(\Omega)]^2} + \left| \int_\Omega \phi_i \, dD^2 v - \int_\Omega \phi_i \, dD^2 u \right| < \epsilon, \quad i \in F \Bigg\},$$

where $v \in BH(\Omega)$, F is finite, $\epsilon > 0$ and $\phi_i \in C_0(\Omega)$. We do not imply here that $BH(\Omega)$ is the dual space of a Banach space, but we name this convergence *weak** to show the correspondence with the weak* convergence in $BV(\Omega)$. We also have the following corresponding compactness result:

Theorem 5.5.2 Compactness in $BH(\Omega)$ *Suppose that the sequence $(u_k)_{k\in\mathbb{N}}$ is bounded in $BH(\Omega)$. Then there exists a subsequence $(u_{k_\ell})_{\ell\in\mathbb{N}}$ and a function $u \in BH(\Omega)$ such that u_{k_ℓ} converges to u weakly* in $BH(\Omega)$.*

Proof From the compact embedding of $BH(\Omega)$ into $W^{1,1}(\Omega)$ and the fact that the sequence $(\nabla u_k)_{k\in\mathbb{N}}$ is bounded in $[BV(\Omega)]^2$, we have that there exists a subsequence $(u_{k_\ell})_{\ell\in\mathbb{N}}$, a function $u \in W^{1,1}(\Omega)$ and a function $v \in [BV(\Omega)]^2$ such that u_{k_ℓ} converges to u in $W^{1,1}(\Omega)$ and ∇u_{k_ℓ} converges to v weakly* in $[BV(\Omega)]^2$ as ℓ goes to infinity. Then $\nabla u = v$, $u \in BH(\Omega)$ and u_{k_ℓ} converges to u weakly* in $BH(\Omega)$. ∎

Definition 5.5.3 Strict convergence in BH Let $(u_k)_{k\in\mathbb{N}}$, u belong to $BH(\Omega)$. We say that u_k converges to u strictly in $BH(\Omega)$ if

$$u_k \to u, \quad \text{in } L^1(\Omega),$$

and

$$|D^2 u_k|(\Omega) \to |D^2 u|(\Omega), \quad \text{as } k \to \infty.$$

It is easily checked that the function

$$d(u,v) = \int_\Omega |u-v| \, dx + \big| |D^2 u|(\Omega) - |D^2 v|(\Omega) \big|$$

is a metric and induces the strict convergence in $BH(\Omega)$. The following lemma can be used to compare these two topologies:

Lemma 5.5.4 *Suppose that $(u_k)_{k\in\mathbb{N}}$, u belong to $BH(\Omega)$ and u_k converges to u strictly in $BH(\Omega)$. Then*

$$\|u_k - u\|_{W^{1,1}(\Omega)} \to 0, \quad \text{as } k \to \infty.$$

Proof We recall from (5.57) that there exist positive constants C_1 and C_2 such that

$$\int_\Omega |\nabla u|\, dx \le C_1 |D^2 u|(\Omega) + C_2 \int_\Omega |u|\, dx, \quad \forall u \in BH(\Omega).$$

Since the sequence $(u_k)_{k\in\mathbb{N}}$ is strictly convergent in $BH(\Omega)$, the sequences $(\|u_k\|_{L^1(\Omega)})_{k\in\mathbb{N}}$ and $(|D^2 u_k|(\Omega))_{k\in\mathbb{N}}$ are bounded. Hence, there exists a positive constant C such that

$$\int_\Omega |\nabla u_k|\, dx < C, \quad \forall k \in \mathbb{N},$$

which implies that the sequence $(u_k)_{k\in\mathbb{N}}$ is bounded in $BH(\Omega)$. From the compact embedding of $BH(\Omega)$ into $W^{1,1}(\Omega)$, we get that there exists a subsequence $(u_{k_\ell})_{\ell\in\mathbb{N}}$ and a function $v \in W^{1,1}(\Omega)$ such that u_{k_ℓ} converges to v in $W^{1,1}(\Omega)$. In particular, u_{k_ℓ} converges to v in $L^1(\Omega)$, so $v = u$ and thus u_{k_ℓ} converges to u in $W^{1,1}(\Omega)$. Since every subsequence of $(u_k)_{k\in\mathbb{N}}$ is bounded in $BH(\Omega)$, we can repeat the same argument and deduce that for every subsequence of $(u_k)_{k\in\mathbb{N}}$, there exists a further subsequence which converges to u in $W^{1,1}(\Omega)$. This proves that the initial sequence $(u_k)_{k\in\mathbb{N}}$ converges to u in $W^{1,1}(\Omega)$ ■

Corollary 5.5.5 *Strict convergence implies weak* convergence in $BH(\Omega)$.*

We are now ready to present the well-posedness analysis for (5.56).

Existence of Solutions

Theorem 5.5.6 *Assuming that $\alpha > 0$, $\beta > 0$, then the minimisation problem* (5.56) *has a solution $u \in BH(\Omega)$.*

Proof Let $(u_n)_{n\in\mathbb{N}}$ be a minimising sequence for (5.56), and let $C > 0$ be an upper bound for $(\mathcal{J}(u_n))_{n\in\mathbb{N}}$. Then we have that

$$|Du_n|(\Omega) = \int_\Omega |\nabla u_n|\, dx < C \quad \text{and} \quad \frac{1}{2}\int_\Omega \mathbb{1}_{\Omega\setminus D}(g - u_n)^2 dx < C \qquad (5.57)$$

for every $n \in \mathbb{N}$. We show that the sequence $(u_n)_{n\in\mathbb{N}}$ is bounded in $L^2(\Omega)$, following essentially the proof for existence of a TV inpainted image in

Theorem 4.3.3. By the Poincaré-Wirtinger inequality (cf. Theorem B.11.8 in Appendix B), there exists a positive constant C_1 such that for every $n \in \mathbb{N}$,

$$\begin{aligned}\|u_n\|_{L^2(\Omega)} &= \left\| u_n - \mathbb{1}_\Omega \frac{1}{|\Omega|} \int u_n \, dx + \mathbb{1}_\Omega \frac{1}{|\Omega|} \int u_n \, dx \right\|_{L^2(\Omega)} \\ &\leq C_1 |Du_n|(\Omega) + \left| \int_\Omega u_n \, dx \right| \\ &\leq C + \left| \int_\Omega u_n \, dx \right|.\end{aligned}$$

Thus, it suffices to bound $\left|\int_\Omega u_n \, dx\right|$ uniformly in n. We have for every $n \in \mathbb{N}$,

$$\begin{aligned}\left\| \mathbb{1}_{\Omega \setminus D} \frac{1}{|\Omega|} \int u_n \, dx \right\|_{L^2(\Omega)} &\leq \left\| \mathbb{1}_{\Omega \setminus D} \frac{1}{|\Omega|} \int u_n \, dx - \mathbb{1}_{\Omega \setminus D} u_n \right\|_{L^2(\Omega)} \\ &\quad + \left\| \mathbb{1}_{\Omega \setminus D}(u_n - g) \right\|_{L^2(\Omega)} + \|\mathbb{1}_{\Omega \setminus D} g\|_{L^2(\Omega)} \\ &\leq \|\mathbb{1}_{\Omega \setminus D}\| \left\| u_n - \mathbb{1}_\Omega \frac{1}{|\Omega|} \int u_n \, dx \right\|_{L^2(\Omega)} \\ &\quad + \left\| \mathbb{1}_{\Omega \setminus D}(u_n - g) \right\|_{L^2(\Omega)} + \|\mathbb{1}_{\Omega \setminus D} g\|_{L^2(\Omega)} \\ &\leq C_1 \|\mathbb{1}_{\Omega \setminus D}\| |Du_n|(\Omega) + \sqrt{2C} + \|\mathbb{1}_{\Omega \setminus D} g\|_{L^2(\Omega)} \\ &\leq C_1 \|\mathbb{1}_{\Omega \setminus D}\| C + \sqrt{2C} + \|\mathbb{1}_{\Omega \setminus D} g\|_{L^2(\Omega)} := C'.\end{aligned}$$

It follows that

$$\left| \int_\Omega u_n \, dx \right| \|\mathbb{1}_{\Omega \setminus D}\|_{L^2(\Omega)} \leq C' |\Omega|,$$

and thus,

$$\left| \int_\Omega u_n \, dx \right| \leq \frac{C' |\Omega|}{\|\mathbb{1}_{\Omega \setminus D}\|_{L^2(\Omega)}},$$

since $\int_\Omega \mathbb{1}_{\Omega \setminus D} \, dx \neq 0$.

Since the sequence is bounded in $L^2(\Omega)$ and Ω is bounded, we have that the sequence is bounded in $L^1(\Omega)$ and also it is bounded in $BH(\Omega)$. From Theorem 5.5.2, we obtain the existence of a subsequence $(u_{n_\ell})_{\ell \in \mathbb{N}}$ and $u \in BH(\Omega)$ such that u_{n_ℓ} converges to u weakly* in $BH(\Omega)$. Trivially, the functional (5.55) is lower semi-continuous with respect to that convergence, which implies that

$$u = \min_{v \in BH(\Omega)} \mathcal{J}(v). \qquad \blacksquare$$

Let us note here that in this proof we needed $\alpha > 0$ in order to get an a priori bound in the L^1 norm of the gradient (for $\beta = 0$, compare Theorem 4.3.3). For $\alpha = 0$, existence can still be proven deriving an L^1 bound on the gradient by integration by parts and the Poincaré-Wirtinger inequality (see [Pap14]).

To complete the well-posedness picture for (5.56), it remains to analyse the stability of the method. More precisely, we want to know the effect that deviations in the data g have on a corresponding minimiser of (5.55). A similar error and stability analysis as we presented for TV-H^{-1} inpainting in Section 5.3 can be adapted to inpainting with bounded Hessian functions (5.56). We refer readers to [PS14 and Pap14] for a full description of this analysis for TV^2 inpainting.

5.5.2. Inpainting Examples

Numerical Solution For the numerical solution of (5.55), we employ the Split-Bregman algorithm [GO09]. For a detailed description of the resulting discrete algorithm for (5.55), we refer readers to the article [PSS13] and the demonstration on IPOL (http://www.ipol.im).

General Examples In this section we give some inpainting examples. Mainly, we want to emphasise the differences in practice between pure first-order ($\beta = 0$) and pure second-order ($\alpha = 0$) TV inpainting.

Figure 5.19 provides a first example where we try to remove a large-font text from a natural image. The TV^2 result seems more pleasant to the human

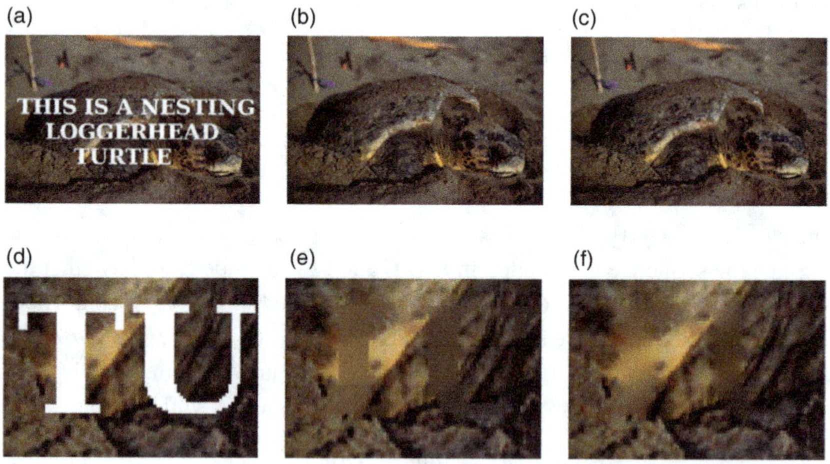

Figure 5.19. Removing large-font text from an image. The PSNR inside the inpainting domain for the three channels is (34.19, 34.97, 36.99) for the TV inpainting and (34.74, 35.34, 37.20) for the TV^2 inpainting. (a) Original image with text. (b) Pure TV inpainting, $\alpha = 0.001$. (c) Pure TV^2 inpainting, $\beta = 0.001$. (d) Detail of original image. (e) Detail of TV inpainting. (f) Detail of TV^2 inpainting. Photo courtesy of Kostas Papafitsoros.

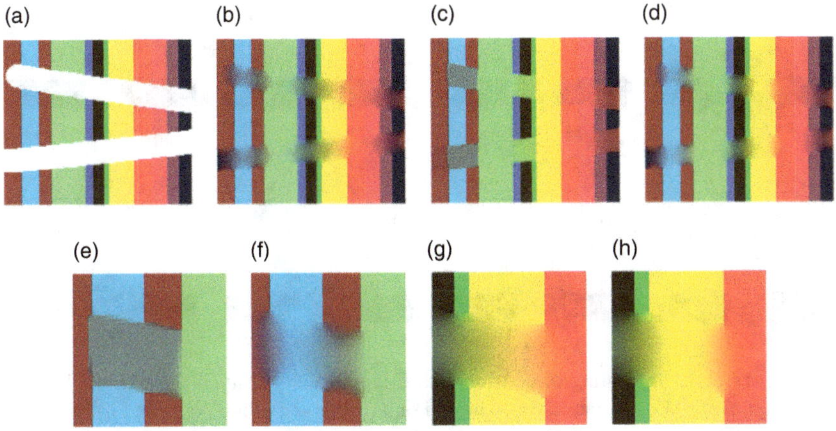

Figure 5.20. Coloured stripes inpainting. Observe the difference between TV and TV^2 inpainting in the light blue stripe at the left (e and f), where in the TV^2 case the blue colour is propagated inside the inpainting domain. (a) Original image. (b) Harmonic inpainting, $\alpha = 0.001$. (c) Pure TV inpainting, $\alpha = 0.001$. (d) Pure TV^2 inpainting, $\beta = 0.001$. (e) Detail of TV inpainting. (f) Detail of TV^2 inpainting. (g) Detail of harmonic inpainting. (h) Detail of TV^2 inpainting.

eye (subjective) as TV inpainting produces piecewise constant parts in the inpainting domain. This is confirmed quantitatively by the PSNR values inside the inpainting domain, as these are higher for the TV^2 inpainting.

Figure 5.20 presents a second example with coloured stripes. Here we also include a harmonic inpainting example. As expected, in the case of TV inpainting, the image is piecewise constant inside the inpainting domain (see Figure 5.20c). This produces a desirable result in stripes whose width is larger than the size of the gap, connecting the gap and preserving the edges there, whereas TV^2 inpainting (Figure 5.20d) adds some additional blur at the edge of the stripe that belongs to the inpainting domain. TV inpainting fails to connect the thin stripes, whereas TV^2 inpainting more reliably propagates the right colour (see Figures 5.20e and f). Finally, notice the difference between harmonic and TV^2 inpainting in terms of connectedness; see, for example, the yellow stripe in Figure 5.20g and h.

What happens with the inpainting result when we continuously move from pure TV to pure TV^2 inpainting? In Figure 5.21, we are making the transition from pure TV to pure TV^2 by varying α and β. The parameters α and β vary as follows:

$$\begin{aligned} \alpha: &\quad 0.01 \to 0.008 \to 0.006 \to 0.005 \to 0.004 \to 0.002 \to 0, \\ \beta: &\quad 0 \quad \to 0.002 \to 0.004 \to 0.005 \to 0.006 \to 0.008 \to 0.01. \end{aligned}$$

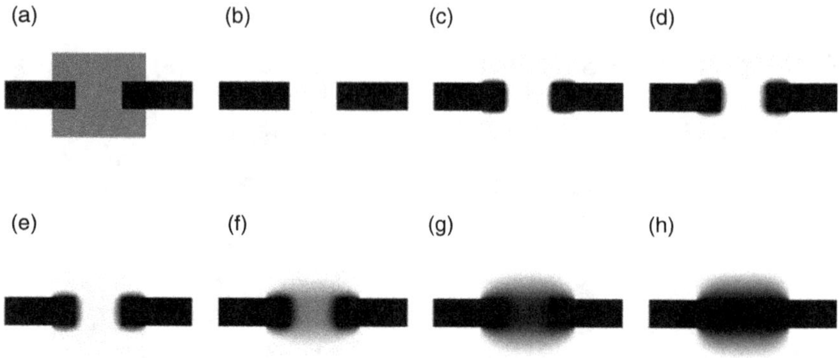

Figure 5.21. Inpainting of a stripe with a large gap. Transition from pure TV to pure TV^2. Connectedness is achieved only for large ratios β/α (g and h). No connectedness is obtained when the weights α and β are equal (e). (a) Inpainting domain. (b) $\alpha = 0$, $\beta = 0.01$. (c) $\alpha = 0.008$, $\beta = 0.002$. (d) $\alpha = 0.006$, $\beta = 0.004$. (e) $\alpha = 0.005$, $\beta = 0.005$. (f) $\alpha = 0.004$, $\beta = 0.006$. (g) $\alpha = 0.002$, $\beta = 0.008$. (h) $\alpha = 0$, $\beta = 0.01$.

As we see in Figure 5.21, the large gap in the stripe is connected only for large values of β and small values of α, that is, for large values of the ratio β/α. In the case where both weights α and β are equal, we observe no connection.

Influence of the Inpainting Domain on Large-Gap Connection Figures 5.18l and 5.21h show that pure TV^2 inpainting has the ability to connect large gaps along the inpainting domain. However, as numerical experiments have shown, the quality of the connection depends on the geometry of the inpainting domain D. In particular, for the broken-stripe example in Figure 5.22, the domain must extend above, underneath, and to the left and right of the gap between the two parts of the stripe. For example, in Figure 5.22e, where the inpainting domain is just the area between the two parts, we do not have connection, whereas in Figure 5.22j, we have. Sufficient conditions on the geometry of D for achieving connectedness are not known in general. Indeed, this could be an interesting topic for future research.

For now – to gain a better intuitive feeling for the nature of this inpainting behaviour – let us discuss the following heuristic considerations for the harmonic and bi-harmonic inpainting results in Figures 5.18 and 5.23. Both these inpainting methods are quadratic and hence possess linear Euler–Lagrange equations, which makes the qualitative study of minimisers easier. In what follows, we assume that both g and the boundary of the inpainting domain ∂D are sufficiently smooth. For harmonic inpainting, the

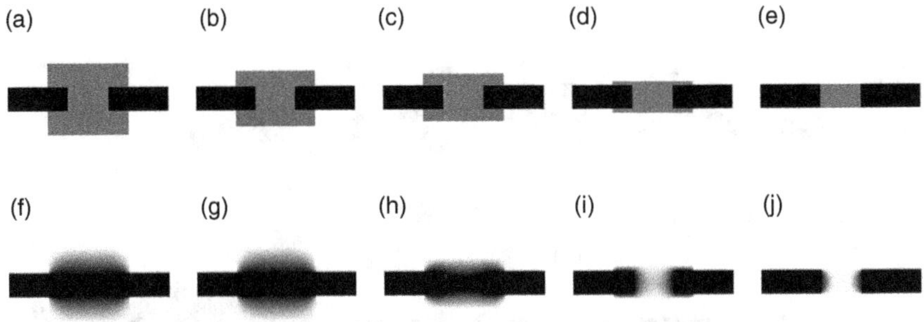

Figure 5.22. Different pure TV^2 inpainting results for different inpainting domains of decreasing height. In all computations, we set $\beta = 0.001$. (a) Domain 1. (b) Domain 2. (c) Domain 3. (d) Domain 4. (e) Domain 5. (f) Domain $1, \mathrm{TV}^2$. (g) Domain $2, \mathrm{TV}^2$. (h) Domain $3, \mathrm{TV}^2$. (i) Domain $4, \mathrm{TV}^2$. (j) Domain $5, \mathrm{TV}^2$.

corresponding Euler–Lagrange equation reads

$$-\alpha \Delta u + 2\mathbb{1}_{\Omega \setminus D}(u-g) = 0, \quad \text{in } \Omega.$$

Taking the limit $\alpha \to 0$ leads to the following boundary-value problem:

$$\begin{aligned} \Delta u &= 0, \quad \text{in } D, \\ u &= g, \quad \text{on } \partial D. \end{aligned} \tag{5.58}$$

Now we can draw the connection to the heat equation. Let $u(x)$ be the temperature in a point $x \in \mathbb{R}^2$. Interpreting the black stripe in Figure 5.18 to be a heating rod that heats up the grey container defined by the inpainting domain D, the inpainting results for small α can be explained as the equilibrium heat distribution inside the grey container, that is, the u that solves (5.58). Whereas in Figure 5.18b the heating rods are only touching the container at the outside of its walls – resulting into a rather modest and almost one-dimensional diffusion of heat in the container – in Figure 5.18c the hot rods are stuck into the grey container, which should be heated up, resulting into a bloblike diffusion of heat in both horizontal and vertical directions. In this case, the inpainting results are almost the same, and no connectedness can be achieved in either case.

Turning our attention to bi-harmonic inpainting

$$\min_u \int_{\Omega \setminus D} (u-g)^2 \, dx + \int_\Omega (\Delta u)^2 \, dx, \tag{5.59}$$

the situation is different. In this case, the corresponding Euler-Lagrange equation reads

$$\alpha \Delta^2 u + 2\mathbb{1}_{\Omega \setminus D}(u-g) = 0, \quad \text{in } \Omega.$$

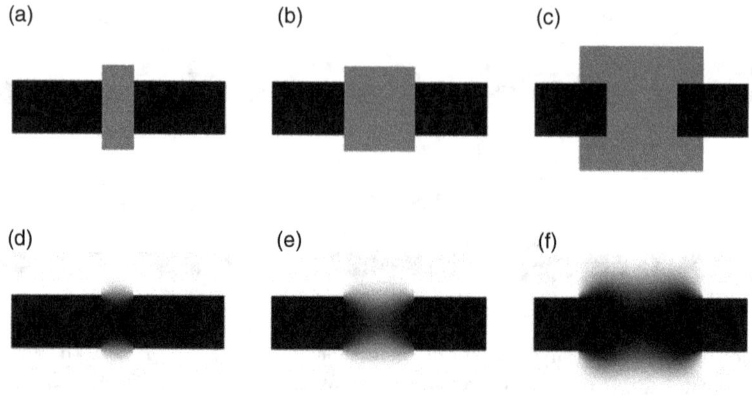

Figure 5.23. Examples of bi-harmonic inpainting. The inpainting domain influences the connection of the stripe in a similar way to pure TV^2 inpainting (Figure 5.22). (a) Small gap. Inpainting domain: grey area. (b) Large gap. Inpainting domain 1: grey area. (c) Large gap. Inpainting domain 2: grey area. (d) Result of bi-harmonic inpainting with $\alpha = 0.05$. (e) Result of bi-harmonic inpainting with $\alpha = 0.05$. (f) Result of bi-harmonic inpainting with $\alpha = 0.05$.

Applying the same argumentation as for harmonic inpainting earlier, we get the following limiting behaviour

$$\begin{aligned} \Delta^2 u &= 0, && \text{in } D, \\ u &= g, && \text{on } \partial D, \\ &+ \text{ an appropriate second boundary condition,} && \text{on } \partial D. \end{aligned} \tag{5.60}$$

The bi-harmonic equation (5.60) models the bending of an elastic flat plate enforced by fixing it to given values on the boundary ∂D. Here $u(x)$ denotes the displacement of the plate from its initial flat state. Taking the plate model (5.60) as a good approximation of the bi-harmonic inpainting model for small α, we can now attempt to interpret the inpainting results in Figure 5.23e and f. In the first setup in Figure 5.23b, the plate is fixed to 0, that is, the grey value is black, on two parts of the vertical boundaries only and fixed to 1 (white) on the rest of the boundary. The resulting 'plate' has a saddle-point shape, failing to connect the two ends of the black stripe. However, in Figure 5.23c the plate is fixed to 0 on parts of the vertical and horizontal boundaries, where the latter presses the saddle point down to a minimum point at 0 such that the ends of the stripe are (although not through a perfect line) connected this time. Indeed, in contrast to harmonic inpainting, the bi-harmonic inpainting result seems to depend heavily on the geometry of the inpainting domain D.

Going back to the numerical examples for TV^2 inpainting in Figure 5.22, a similar behaviour to that of bi-harmonic inpainting can be observed. As before, we see that in order to achieve perfect connectedness, it seems crucial that the inpainting domain is larger than the gap between the two stripes. In contrast to bi-harmonic inpainting, though, the non-linearity within the fourth-order diffusion of the corresponding Euler-Lagrange equation introduces less blur and results in a straighter connection of the line.

5.6. Further Reading and Some Extensions

In this chapter we have discussed the idea of introducing higher-order derivatives in PDE inpainting methods to produce inpaintings that are closer to Kanizsa's good continuation principle from Chapter 3. In particular, we have discussed the idea of curvature penalisation in inpainting with Euler elastica curves in Section 5.2 and fourth-order non-linear diffusion inpainting with Cahn-Hilliard, higher-order TV and low curvature image simplifiers in Sections 5.3 through 5.5. These are, of course, not the only attempts to develop higher-order inpainting methods. Let us give credit to some more of them, very briefly, in this concluding section.

The first work connecting image inpainting to a third-order PDE is the transport process of Bertalmio et al. [BSCB00], which we will discuss in more detail in Chapter 6. A diffusion-based third-order approach to image inpainting is curvature-driven diffusion (CDD) [CS01b], which we have already touched on briefly in Section 4.5. Although solving the problem of connecting level lines over large distances (connectedness principle), the level lines are still interpolated linearly, and thus, curvature is not preserved. Other higher-order inpainting algorithms are inpainting with combinations of second- and higher-order methods [LLT03, LT06, TOH07, HTBB11], similar to TV-TV^2 inpainting from Section 5.5, and a higher-order version of the free-discontinuity approach of Mumford-Shah in [ES02], which will be discussed in more detail in Section 7.2.

Let us mention one more higher-order approach that is interesting because it raises the order of existing approaches by one. In [LMS13], a third-order anisotropic TV regulariser – resulting into a sixth-order corresponding PDE – was proposed for the interpolation of surfaces. In this approach, both the third order and the anisotropy introduced by a vector field are crucial for the inpainting to carry over any non-smoothness of given level lines to the interpolant. There the authors considered the problem of reconstructing an unknown two-dimensional height map $u\colon \Omega \to \mathbb{R}$ on a two-dimensional

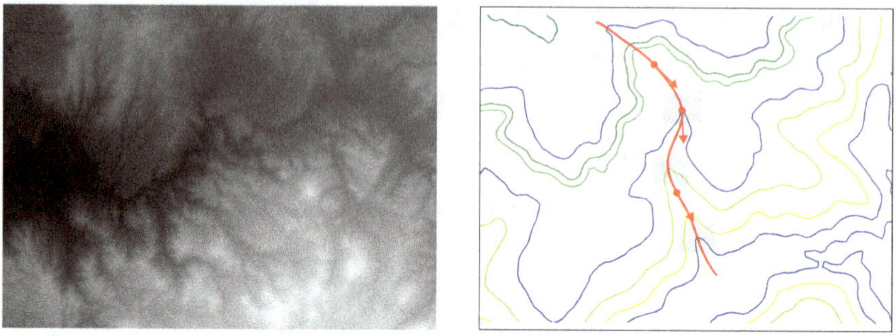

Figure 5.24. Surface interpolation for digital elevation maps (DEMs). Left: Exemplary digital elevation map. DEMs have a unique structure which requires careful consideration when choosing a regulariser to avoid removing important features. Right: Surface interpolation problem. Based on the given level lines (blue), the task is to reconstruct the surface between the level lines. A particular difficulty is that level lines can have points of high curvature or even be non-smooth (marked rectangular regions), whereas there are generally no non-differentiabilities when *crossing* the contours along a path that associates similar points (red). The proposed approach relies on a vector field v (red arrows) that approximates the tangents to such paths in conjunction with a suitable anisotropic regulariser.

domain $\Omega \subseteq \mathbb{R}^2$ based on the values of u on a small number of level lines: $u(x) = l_i$ for $x \in C_i$, $i = 1, \ldots, N$, where $C_i = u^{-1}(\{l_i\})$ are the known level lines (or only parts of level lines – or even only the values on a set of disjoint points). This is a problem that often appears in connection with digital elevation maps (DEMs), such as in DEM reconstruction from sparse measurements or tidal coastline data. The author, proposed the following variational problem for this task:

$$\min_u \{\|D^3 u(v, \cdot, \cdot)\|, \quad u(x) = u_0(x) \quad \text{for } x \in C\}, \tag{5.61}$$

where $D^3u(v,\cdot,\cdot)$ denotes the directional derivative of the second-order derivative operator of u in the direction of v, v is an *auxiliary vector field* and $C \subseteq \Omega$ is the set on which the data are known. The assumption underlying the derivation of (5.61) in [LMS13] is that the level lines of u can be non-smooth but are generally 'similar' to each other; that is, the points on two sufficiently close level lines can be associated with each other (Figure 5.24, right). Therefore, an inpainting method for reconstructing a surface from sparse elevation data should (1) *coincide with the given data* on the set C, (2) the interpolated level lines should *preserve the geometry of the given level lines* – in particular, non-differentiabilities – as accurately as possible and

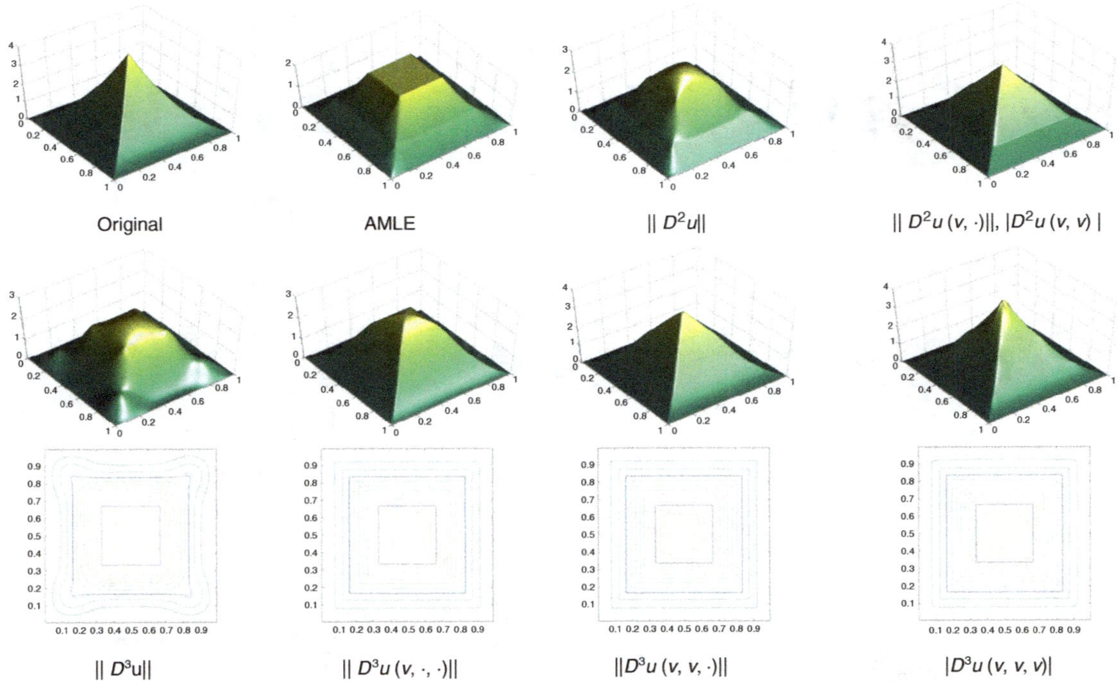

Figure 5.25. Comparison of different notions of anisotropy with known directions v. The input consists of the contour lines marked in blue, including the boundary of the domain, and contains a level line around a region with a local maximum. In consequence, this is an example of a problem that cannot be solved by pointwise interpolation between contour lines. AMLE does not extrapolate the tip and introduces kinks along the given contours. The non-directional approaches result in smoothed-out contours. In contrast, the directional methods do not smooth the level lines, but they still regularise – as desired – the *spacing* of the contours to a varying degree.

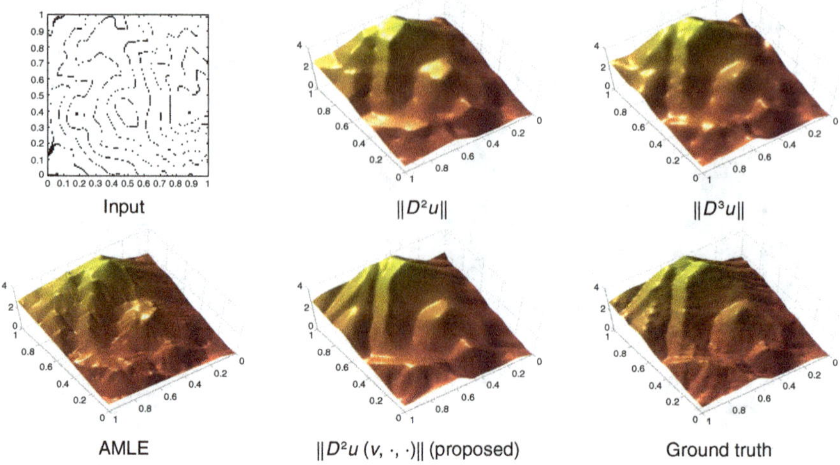

Figure 5.26. Reconstruction of real-world digital elevation maps. Top row: Input contours, second-order isotropic total variation, third-order isotropic total variation. Bottom row: AMLE, proposed method using $\|D^3u(v,\cdot,\cdot)\|$, ground truth. AMLE does not correctly recover the small peak and tends to hallucinate features. The proposed method correctly recovers the mountaintops and ridges (top left corner).

(3) the interpolated surface should define a *smooth transition* – at least continuity of the gradient – *across* level lines (e.g., along the red path in Figure 5.24).

It is demonstrated in [LMS13] that for an interpolation algorithm to fulfil all these requirements, a *third-order* regulariser R is needed, and moreover, it is necessary to include *directional* information, that is, anisotropy, in the form of an *auxiliary vector field* v that incorporates information about the relation between adjacent level lines. This unknown vector field v models the direction in which the *normals of the level lines change least.* In [LMS13], the authors proposed an efficient method – starting with a random initialisation – of alternatingly computing u and v. Figures 5.25 and 5.26 provide an illustration of (5.61) and its comparison with other diffusion-based inpainting methods such as AMLE inpainting from Section 4.4 and TV^2 from Section 5.5.

6
Transport Inpainting

In this chapter we consider the generic transport model

$$u_t = \vec{c} \cdot \nabla u \tag{6.1}$$

with direction field $\vec{c}$ for image inpainting. In contrast to the previous two chapters, where the main mechanism was diffusion or – as we will see in Chapter 8 – a combination of diffusion and transport, in this chapter we discuss two particular partial differential equation (PDE) – based inpainting methods that are solely based on transport dynamics. In Section 6.1 we start with inpainting model of Bertalmio et al. [BSCB00], which interpolates by transporting the known image structures along the level lines of the interpolant. Section 6.2 is dedicated to one of the most successful PDE inpainting approaches, that proposed by Bornemann and März [BM07], which interpolates by coherence transport. Note that in this chapter g denotes either a nonlinear diffusivity function (in Section 6.1) or a guidance vector field (in Sections 6.2 and 6.3). The given image with inpainting domain D is donated by f.

6.1. Inpainting by Transport Along Level Lines

Digital image restoration received its synonym 'inpainting' in the pioneering work by Bertalmio et al. [BSCB00]. Their work constitutes the first automatic inpainting algorithm for digital images, which used a discrete PDE model that was motivated by the work of art conservators. Further, their algorithm introduced the important concept of propagating both the gradient direction and the grey values into the inpainting domain.

Consulting conservators at the Minneapolis Institute of Art their first conclusion was that inpainting is a subjective procedure which differs for each art piece and for each conservator. Nonetheless, the authors of [BSCB00] gathered the underlying methodology of inpainting in the following four (chronological) points:

1. The restoration process should preserve the unity of the art piece. Hence, the respective inpainting procedure is determined in dependence on the global impression of the work.
2. The actual inpainting of the picture starts with the prolongation of structural information from the surroundings of the gap into its inner part. This means that contour lines arriving at the boundary of the gap are continued into the gap 'smoothly'.
3. From reconstruction of the contour lines, different regions inside the gap form. These regions are filled in with colour given by adjacent parts of the gap's boundary.
4. In a final step, small details, that is, texture, are added.

The algorithm proposed in [BSCB00] simultaneously performs steps 2 and 3 of this methodology in an iterative manner. The method is motivated by the PDE

$$u_t = \nabla^{\perp} u \cdot \nabla \Delta u. \tag{6.2}$$

This equation convects the image intensity u along level lines Δu, which were chosen as a 'smoothness' estimator of the image function. Thinking about the inpainted image as a steady state of this equation, the inpainted image is the one whose level lines in the direction of $\nabla^{\perp} u$ are parallel to the level lines of Δu, that is,

$$\nabla^{\perp} u \cdot \nabla \Delta u = 0. \tag{6.3}$$

This fits with the observation that $\nabla^{\perp} u \cdot \nabla \Delta u = -\nabla^{\perp} \Delta u \cdot \nabla u$, which lends itself to another interpretation of the inpainting process in terms of (6.1): the transport of the image intensity u that is convected by the field $\vec{c} = -\nabla^{\perp} \Delta u$.

To avoid the crossing of level lines, the 'transport equation' (6.2) is extended by a small anisotropic diffusion term, added to the inpainting process with a small weight parameter $\nu > 0$. More precisely, the authors in [BSCB00] considered a discrete approximation of the PDE

$$u_t = \underbrace{\nabla^{\perp} u \cdot \nabla \Delta u}_{\text{Smoothness transport}} + \underbrace{\nu \nabla \cdot (g(|\nabla u|) \nabla u)}_{\text{Anisotropic diffusion}}. \tag{6.4}$$

Here the diffusion function g is monotonically decreasing and depends on the size of the gradient of the image; for example, $g(|\nabla u|) = 1/(1 + |\nabla u|^2)$. In this way, the diffusion is stronger in homogeneous areas of the image and weaker in the presence of edges (large $|\nabla u|$) and as such is edge preserving. The inpainted image is computed as a steady-state solution of (6.4).

Fluid Dynamics Interpretation Another interpretation of the inpainting process of (6.4) is given in [BBS01] in terms of an incompressible fluid flow. Incompressible Newtonian fluids are described by the Navier-Stokes equations for the divergence-free velocity field $v = (v^1, v^2)^t \in \mathbb{R}^2$ of the fluid and its pressure $p \in \mathbb{R}$

$$\begin{cases} v_t + v \cdot \nabla v = -\nabla p + \nu \Delta v, \\ \nabla \cdot v = 0. \end{cases} \tag{6.5}$$

The second equation in (6.5) expresses the divergence-free property of the velocity field. This can be encoded in choosing a velocity field v that points in the direction of the level lines of a stream function Ψ, that is, choosing $v = \nabla^{\perp}\psi$, since then $\nabla \cdot v = \partial_{xy}\Psi - \partial_{xy}\Psi = 0$. Moreover, we can define the vorticity function $\omega = \nabla \times v = v_x^2 - v_y^1$ for v. Then, taking the *curl* in the first equation of (6.5) with $\nu = 0$, we get an equation for the vorticity ω:

$$\begin{aligned} (\nabla \times v)_t + \nabla \times \left(v^1 v_x^1 + v^2 v_y^1, v^1 v_x^2 + v^2 v_y^2\right)^t &= -(p_{yx} - p_{xy}), \\ (\nabla \times v)_t + v^1(v_{xx}^1 - v_{yx}^2) + v^2(v_{yx}^1 - v_{yy}^2) + (v_x^1)^2 - (v_y^2)^2 &= 0, \\ \omega_t + v \cdot \nabla \omega &= 0, \end{aligned}$$

where the last equivalence follows from the fact that v is divergence free. Then the vorticity equation for $\nu = 0$ is the Euler equation for inviscid flow and reads

$$\omega_t + v \cdot \nabla \omega = 0. \tag{6.6}$$

In terms of the stream function ψ, where $\omega = \Delta\psi$, a stationary solution of (6.6) fulfils

$$\nabla^{\perp}\psi \cdot \nabla\Delta\psi = 0.$$

This makes the analogy to inpainting obvious. The stream function for an inviscid flow in two dimensions satisfies the same equation as the steady state of (6.3) for the image intensity function u. In this flavour, the authors of [BBS01] proposed the following solution of the inpainting task: For a given image $f \in c^2(\Omega)$, solve for ω and u in D

$$\omega_t + v \cdot \nabla\omega = \nu \nabla \cdot (g(|\nabla\omega|)\nabla\omega), \quad \omega|_{\partial D} = \Delta f, \tag{6.7}$$

$$\omega = \Delta u, \qquad u|_{\partial D} = f. \tag{6.8}$$

Note that this inpainting approach also resembles the transport equation (6.4) in terms of the conditions imposed on the boundary of the inpainting domain D. In [BBS01], the boundary conditions for (6.4) were calculated from a small sub-band around the inpainting domain D (i.e., D is dilated to D_ϵ by small balls of radius ϵ, and the sub-band is defined as $D_\epsilon \setminus D$). Fixing both the grey values u and the gradient of the image function within this sub-band, the inpainting equation (6.4) transports both the grey values and the direction of the level lines into D.

For image inpainting, the fluid interpretation of smoothness transport is interesting because it opens up a completely new door to analytical results for the inpainting model. Existence and uniqueness for solutions of (6.7) with linear diffusion, that is, $g = 1$, is well established (cf., e.g., [MB02]). For other diffusion functions g, that is, in the case of non-linear diffusion, the situation is not so clear. In fact, non-linear diffusion processes are known to be ill posed for a wide class of functions g. However, as pointed out in [BBS01], this ill-posedness can be removed by smoothing the image gradient inside the argument of g (cf., e.g., [CLMC92]).

The next paragraph describes the inpainting algorithm from [BSCB00] in its discrete formulation.

The Algorithm The algorithm in [BSCB00] was proposed in a discrete setting. Let $\Omega = [0,M] \times [0,N]$ be the image domain and $F = F(i,j) : \Omega \to \mathbb{R}$ a grey value image which is damaged inside $D \subset \Omega$, the inpainting domain. Then a sequence of images $U = U^n(i,j) : \Omega \times \mathbb{N} \to \mathbb{R}$ is constructed such that $U^0(i,j) = F(i,j)$ and $\lim_{n\to\infty} U^n(i,j) = U^*(i,j)$, the inpainted image. The images U^n are computed by the iteration

$$U^{n+1}(i,j) = U^n(i,j) + \Delta t U_t^n(i,j), \quad \forall (i,j) \in D,$$

where Δt is the step size of the iteration, whose update is given by U_t^n. The latter is a discrete approximation of $\nabla^\perp u \cdot \nabla \Delta u$ in (6.4). It reads

$$U^{n+1}(i,j) = U^n(i,j) + \Delta t \vec{\delta \Delta U^n}(i,j) \cdot \nabla^\perp U^n(i,j), \quad \forall (i,j) \in D, \tag{6.9}$$

where $\vec{\delta L}$ is a measure of the change in L (the discrete approximation of the continuous gradient ∇L).

To ensure the correct evolution of the direction field, in particular, to avoid the crossing of level lines, the preceding discrete inpainting process is intervened with intermediate steps of anisotropic diffusion that approximate the second term in (6.4). This reads

$$U^{n+1}(i,j) = U^n(i,j) + \Delta t \mathbb{1}_\epsilon(i,j) \cdot \nabla \cdot \left(\frac{\nabla U^n}{|\nabla U^n| + \epsilon} \right)(i,j) \cdot |\nabla U^n|(i,j), \quad \text{in } D^\epsilon, \tag{6.10}$$

where D^ϵ is a dilation of D with a ball of radius ϵ, and $\mathbb{1}_\epsilon$ is a smooth function that is fixed to $\mathbb{1}_\epsilon = 0$ on ∂D^ϵ and $\mathbb{1}_\epsilon = 1$ in D. The role of $\mathbb{1}_\epsilon$ is to impose Dirichlet boundary conditions on the stripe $D^\epsilon \setminus D$.

Transport Inpainting Algorithm [BSCB00]

Solve T times:
  Solve M times:
  (6.9): $U^{n+1}(i,j) = U^n(i,j) + \Delta t \delta \vec{\Delta U}^n(i,j) \cdot \nabla^\perp U^n(i,j)$, in D
  end

  Solve N times:
  (6.10): $U^{n+1}(i,j) = U^n(i,j) + \Delta t \mathbb{1}_\epsilon(i,j) \cdot \nabla \cdot \left(\frac{\nabla U^n}{|\nabla U^n|}\right)(i,j) \cdot |\nabla U^n|(i,j)$, in D^ϵ
  end
end

In what follows, we discuss the implementation details of the inpainting algorithm (6.4) as proposed in [BSCB00] and described in detail in Bertalmio's thesis [Ber01].

The Implementation Let us start with implementation of the transport equation (6.9)

$$U^{n+1}(i,j) = U^n(i,j) + \Delta t U_t^n(i,j), \quad \text{in } D.$$

Here U_t^n is computed as

$$U_t^n(i,j) = \left(\delta \vec{\Delta U}^n(i,j) \cdot \frac{\nabla^\perp U^n(i,j)}{|\nabla^\perp U^n(i,j)|}\right) |\nabla U^n(i,j)|_{sl}, \tag{6.11}$$

where the original update term in (6.9) has been multiplied by a slope-limited version of the norm of the gradient of U, that is, by $|\nabla U^n(i,j)|_{sl}$, in order to stabilise the numerical scheme (see [OS88] for details). The single terms that constitute the update (6.11) are computed as follows:

$$\begin{aligned}
L^n(i,j) &:= \Delta U^n(i,j) = U^n_{xx}(i,j) + U^n_{yy}(i,j),\\
\delta \vec{L}^n(i,j) &:= (L^n(i+1,j) - L^n(i-1,j), L^n(i,j+1) - L^n(i,j-1)),\\
\vec{N}^n(i,j) &:= \nabla^\perp U^n(i,j),\\
\frac{\vec{N}^n(i,j)}{|\vec{N}^n(i,j)|} &:= \frac{(-U^n_{yc}(i,j), U^n_{xc}(i,j))}{\sqrt{(U^n_{xc}(i,j))^2 + (U^n_{yc}(i,j))^2}},\\
\beta^n(i,j) &:= \delta \vec{L}^n(i,j) \cdot \frac{\vec{N}^n(i,j)}{|\vec{N}^n(i,j)|},
\end{aligned}$$

and

$$|\nabla U^n(i,j)|_{sl} := \begin{cases} \Big[\min^2(U^n_{xb},0) + \max^2\big(U^n_{xf},0\big) \\ \quad + \min^2\big(U^n_{yb},0\big) + \max^2\big(U^n_{yf},0\big)\Big]^{1/2}, & \beta^n(i,j) > 0, \\ \Big[\max^2(U^n_{xb},0) + \min^2\big(U^n_{xf},0\big) \\ \quad + \max^2\big(U^n_{yb},0\big) + \min^2\big(U^n_{yf},0\big)\Big]^{1/2}, & \beta^n(i,j) \le 0, \end{cases}$$

where the subindices c,b and f denote central, backward and forward difference, respectively. The Laplacian L^n is discretised by finite differences based on the five-point rule. Moreover, whenever a division by the norm of the gradient takes place, $|\nabla U|$ is replaced by its regularised version $|\nabla U|_\epsilon = \sqrt{U_x^2 + U_y^2 + \epsilon}$ for a small $0 < \epsilon \ll 1$ to avoid division by zero.

As already indicated in the presentation of the algorithm inside the preceding box, the anisotropic diffusion equation (6.10) is approximated by the explicit time-stepping scheme

$$U^{n+1}(i,j) = U^n(i,j) + \Delta t \mathbb{1}_\epsilon(i,j) \cdot \nabla \cdot \left(\frac{\nabla U^n}{|\nabla U^n|}\right)(i,j) \cdot |\nabla U^n|(i,j), \quad \text{in } D^\epsilon. \tag{6.12}$$

After applying the divergence operator to the terms inside the parentheses and multiplying them by the norm of the gradient of U^n, the update term in the preceding equation reads

$$\nabla \cdot \left(\frac{\nabla U^n}{|\nabla U^n|}\right) \cdot |\nabla U^n| = \left(U_y^2 U_{xx} + U_x^2 U_{yy} - 2U_x U_y U_{xy}\right) \frac{1}{|\nabla U|^2}.$$

As before, $|\nabla U|$ is replaced by its regularised version $|\nabla U|_\epsilon = \sqrt{U_x^2 + U_y^2 + \epsilon}$ for a small $0 < \epsilon \ll 1$.

Note that both (6.11) and (6.12) are solved on an extension D_ϵ of the inpainting domain D – similar to the setup in Section 5.2.5 and illustrated in Figure 5.4. No explicit boundary conditions are imposed on the solution, but after every outer iteration of transport and anisotropic diffusion iterations, the solution U^{n+1} in $D_\epsilon \setminus D$ is set equal to the given image F.

Numerical Results For the numerical examples in Figures 6.1 and 6.2, we used $\Delta t = 0.1$ and a total number of iteration steps MN between 1,000 and 10,000. In Figure 6.1 we used $M = 15$ steps of transport and $N = 2$ steps of anisostropic diffusion. This amount of diffusion turned out to be too much for the example of the sleeping dog in Figure 6.2, where we switched to $M = 100$ and $N = 1$. The inpainting result in Figure 6.3 was obtained with $M = 50$ and $N = 1$. The inpainting domain in Figure 6.4 consists of small letters

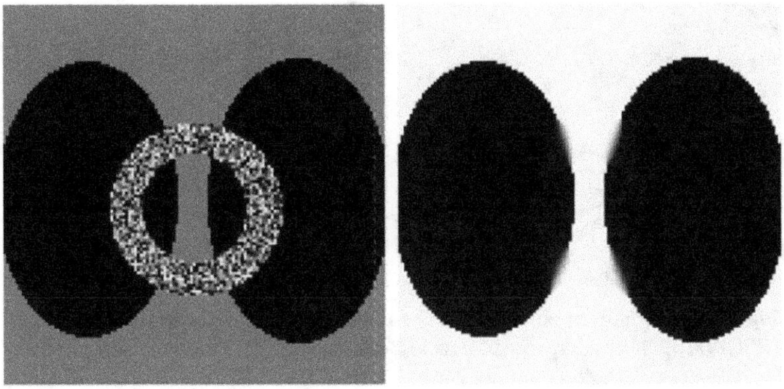

Figure 6.1. Transport inpainting (6.4) for two rings.

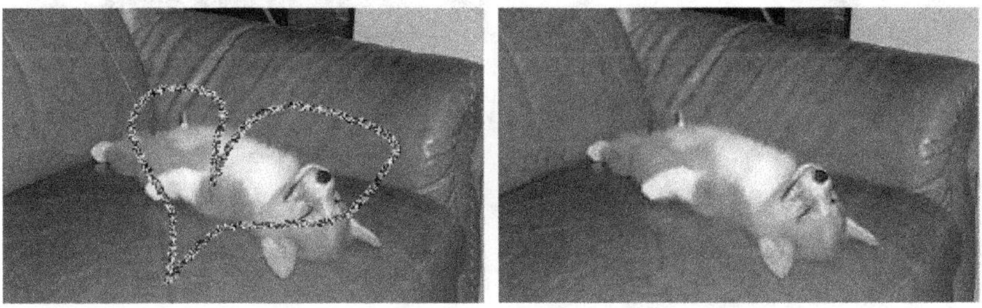

Figure 6.2. Transport inpainting (6.4) for the sleeping dog example.

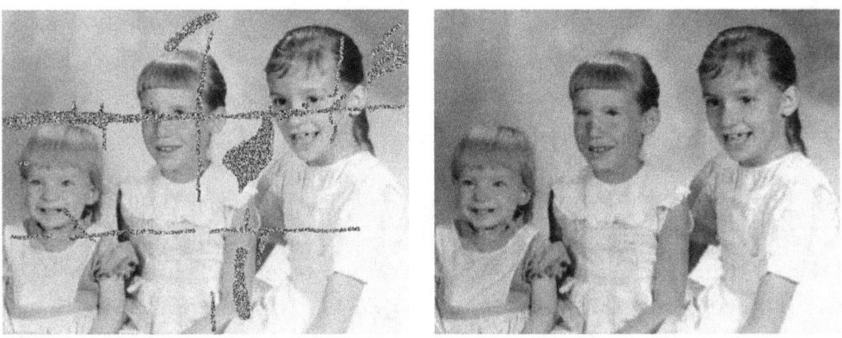

Figure 6.3. Transport inpainting (6.4) for the restoration of an old photograph (see [BSCB00]). Photo courtesy of Marcelo Bertalmio.

that have been superimposed on the photograph. Because each letter is very small, the inpainted image can be computed relatively quickly (with only 1,000 iterations). Here we chose $M = 30$ and $N = 1$. Figures 6.5 and 6.6 demonstrate

Figure 6.4. Transport inpainting (6.4) for the removal of superimposed text (see [BSCB00]). Photo courtesy of Marcelo Bertalmio.

Figure 6.5. Good continuation with transport inpainting (6.4).

Figure 6.6. Good continuation with transport inpainting (6.4).

the successful continuation of a straight line, even across large differences, as in Figure 6.6.

6.2. Inpainting with Coherence Transport

For presentation of the discrete transport approach, we closely follow Thomas März [Mär10, Mär11], who discussed this approach in his doctoral thesis [Mär10]. The main algorithm was proposed by Bornemann and März in 2007 [BM07] and developed further by März [Mär11]. Moreover, in [Mär13 and Mär10], März gives a full account of the analysis in the continuum of the PDE that arises from the coherence transport inpainting approach. Let us start this section by first presenting the idea of coherence transport for inpainting as it was motivated in the original paper [BM07].

There are two inpainting methods that constitute the main sources of inspiration for the proposal of Bornemann and März: the inpainting approach of Bertalmio et al. presented in the preceding section for its high-quality inpainting results and Telea's inpainting algorithm [Tel04] for its efficient computation. The latter is an inpainting method that inpaints an image by transporting known image values from the boundary of the inpainting domain inwards in the direction given by the fast marching method (orthogonal to the signed distance function of ∂D). This can be computed extremely quickly. Telea proposes a non-iterative algorithm that is able to compute the inpainted image in a single pass through all pixels in the inpainting domain. This is in great contrast to all the inpainting methods we have discussed so far. Those are based on the computation of discrete solutions to a PDE and are therefore all iterative in nature (note that a direct solution of the system of linear or non-linear equations that arises from the discretisation of the respective PDE is in general computationally prohibitive because of its size and/or bad conditioning). The inpainted image is then the fixed point of this iterative procedure, whose computation might need a large number of iterations depending on the design of the iterative method (e.g., the time discretisation if a steepest-descent approach is employed) and the size of the inpainting domain. Therefore, Telea's approach seems very promising. However, as was already pointed out in [BSCB00], the propagation direction chosen for Telea's inpainting method is rather suboptimal. The geometry of the inpainting domain in general does not correlate with the geometry of the image structures that should be inpainted. Hence, basing the propagation direction solely on ∂D seems unnatural and might force level lines arriving at ∂D to curve in order to meet the enforced direction inside D. To choose the tangent to the level lines as the propagation direction – as in the Bertalmio et al. inpainting approach in the preceding section – seems much more intuitive and indeed gives improved inpainting results.

The pioneering contribution of [BM07] is the derivation of a single-pass algorithm à la Telea to solve inpainting along level lines à la Bertalmio et al. and to further enhance the inpainting method by replacing the tangents to the level lines as the transport direction by the coherence direction proposed by Weickert [Wei98] for image smoothing and used by Tschumperlé for image inpainting [Tsc06]. The resulting inpainting approach is called *inpainting by coherence transport*. In what follows, we explain this inpainting approach in more detail mainly following the exposé of März [Mär10].

Definitions The algorithm in [BM07] was proposed in the discrete setting. The continuum limit transport equation was only discussed afterwards asymptotically. We define the following quantities:

- $\Omega_h = \{x_{i,j}\}_{\substack{i=1,\dots,M \\ j=1,\dots,N}}$ is the discrete image domain, and $u_h : \Omega_h \to [0,255]$ is the discrete restored image.

- $D_h \subset \Omega_h$ is the inpainting domain, and $\Omega_h \setminus D_h$ defines the intact part of the image. The given damaged image f_h is defined in the intact part only; that is, $f_h : \Omega_h \setminus D_h \to [0,255]$.
- Analogous to the continuum case, where we assume that $\partial\Omega \cap \partial D = \emptyset$, we assume here that the distance of boundary pixels of D_h to boundary pixels of Ω_h is at least one pixel.
- Finally, for an $\epsilon > 0$, a discrete ϵ-neighbourhood of a point $x \in \Omega_h$ is defined by

$$B_{\epsilon,h}(x) = \{y \in \Omega_h : |y - x| \leq \epsilon\}.$$

The Generic Discrete Algorithm The algorithm starts with an ordering of the pixels inside the inpainting domain, determining the order in which pixels are going to be inpainted. In general, this imposes the evolution of the inpainting mechanism in an onion-peel manner. With this we define $D_h = (x_1, \ldots, x_L)$, the ordered pixels inside the inpainting domain, and

$$B^{<}_{\epsilon,h}(x_l) = B_{\epsilon,h}(x_l) \setminus \{x_l, \ldots, x_L\}, \quad l = 1, \ldots, L,$$

the ϵ-neighbourhood of x_l that contains only known or already inpainted pixels. Further, we assign non-negative weights $w(x,y)$ to the dependence of pixels x and y, and we assume that for any pixel x_l inside the inpainting domain there is at least one known or already inpainted pixel within an ϵ-neighbourhood of x_l that has a positive dependence on x_l, that is,

$$\sum_{y \in B^{<}_{\epsilon,h}(x_l)} w(x_l, y) > 0, \quad x_l \in D_h.$$

With this, we can write down the following generic form of a discrete transport algorithm:

Discrete Transport Algorithm

$$u_h|_{\Omega_h \setminus D_h} = f_h$$

$$u_h(x_l) = \frac{\sum_{y \in B^{<}_{\epsilon,h}(x_l)} w(x_l,y) u_h(y)}{\sum_{y \in B^{<}_{\epsilon,h}(x_l)} w(x_l,y)}, \quad l = 1, \ldots, L. \tag{6.13}$$

To specify the details of this algorithm, we are going to discuss choices for pixel ordering and the weights w.

Ordering of the Pixels First of all, we fix the ordering considered here to the one that is defined in an onion-peel manner. This means that we start on the boundary of the inpainting domain and work inwards in the order that monotonically increases a certain 'distance map' from the boundary. To this end, we define the onion-peel map $P_h : D_h \to [0, P_{\max}]$ with $P_h(x_l) = 0$ for x_l on the boundary of D_h, inducing the following ordering rule on pixels inside D_h:

$$\text{If } P_h(x_k) < P_h(x_l), \quad \text{then } k < l.$$

For pixels that lie on the same onion peel, that is, $P_h(x_l) = P_h(x_k)$, we assume that the inpainting of x_l and x_k is independent of each other. Hence, the order in which such pixels are inpainted does not influence the inpainting result. Then the ϵ-neighbourhood $B^{<}_{\epsilon,h}(x_l)$ enjoys the following alternative definition in terms of the map P_h:

$$B^{<}_{\epsilon,h}(x_l) = \{x \in B_{\epsilon,h}(x_l) : P_h(x) < P_h(x_l) \text{ if } x \in D_h\}.$$

So how do we choose the ordering map P_h? The classical choice proposed in Telea's work [Tel04] is to choose $P_h(x)$ as the distance from x to the boundary ∂D_h, that is,

$$P_h(x) = dist(x, \partial D), \quad \text{for } x \in D_h.$$

The same choice was used in [BM07]. The advantage of this distance map is that it can be computed very efficiently by the so-called fast-marching method. In a very interesting contribution from März [Mär11], the author discussed a variety of other choices for P_h, that could be tuned to the specific inpainting setting and as such are able to improve the inpainting result in many cases.

The Transport Direction The next building block of discrete transport inpainting (6.13) is the direction of the transport, which is determined by the choice of the weights $w(x,y)$. In general, we consider weights of the form

$$w(x,y) = \frac{1}{|x-y|} K_{(x-y)/\epsilon}(x),$$

with kernels $K_\eta(x)$. Having fixed the onion-peel map P_h in a pre-processing step, different directions of transport are imposed by specifying the kernel functions.

- Telea [Tel04] proposed inpainting in normal direction $\vec{n}$ to the boundary of D_h. More precisely,

  $$K_{(x-y)/\epsilon}(x) = |(\vec{n}(x), x-y)|, \qquad \text{where} \quad \vec{n}(x) = \frac{\nabla P_h(x)}{|\nabla P_h(x)|}.$$

  In [Tel04], the ordering map $P_h = d$. Hence, the weight $w(x,y)$ is maximal for pixels x, y, where the direction x–y aligns with the gradient of the distance function $\text{dist}(x, \partial D_h)$. Hence, pixels are inpainted based on the known (or already estimated) pixel grey values that are on the line which is normal to the boundary ∂D_h.

- Another choice was considered by Bornemann and März [BM07], which they called *coherent transport*. There the kernel they used was given by

$$K_\eta(x) = \sqrt{\frac{\pi}{2}}\mu e^{-\mu^2/2(g^\perp(x),\eta)^2}, \tag{6.14}$$

where $\eta = (x-y)/\epsilon$, g is a given guidance field for coherence transport, and the parameter $\mu \geq 0$ models the strength of influence of g. The guidance vector is modelled by the orientation of the least change in grey value; that is, it is defined to approximate the orientation of the tangent $\nabla^\perp u$. This approximation is done by means of the structure tensor $\mathcal{J}_\rho$ [Wei98]. The structure tensor $\mathcal{J}_\rho$ of an image u is defined to be

$$\mathcal{J}_\rho(\nabla u_\sigma) := G_\rho * (\nabla u_\sigma \otimes \nabla u_\sigma), \quad \rho > 0.$$

Here G_ρ is a Gaussian kernel with variance ρ, u_σ is the image u convolved with G_σ and $\nabla u_\sigma \otimes \nabla u_\sigma := \nabla u_\sigma \cdot \nabla u_\sigma^\perp$. For its application here, the convolution is defined over the union of the known and already inpainted pixels only. These are characterised by means of the ordering map; that is, for every $x \in D$, let $D(x) = \{y \in D : P_h(y) < P_h(x)\} \cup \Omega \setminus D$. More precisely,

$$u_\sigma(y,x) = \frac{1}{\int_{D(x)} G_\sigma(y-z)\, dz} \int_{\Omega(x)} G_\sigma(y-z)u(z)\, dz,$$

and

$$\mathcal{J}_\rho(\nabla u_\sigma)(x) = \frac{1}{\int_{D(x)} G_\sigma(x-y)\, dy} \int_{D(x)} G_\rho(x-y)\nabla u_\sigma(y,x) \cdot \nabla u_\sigma^\perp(y,x)\, dy.$$

The use of $\nabla u_\sigma \otimes \nabla u_\sigma$ as a structure descriptor aims at making the tensor insensitive to noise and sensitive to change in orientation only; that is, the sign of the gradient should not be taken into account. Then the guidance vector $g(x)$ is defined as the eigenvector of $\mathcal{J}_\rho(\nabla u_\sigma)(x)$ that corresponds to the minimal eigenvalue, encoding the coherent direction and its strength.

The Continuum Model In [BM07 and Mär10], the continuum ($h \to 0$) and vanishing viscosity ($\epsilon \to 0$) limit of the discrete transport algorithm (6.13) is derived. For the two choices of $w(x,y)$ discussed earlier, the limiting equations for $h \to 0$ are transport and advection PDEs defined on $D \setminus \Gamma$. The set Γ is the so-called stopping set of the algorithm. It depends on the choice of the ordering map P. More precisely, it is defined as the set of points where $\vec{n}(x)$ is singular, that is, the central onion peel of the ordering map. With this, the resulting continuum equation for the normal transport reads

$$(\vec{n}(x), \nabla u(x)) = 0 \text{ in } D \setminus \Gamma, \quad u = f \text{ on } \partial D,$$

and for the coherent transport gives

$$(c_\mu(x), \nabla u(x)) = 0 \text{ in } \Omega \setminus \Gamma, \quad u = f \text{ on } \partial D, \quad (c_\mu(x), \vec{n}(x)) > 0. \tag{6.15}$$

Here c_μ is a transport field that depends on the kernel (6.14) and has to fulfil a side condition, the last inequality in the preceding definition. In [BM07], the authors proved that c_μ asymptotically for $\mu \to \infty$ is mainly determined by the guidance vector g. Namely, it behaves in the following way:

$$\lim_{\mu \to \infty} c_\mu(x) = \begin{cases} g(x), & (g(x), \vec{n}(x)) > 0, \\ -g(x), & (g(x), \vec{n}(x)) < 0, \\ \vec{n}(x), & (g(x), \vec{n}(x)) = 0. \end{cases}$$

For a thorough analysis of the preceding limiting equations and their well-posedness, compare [Mär10]. The natural setting for the analysis of these transport equations is – similar to the study of second-order diffusion inpainting in Section 4.1 – the notion of viscosity solutions (compare Appendix B.12).

Numerical Discussion For more details on the numerical implementation of (6.13) with coherence transport, we refer readers to [BM07]. A gimp-plugin for inpainting with coherence transport is available at http://inpaintgimpplugin.github.io. In what follows, let us finish this chapter by discussing some representative numerical results that have been computed with the C-code of Tom März. Our first example in Figure 6.7 shows the difference between diffusion-based inpainting, such as TV-H^{-1} inpainting, and inpainting by transport. On the one hand, the TV-H^{-1} inpainting result is more fuzzy than the transport inpainted one. On the other hand, transport inpainting might introduce artefacts inside the inpainting domain, possibly in the surroundings of the stopping set. In Figure 6.8, we consider the coloured-stripe example. Here the parameters have been set to $[\epsilon, \mu, \sigma, \rho] = [5, 50, 0.1, 10]$. The result seems almost perfect. The few artifacts are probably due to the choice of the distance function (also see the later work of Tom März [Mär11] for alternative choices). Note that coherence transport inpainting is not rotationally invariant. In fact, a rotation of the image in Figure 6.8 by 14 degrees gives a completely

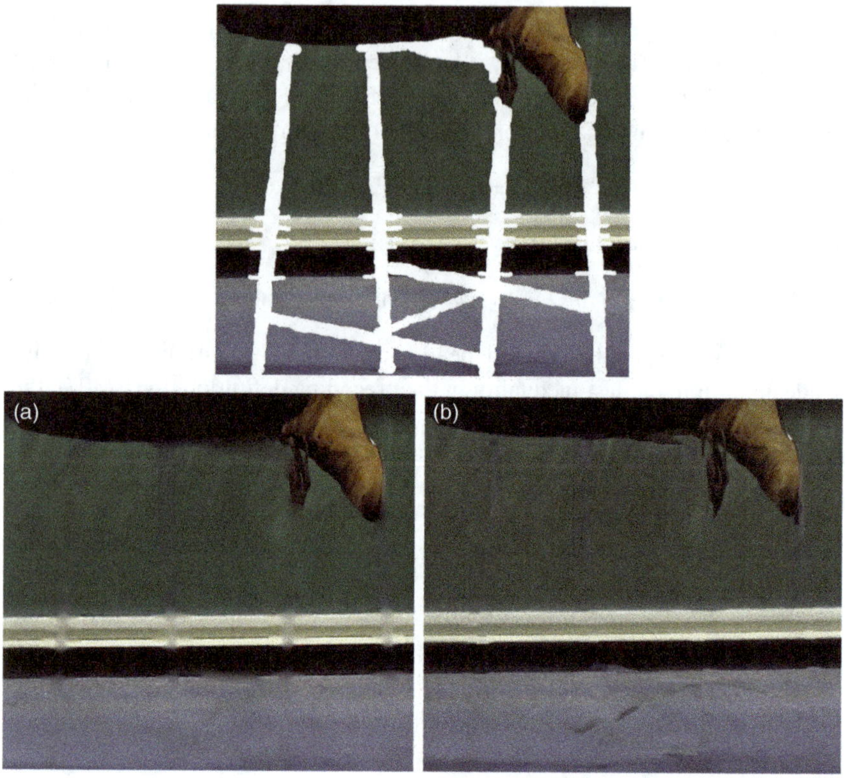

Figure 6.7. Comparison of coherence transport inpainting (6.13) and TV-H^{-1} inpainting (5.20) for a detail of the EPSRC winner photo in Figure 1.1. (a) TV-H^{-1} inpainted image. (b) Coherence transport inpainted image. Photo courtesy of Joana Grah, Kostas Papafitsoros and the author.

different inpainting result. With the same choice of parameters as before, coherence transport now fails to connect the coloured stripes in Figure 6.9. In this case, tweaking the parameters or choosing another distance will not improve the inpainting result significantly. We refer to the next section for a discussion of this phenomenon and for an alternative transport inpainting approach that resolves this issue.

6.3. GuideFill: Fast Artist-Guided Transport Inpainting

This section was contributed by my student Rob Hocking, who has developed this new transport-based inpainting technique as part of his Ph.D. work.

While powerful, coherence transport has limitations, namely:

Figure 6.8. Good continuation with coherence transport inpainting (6.13).

Figure 6.9. When we rotate the image in Figure 6.8 by 14 degrees, coherence transport inpainting (6.13) fails.

1. Users may need to tune parameters in order to obtain a good result.
2. The user may need to semi-manually specify a suitable distant map P_h in order to avoid artefacts related to pixel ordering [Mär11].
3. The discrete solution may exhibit grid artefacts not present in the continuum limit (see Figure 6.11).
4. The method is a black box with no artist control (aside from the ability to influence the ordering of the pixels in [Mär11]).

GuideFill is a more recent inpainting method that aims to overcome these limitations while maintaining the same high speeds as coherence transport. Just as the latter was inspired by Telea's algorithm [Tel04] and designed to improve on it, GuideFill adopts the basic framework of coherence transport while adjusting the details. It has three features in common with coherence transport:

1. The method is non-iterative in the sense that each pixel is filled only once, in an order determined by a distance function P_h.
2. The inpainted image u_h is given by a weighted sum similar to (6.13).
3. The weights are assigned based on a 'guide direction' g in such a way as to encourage transport parallel to g.

Despite high-level similarities, the details of the methods are very different, centering around three novel features.

The first of these, and the main idea of GuideFill, is to pre-compute a *guide field*

$$g : D_h \to \mathbb{R}^2$$

prior to inpainting, rather than computing g and u_h concurrently, as is done in coherence transport. The guide field is generated based on a small set of splines carrying information about how key edges in $\Omega_h \backslash D_h$ should be continued into D_h.

The method begins with the user either drawing the desired edges directly onto the image as Bézier splines using a GUI or else by having a set of splines automatically generated for them based on the output of a suitable edge-detection algorithm. In the latter case, the user may either accept the result or use it as a starting point which the user may improve on by editing and/or removing existing splines, as well as drawing new ones. This is illustrated in Figure 6.10a and b.

Next, the idea is to choose $g(x)$ to be 0 when x is far away from any splines and 'parallel' to the splines when x is close. More precisely, we define a *spline distance map* $P_h^B(x)$ which gives the Euclidean distance from x to the closest point y on the closest spline to x. Then the direction of $g(x)$ is given by $\nabla^\perp P_h^B(x)$, whereas the magnitude is a rapidly decreasing function of $P_h^B(x)$. This ensures that the inpainting will tend to follow the splines wherever they are present (see Figure 6.10c).

The primary motive for constructing the guide field in this manner is to put the user in the loop by providing him or her with a mechanism for control. However, it also has two other consequences.

The first consequence is also the second novel feature of GuideFill. Knowing g ahead of time facilitates the automatic computation of a modified distance map P_h adapted to g in such a way as to give an improved ordering of the pixels, leading to a higher-quality visual result. Adapted distance functions are also considered in [Mär11], but their computation is not automatic.

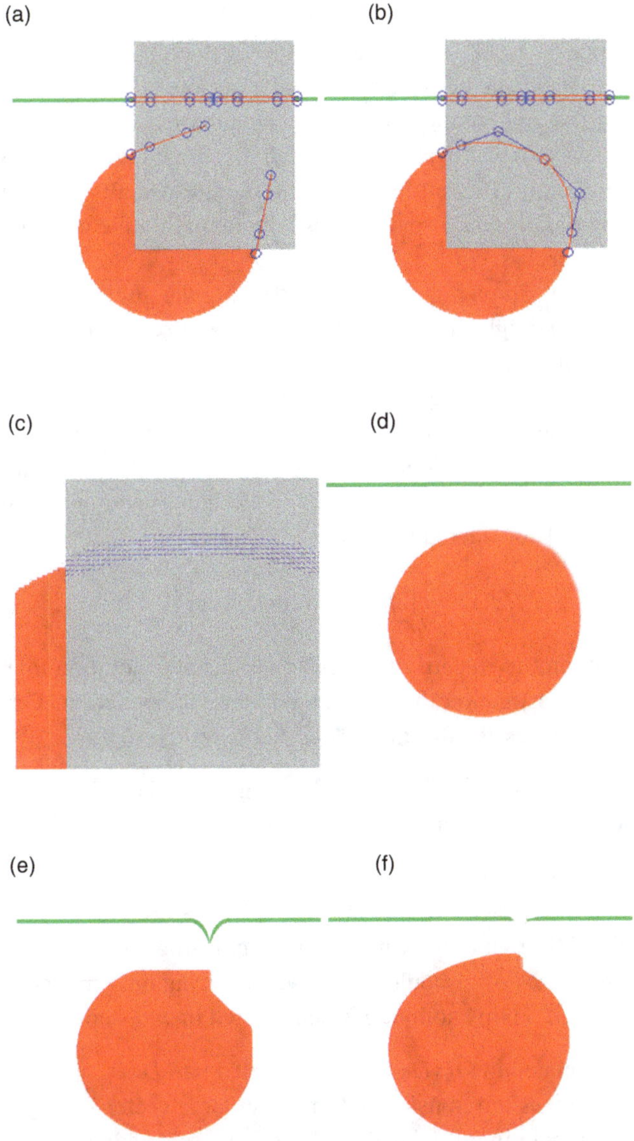

Figure 6.10. Using GuideFill and coherence transport to inpaint a circle. (a) GuideFill's automatically detected splines. (b) The user adjusts two of the splines to close the circle. (c) Section of the resulting guide field. (d) The result obtained with GuideFill $(\epsilon,\mu,\rho,\sigma) = (3,50,2,2)$. (e) Coherence transport with $(\epsilon,\kappa,\rho,\sigma) = (5,125,5,2)$. (f) Coherence transport with $(\epsilon,\kappa,\rho,\sigma) = (6,125,24,12)$.

A second consequence is that the guide field g has to be computed using the values of u_h in $x \in \Omega_h \backslash D_h$ without using any of the inpainted pixels $x \in D_h$. This a potential drawback because coherence transport, which includes freshly filled pixels in the computation of g, is in some cases able to use this extra information to close or nearly close broken curves (see Figure 6.10f). However, in practice, this requires careful parameter tuning, especially of the pre- and post-smoothing parameters σ and ρ. This is potentially time consuming with no guarantee of success.

GuideFill takes a simpler approach, constructing the guide field g based on splines which are straight lines by default. This approach has the advantage that default parameters are rarely, if ever, inadequate, effectively eliminating the need for parameter tuning. At the same time, the ability to close broken curves is retained by having the user simply edit the splines as in Figure 6.10b.

The third novelty of GuideFill arises from a careful analysis of the relationship between the discrete algorithm (6.12) and its continuum limit. The analysis in [Mär10] considered the limit

$$u(x) = \lim_{\mu \to \infty} \left[\lim_{\epsilon \to 0} \left[\lim_{h \to 0} u_h(x) \right] \right]$$

but made no attempt to quantify how close the discrete solution u_h is to this triple limit. An investigation thereof explains the phenomena in Figure 6.11 and also suggests a way to adjust (6.12) in such a way as to

- Retain the same continuum limit u as coherence transport.
- Reduce the discrepancy between u_h and u.

See [HSMM14] for details.

The Guide Field The guiding principle in the construction of the guide field is that non-zero values of g should be sparse, occurring only in the vicinity of image edges. Based on this notion, g is constructed in two steps:

1. Edges are estimated in $\Omega_h \backslash D_h$ and extrapolated into D_h.
2. Extrapolated edges are transformed into a suitable vector field g such that the resulting inpainting u_h is likely to have these same edges.

By 'edges', we mean, in the continuum limit, jump discontinuities of u. In the discrete case, edges may be thought of as long and thin chains of pixels across which there is a sharp or abrupt change in u_h.

As we have already mentioned, extrapolated edges take the form of Bézier splines which may be supplied either by the user or else detected automatically. Here we describe the automatic detection process in more detail. We also provide more details regarding the how g itself is computed once the splines are in place.

The first step in automatic edge estimation is to extend the distance map P_h (giving the Euclidean distance from ∂D_h) defined on D_h only into a signed distance map on Ω_h. P_h need not be defined everywhere – all that is required is that the set $\{x : P_h(x) \geq -\Delta\}$ be defined for some sufficiently large threshold Δ (set by the user). The next step is to run a suitable edge-detection algorithm on u_h in the region $-\Delta \leq P_h \leq 0$. The authors use Canny edge detection [Canny86] because it is particularly robust and has the property that detected edges are a single pixel wide, but any method with the latter property will do. The output of Canny edge detection is a map

$$E : \{x \in \Omega_h : -\Delta \leq P_h(x) \leq 0\} \to \{0, 1\}$$

such that $E(x) = 1$ indicates the presence of an edge, and vice versa.

Defining the dilated inpainting domain $D_{h,\delta}$ by

$$D_{h,\delta} = \{x : P_h(x) \geq -\delta\}, \qquad \text{for} \quad 0 < \delta \leq \Delta,$$

the next step is to compute the intersection

$$\Lambda = \{x : E(x) = 1\} \cap \partial D_{h,\delta}.$$

Each $x \in \Lambda$ is the point of origin of a potential spline.

Similarly to coherence transport, we then estimate the local coherence direction by evaluating the structure tensor $\mathcal{J}_\rho(\nabla u_\sigma)$ [Wei98] at each $x \in \Lambda$. The dilation of D_h is helpful in ensuring that there are enough uncorrupted pixels in the vicinity of x to get a good estimate of $\mathcal{J}_\rho(\nabla u_\sigma)(x)$. The algorithm sets $\delta = \max(\rho, \sigma)$ by default.

We denote the eigenvalues of $\mathcal{J}_\rho(\nabla u_\sigma)(x)$ by $\lambda_2(x) \geq \lambda_1(x) \geq 0$ and corresponding (unit) eigenvectors by $v_1(x)$, $v_2(x)$. We first eliminate edges which are parallel or nearly parallel to ∂D_h by requiring

$$\frac{|v_1(x) \cdot \nabla P_h(x)|}{\|\nabla P_h(x)\|} \leq \cos\phi_0, \qquad \text{for } x \in \Lambda$$

($\phi_0 = 10°$ by default). [Although we should have $\|\nabla P_h(x)\| = 1$ in theory, we leave it in the denominator for the sake of robustness against numerical error.] After removing those points which fail, to each remaining $x \in \Lambda$ we associate the line

$$L_{x,v_1(x)} = \begin{cases} \{x + t v_1(x) : 0 \leq t \leq \ell\}, & \text{if } v_1(x) \cdot \nabla P_h(x) > 0, \\ \{x - t v_1(x) : 0 \leq t \leq \ell\}, & \text{otherwise,} \end{cases}$$

where ℓ is a parameter that controls the distance that edges are extrapolated into D_h. The two cases are to ensure that $L_{x,v_1(x)}$ points into D_h instead of away from it.

After this, it is helpful to check that $\{y : E(y) = 1\}$ does not deviate too much from the line $L_{x,v_1(x)}$, at least in some small neighbourhood of x. To that end, we check that no point in the connected component of

$$\{y : E(y) = 1\} \cap \{y : -\delta \leq P_h(y) \leq 0, \|y - x\| \leq \delta'\}$$

containing x deviates from $L_{x,v_1(x)}$ by more than some tolerance ($\delta' = 2\delta$ by default). The lines which pass this second round of testing are then converted into cubic Bézier splines which can be edited by the user as in Figure 6.10a and b.

Once this has been done for every $x \in \Lambda$ and following the optional pass of user adjustment, we are left with a set of Bézier splines

$$\{B_i\}_{i=1}^N$$

where $N \in \mathbb{N}$ and each B_i is a function on $[0, 1]$ taking values in $\mathbb{R}^2$.

It remains to convert this family into a suitable guide field g, as in Figure 6.10c. This is done based on two assumptions.

Firstly, we assume that the curves B_i are integral curves of the as yet unknown vector field g. That is, we assume that each curve B_i originating at $x_i \in \Lambda$ solves the ODE initial-value problem

$$\frac{d}{dt}B_i(t) = g(B_i(t)), \quad B_i(0) = x_i.$$

Next, we assume that g drops to 0 a short distance ζ away from each curve.

Conceptually, such a g can be found as follows: define the distance map P^B from $\{B_i\}$ by

$$P^B(x) = \inf\{\|x - B_i(t)\| : 1 \leq i \leq N \text{ and } t \in [0, 1]\}.$$

Then

$$g(x) = \exp\left(\frac{-P^B(x)^2}{2\zeta^2}\right) \frac{\nabla^\perp P^B(x)}{\|\nabla^\perp P^B(x)\|}$$

has the desired properties. However, in practice, g is computed via a more efficient approach (see [Hocking2014] for details).

The Modified Discrete Algorithm Figure 6.12 shows coherence transport used to connect a broken line (making an angle of 73 degrees with the horizontal). In this case, the line 'refracts' at the interface ∂D_h and continues into D_h at a new angle of 90 degrees, thereby failing to connect.

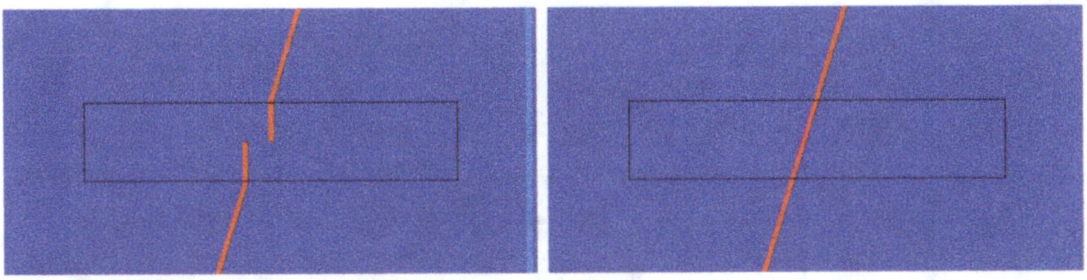

Figure 6.11. Connecting a broken line using coherence transport (left) and GuideFill (right). In this case, each method is provided $g \equiv (\cos(73°), \sin(73°))^T$ directly. Each method uses $\epsilon = 3$ pixels and $\mu = 50$. The inpainting domain D_h is outlined in black.

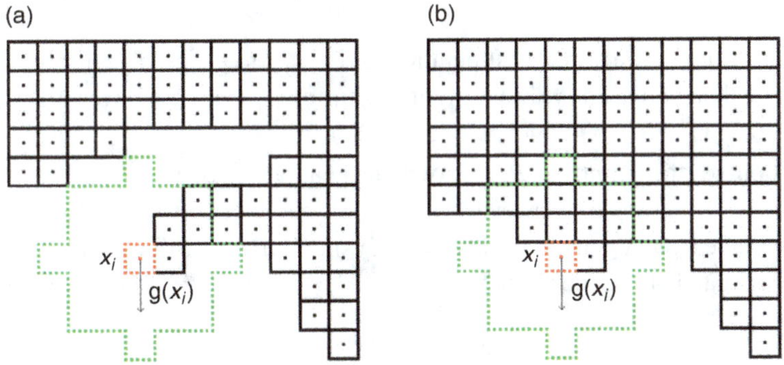

Figure 6.12. The value u_h at a pixel x_i due to be filled is given as a weighted sum of $\tilde{u}_h$ at $x_j \in \tilde{B}^<_{\epsilon,h}(x_i)$. Here x_i is outlined in red, already filled pixels are drawn in black and $\tilde{B}_{\epsilon,h}(x_i)$ is the area enclosed within the green dotted line ($\epsilon = 3$ pixels). The guide direction $g(x_i)$ is visualised as an arrow originating at x_i. (a) If $\tilde{B}^<_{\epsilon,h}(x_i)$ contains no pixels in the directions $\pm g(x_i)$, we do not expect a good inpainting. In this case, it is better to delay filling x_i until more of its neighbours have been filled. (b) Now we can expect a good result since $\tilde{B}^<_{\epsilon,h}(x_i)$ contains several pixels in the direction $-g(x_i)$.

To understand this behaviour, note that the continuum limit (6.15) is derived based on the assumption that (6.12) may be replaced by the following 'formal' limit:

$$u(x) = \frac{\int_{y \in B^<_\epsilon(x)} w(x,y)u(y)dy}{\int_{y \in B^<_\epsilon(x)} w(x,y)dy} \tag{6.16}$$

for some suitable set $B^<_\epsilon(x) \subset B_\epsilon(x)$. In other words, it is assumed that we may replace the discrete sums in the numerator and denominator by their limiting integrals. However, for the typical parameter values used by the method ($\epsilon = 5$ pixels is the recommended default), $B_{\epsilon,h}(x)$ does not contain very many points (just eighty-one in this case). Hence the sum may not be an accurate approximation of the integral.

In fact, for large μ, the success with which (6.12) approximates (6.16) depends not only on the number of pixels inside $B^{<}_{\epsilon,h}(x)$ but also on their position relative to the line $L_{x,g(x)}$ passing through x in direction $g(x)$. As μ increases, the mass of $w(x,y)$ concentrates more and more closely around this line, and if there are no pixel centers on or close to it, then neither sum will be well approximated by its corresponding integral. Thus, we expect to get problems when $L_{x,g(x)}$ 'misses' pixel centers in $B^{<}_{\epsilon,h}(x)$.

To alleviate this problem, GuideFill introduces a rotated ball

$$\tilde{B}_{\epsilon,h}(x) = R(B_{\epsilon,h}(x)),$$

where R is a rotation about x taking the line $L_{x,(1,0)^{\mathrm{T}}}$ to $L_{x,g(x)}$. The discrete sum (6.12) is then replaced with a weighted sum of u_h evaluated over a suitable subset $\tilde{B}^{<}_{\epsilon,h}(x) \subseteq \tilde{B}_{\epsilon,h}(x)$ which we define shortly.

Of course, the points in $\tilde{B}_{\epsilon,h}(x)$ will in general lie *between* pixel centers, where u_h is unknown. However, this is not a serious problem, as we can approximate u_h at a general point $y = (y_1, y_2) \in \Omega$ using bi-linear interpolation of u_h evaluated at the four lattice points $y^{(0,0)}$, $y^{(1,0)}$, $y^{(0,1)}$, $y^{(1,1)} \in \Omega_h$, where

$$y^{(i,j)} \equiv (h\lfloor y_1/h \rfloor + ih, h\lfloor y_2/h \rfloor + jh),$$

provided that u_h is known at these points.

With this in mind, we define

$$\tilde{B}^{<}_{\epsilon,h}(x) \equiv \{y \in \tilde{B}_{\epsilon,h}(x) : T_h(y^{(i,j)}) < T_h(x) \text{ for all } 0 \le i,j \le 1\}.$$

and replace (6.12) with

$$u_h(x_i) = \frac{\sum_{y \in \tilde{B}^{<}_{\epsilon,h}(x_i)} w(x_i,y)\tilde{u}_h(y)}{\sum_{y \in \tilde{B}^{<}_{\epsilon,h}(x_i)} w(x_i,y)} \tag{6.17}$$

where $\tilde{u}_h : \Omega \to \mathbb{R}^3$ is the bi-linear interpolant of u_h. We can think of (6.17) in terms of a sum of u_h evaluated on a local grid of virtual pixels aligned with the local guide direction $g(x_i)$.

Computing a Good Pixel Order In [Mär11], März suggested a number of methods for generating a modified distance function P_h tailored to the problem at hand, in many cases leading to a much nicer restoration. Whereas all these methods require some form of user input – for example, a 'stopping set' consisting of a curve of pixels to be filled last – GuideFill uses a simple heuristic to generate P_h automatically and concurrently with the inpainting itself.

The basic idea is as follows: let x_i be a pixel due to be filled. Since $u_h(x_i)$ is given as a weighted sum (6.17) of $\tilde{u}_h(x_j)$ for $x_j \in \tilde{B}^{<}_{\epsilon,h}(x_i)$ with weights depending on $g(x_i)$, we can expect the 'quality' of $u_h(x_i)$ to depend on *how many* x_j are in this set and *where they are* in relation to $g(x_i)$. In particular, if there are no such x_j on or near the line $L_{x_i,g(x_i)}(x_i)$ passing through x_i in direction $g(x_i)$, then we do not expect to get a good result. This is illustrated in Figure 6.12.

We formalise this intuition as follows: to each $x \in D_h$ we assign a non-negative measure m_x defined by

$$m_x(A) = \sum_{y \in A} w(x,y)$$

where $A \subset \Omega$ is a finite set and $w(x,y)$ is the weight function with kernel (6.14) depending implicitly on g.

Our goal is to construct P_h in such a way that

$$m_{x_i}(\tilde{B}^{<}_{\epsilon,h}(x_i)) \geq c\, m_{x_i}(\tilde{B}_{\epsilon,h}(x_i)) \tag{6.18}$$

for all $x_i \in D_h$, where $c \in (0,1)$ is a small user-specified constant ($c = 0.05$ by default). In other words, we are insisting that the total weight of the partially filled ball $\tilde{B}^{<}_{\epsilon,h}(x_i)$ of available pixels is always at least some constant fraction c of the weight of the full ball.

To accomplish this, we begin by initialising P_h to the Euclidean distance from ∂D_h as before. After serialising the pixels in terms of P_h as $D_h = (x_1, \ldots, x_L)$, we visit them one at a time in order as usual. However, in our case, a given pixel x with serialisation x_i is only filled if (6.18) is satisfied. If (6.18) is *not* satisfied, filling is delayed by updating both the distance function $P_h(x)$ and the serialisation of x so as to move x further down the list. To prevent an infinite loop, each pixel is given a maximum number of deferrals before it must be filled, even if (6.18) is not satisfied.

The following pseudocode for GuideFill formalises this discussion:

GuideFill

Compute the guide field $g(x)$
Initialise P_h to the Euclidean distance from ∂D_h
Initialise the number of visits $V_h : D_h \to \{0,1,\ldots\}$ to zero everywhere
Serialise the pixels as $D_h = (x_1, x_2, \ldots, x_L)$ based on P_h
$-i \leftarrow 1$
while $i \leq L$
  if $m_{x_i}(\tilde{B}^{<}_{\epsilon,h}(x_i)) \geq c\, m_{x_i}(\tilde{B}_{\epsilon,h}(x_i))$ or $V_h(x_i) = V_{\max}$
    Compute $u_h(x_i)$ using (6.17)
    $i \leftarrow i+1$

else
  $P^* \leftarrow \min(\{P_h(x_j) : P_h(x_j) > P_h(x_i)\})$
  Delete x_i from position i of the list and reinsert it between
    x_r and x_{r+1}, where $P_h(x_r) = P^*$ and $P_h(x_r) < P_h(x_{r+1})$.
  $P_h(x_i) \leftarrow P^*$
  $V_h(x_i) \leftarrow V_h(x_i) + 1$

Figure 6.13. A damaged piece of calligraphy is restored by having the user draw the splines by hand. In this, $\epsilon = 3$ pixels, $\mu = 50$ and c was increased from its default value $c = 0.05$ up to $c = 0.1$. Photo courtesy of Rob Hocking.

Numerical Results Figure 6.12 showed a side by side comparison of coherence transport and GuideFill applied to the problem of connecting a broken line. In order to isolate the effect of using the modified sum (6.17) in place of (6.12), we have by-passed the computation of g and instead provided the correct value to each method directly. In the case of coherence transport, the line fails to connect, whereas with GuideFill, the line is connected faithfully, at the cost of some blurring.

As a second example, we consider the problem of restoring the damaged painting shown in Figure 6.13. In this case, which would be next to impossible for a fully automatic method, we are able to obtain a satisfactory result by having the user draw the splines directly.

7
The Mumford-Shah Image Model for Inpainting

Mumford and Shah [MS89] introduced in 1989 a segmentation model that is based on the idea of decomposing an image into piecewise smooth parts that are separated by an edge set Γ. The idea to use this model for image inpainting originates in the works [TYJW01, SC02 and ES02]. In this chapter we discuss the original Mumford-Shah inpainting approach with a length penalty on the edge set Γ in Section 7.1, along with a higher-order extension of this approach by an Euler elastica regularisation on the edge set in Section 7.2.

7.1. Inpainting with Mumford-Shah

As before, let $\Omega \subset \mathbb{R}^2$ be a rectangular domain and g a given (possibly noisy) image with an inpainting domain $D \subset \Omega$. Further, define an edge set Γ to be a relatively closed subset of Ω with finite one-dimensional Hausdorff measure (compare Definition B.7.2). We search for a pair (u, Γ) minimising

$$\mathcal{E}(u,\Gamma) = \frac{1}{2}\int_{\Omega\setminus D} (u-g)^2\, dx + \mathcal{J}(u,\Gamma), \tag{7.1}$$

with

$$\mathcal{J}(u,\Gamma) = \frac{\alpha}{2}\int_{\Omega\setminus\Gamma} |\nabla u|^2\, dx + \beta\mathcal{H}^1(\Gamma). \tag{7.2}$$

Here α and β are non-negative constants, and $\mathcal{H}^1(\Gamma)$ is the one-dimensional Hausdorff measure of Γ (which is the length of Γ if Γ is regular). The Mumford-Shah model is inspired by former statistical approaches (e.g. [GG84, BZ87]). It aims to decompose a given image into its piecewise smooth part u and its edge set Γ, where the former is measured with the H^1 norm and the latter by its length or, more generally, by the one-dimensional Hausdorff dimension $\mathcal{H}^1(\Gamma)$.

Well-Posedness of the Scheme: The existence and uniqueness theory for the Mumford-Shah functional is only partly understood. The difficulty in studying $\mathcal{J}$ is that it involves two unknowns u and Γ of different natures: u is a function defined on a two-dimensional space, whereas Γ is a one-dimensional set. A particular difficulty is that the Hausdorff measure is not semi-continuous with respect to any compact topology (cf. Exercise 13 in Appendix A). Hence, for getting a well-defined imaging model, we shall introduce a relaxed version of (7.2) for which the existence of solutions in the space of so-called special functions of bounded variation $SBV(\Omega)$ [AFP00], defined later, can be proven.

Let us start the well-posedness discussion of (7.1) with a re-phrasing of the problem within a more regular setting (cf. [CS05a]) than that of the relaxed functional with which we will end up. This provides us with a better ground to motivate the idea on which the relaxation is built. Let u be defined as piecewise smooth; that is, u is defined on a partition of $\Omega = \bigcup_{i=1}^{K} \Omega_k \bigcup \Gamma$ with $u|_{\Omega_k} \in H^1(\Omega_k)$ for $k = 1, \ldots, K$ and piecewise C^1 edge set Γ. Then one can define a piecewise continuous vector field $\vec{n}$ for each point $x \in \Gamma$ that is part of a C^1 component and with that the jump of u at x as

$$[u](x) = \lim_{\rho \to 0^+} u(x + \rho \cdot \vec{n}(x)) - u(x - \rho \cdot \vec{n}(x)),$$

where the limit for $\rho \to 0^+$ is taken as the trace sense. With this, we can define a vectorial measure on Γ as

$$J_u := [u]\vec{n}\, d\mathcal{H}^1 \quad \text{with} \quad J_u(\gamma) = \int_\gamma [u](x)\vec{n}(x)\, d\mathcal{H}^1, \quad \forall \gamma \in \mathcal{B}(\Gamma).$$

In this setting, one can easily check that the restriction of the distributional derivative Du to the edge set Γ equals J_u. Moreover, on the components Ω_k where u is H^1, one has that $Du = \nabla u \in L^2(\Omega_k)$, and hence,

$$Du = \nabla u|_{\Omega \setminus \Gamma} + J_u|_\Gamma.$$

This is our key observation: instead of defining the edge set Γ separately from u, we rather capture it within the jump set of u, that is, the set in which Du is singular and one-dimensional.

The problem with assuming $\Gamma \in C^1$ (or even piecewise C^1) is that this space does not provide us with any compactness property and, as a consequence, any existence proof. Hence, we have to loosen our assumption on Γ and prove existence in a less restrictive (i.e., larger) function space. This space will turn out to involve the space of functions of bounded variation $BV(\Omega)$. This space provides us with sufficient compactness and semi-continuity properties and gives sense to one-dimensional discontinuities (edges) of functions. The latter becomes clear when we recall some facts about functions of bounded variation such as that the distributional derivative Du of a BV function can be written as the sum of its absolute continuous part $\nabla u\, dx$, its jump part J_u and its Cantor part C_u, that is,

$$Du = \nabla u\, dx + \underbrace{(u^+ - u^-)\vec{n}_u\, \mathcal{H}^1|_{S_u}}_{J_u} + C_u, \tag{7.3}$$

(cf. Appendix B.11). The idea now, similar to the C^1 case discussed earlier, is to identify the edge set Γ with the jump set S_u. Hence, instead of (7.1), we minimise

$$\mathcal{E}(u) = \frac{1}{2}\int_{\Omega\setminus D} (u-g)^2\, dx + \frac{\alpha}{2}\int_{\Omega} |\nabla u|^2\, dx + \beta\mathcal{H}^1(S_u).$$

Solving this would allow us to eliminate the unknown Γ in the minimisation problem. The issue, however, is that we cannot do this in BV. The objectionable element of BV is the Cantor part C_u in the decomposition (7.3). For a function $u \in BV$, this part may contain pathological functions such as the Cantor-Vitali function that make the minimisation problem ill-posed (cf. [Amb89b]). The Cantor-Vitali function is non-constant, continuous but with approximate differential equals zero almost everywhere. For such a function v, we would have

$$\mathcal{E}(v) = \frac{1}{2}\int_{\Omega\setminus D} (v-g)^2\, dx \geq \inf_{u\in BV(\Omega)} \mathcal{E}(u) = 0,$$

because BV functions are dense in L^2. But this means that the infimum cannot be achieved in general. To exclude this case, we consider the space of special functions of bounded variation $SBV(\Omega)$, which is the space of BV functions such that $C_u = 0$. Then our new problem that replaces (7.1) reads

$$\min\{\mathcal{E}(u) : \ u \in SBV(\Omega)\cap L^\infty(\Omega)\}. \tag{7.4}$$

For the relaxed problem (7.4), we have the following existence result:

Theorem 7.1.1 *Let $g \in L^\infty(\Omega)$. Then the minimisation problem* (7.4) *obtains a solution.*

To prove Theorem 7.1.1, we will use the following compactness and closure results for SBV functions (cf. [AFP00, section 4.]):

Theorem 7.1.2 Closure of *SBV* *Let $\Omega \subset \mathbb{R}^d$ open and bounded and $(u_n) \subset SBV(\Omega)$ with*

$$\sup_n \left\{ \int_\Omega |\nabla u_n|^2\, dx + \int_{S_{u_n}} |u_n^+ - u_n^-|\, d\mathcal{H}^{d-1} \right\} < \infty. \tag{7.5}$$

If (u_n) weakly converges in BV to u, then $u \in SBV(\Omega)$, ∇u_n weakly* converges to ∇u in $[L^2(\Omega)]^d$ and $D^j u_n$ weakly* converges to $D^j u$ in Ω. Moreover,*

$$\begin{aligned} \int_\Omega |\nabla u|^2\,dx &\le \liminf_{n\to\infty} \int_\Omega |\nabla u_n|^2\,dx, \\ \int_{S_u} |u^+ - u^-|\,d\mathcal{H}^{d-1} &\le \liminf_{n\to\infty} \int_{S_{u_n}} |u_n^+ - u_n^-|\,d\mathcal{H}^{d-1}. \end{aligned} \tag{7.6}$$

Theorem 7.1.3 Compactness of *SBV* *Let $\Omega \subset \mathbb{R}^d$ open and bounded and $(u_n) \subset SBV(\Omega)$. Assume that u_n satisfies (7.5) and that $|u_n(x)| \le C$ for a.e. $x \in \Omega$ for a constant $C \ge 0$ and all $n \ge 1$. Then there exists a subsequence $(u_{n(k)})$ weakly* converging in $BV(\Omega)$ to $u \in SBV(\Omega)$ with $|u(x)| \le C$ for a.e. $x \in \Omega$.*

Proof of Theorem 7.1.1 Let $(u_n) \in SBV(\Omega) \cap L^\infty(\Omega)$ be a minimising sequence of $\mathcal{E}$. Firstly, we convince ourselves of the fact that we can restrict the minimisation problem to functions u that are essentially bounded by $C = \|g\|_{L^\infty}$. This is so because for $\tilde{u} = \max(\min(u, C), -C)$, the truncated version of u, we have

$$S_{\tilde{u}} \subset S_u, \quad \text{and} \quad \alpha \int_\Omega |\nabla \tilde{u}|^2\,dx + \int_\Omega |g - \tilde{u}|^2\,dx \le \alpha \int_\Omega |\nabla u|^2\,dx + \int_\Omega |g - u|^2\,dx.$$

Then, for such a minimising sequence, we have the uniform bound

$$\mathcal{E}(u_n) = \frac{1}{2}\int_{\Omega\setminus D} (u_n - g)^2\,dx + \frac{\alpha}{2}\int_\Omega |\nabla u_n|^2\,dx + \beta\mathcal{H}^1(S_{u_n}) \le C,$$

for a constant $C \ge 0$ and for all $n \ge 1$. By Theorem 7.1.3, we can find a subsequence $u_{n(k)}$ that weakly* converges in BV to a $u \in SBV(\Omega)$ with $|u(x)| \le C$ for a.e. $x \in \Omega$. Moreover, by Theorem 7.1.2, ∇u_n weakly* converges to ∇u in $(L^2(\Omega))^d$, and $D^j u_n$ weakly* converges to $D^j u$ in Ω. Applying the lower semi-continuity properties (7.6) finishes the existence proof. ∎

Having established the existence theory for the relaxed problem (7.4), the question arises as to what the connection between the relaxed and original formulation (7.1) exactly is. To answer this, we make use of the following theorem from [Amb89a]:

Theorem 7.1.4 *Let $\Gamma \subset \Omega$ be a closed set such that $\mathcal{H}^{d-1}(\Gamma) < \infty$, and let $u \in H^1(\Omega \setminus \Gamma) \cap L^\infty(\Omega)$. Then $u \in SBV(\Omega)$, and $S_u \subset \Gamma \cup P$ with $\mathcal{H}^{d-1}(P) = 0$.*

Hence,

$$\min_u \mathcal{E}(u) \le \inf_{(u,\Gamma)} \mathcal{E}(u, \Gamma).$$

Moreover, for a minimiser of $\mathcal{E}$ it is proven in [MD97, DMMS92 and MS95] that $\mathcal{H}^{d-1}(\Omega \cap (\bar{S_u} - S_u)) = 0$. Then, by choosing $\Gamma = \Omega \cap \bar{S_u}$, we get a solution of the original problem, and

$$\min_u \mathcal{E}(u) = \min_u \mathcal{E}(u, \Omega \cap \bar{S_u}).$$

Following the existence theory, there exists a series of works concerned with the regularity of the edge set Γ (cf., e.g., [MS89 and Bon96]). In practice, mostly the edge set Γ is assumed to be at least Lipschitz continuous, in which case the Hausdorff measure as a regularity measure of Γ is replaced by the length of Γ as defined in (B.2).

7.1.1. Ambrosio-Tortorelli Approximation

For numerical implementation of the minimisation problem for (7.1), we face the problem of non-differentiability and the discretisation of the unknown edge set Γ. Esedoglu and Shen [ES02] further considered this model and its numerical solution via the Ambrosio-Tortorelli approximation [AT90, AT92], an elliptic approximation of (7.2). There a sequence of regular functionals $\mathcal{J}_\epsilon$ is considered which approximates $\mathcal{J}$ in the sense of Γ convergence (cf. Section B.8 in Appendix B) and solves the minimisation problem for $\mathcal{J}_\epsilon$, $\epsilon \ll 1$. That is, the edge set Γ in (7.2) is replaced by an ϵ-blowup χ_ϵ of its indicator function; that is, χ_ϵ is close to 1 a.e. in Ω except on an ϵ-tube Γ_ϵ around Γ, where it is close to 0 (cf. Figure 7.1). Then $1/\epsilon|1-\chi_\epsilon|^p$, $p \geq 1$, is an approximation of the Dirac delta measure δ_Γ that lives on Γ, and

$$\text{Length}(\Gamma) = \int_\Omega \delta_\Gamma(x)\, dx \propto \int_\Omega \frac{|1-\chi_\epsilon|^p}{\epsilon}\, dx.$$

In the Ambrosio-Tortorelli's approximation, χ_ϵ is computed, for a given image u, as the minimiser of

$$\mathcal{J}_\epsilon(u,\chi) = \frac{\alpha}{2}\int_\Omega \chi^2 |\nabla u|^2\, dx + \beta \int_\Omega \left(\epsilon|\nabla \chi|^2 + \frac{(1-\chi)^2}{4\epsilon}\right) dx. \tag{7.7}$$

The latter replaces the regularisation energy (7.2) in the inpainting scheme, which now reads

$$u = \text{argmin}_u \left(\mathcal{J}_\epsilon(u,\chi_\epsilon) + \frac{1}{s}\|u-g\|^s_{L^s(\Omega\setminus D)}\right). \tag{7.8}$$

For the existence of minimisers, we have to change $\mathcal{J}_\epsilon$ slightly to $\tilde{\mathcal{J}}_\epsilon(u,v) = \mathcal{J}_\epsilon(u,v) + h(\epsilon)\int_\Omega |\nabla u|^2\, dx$.

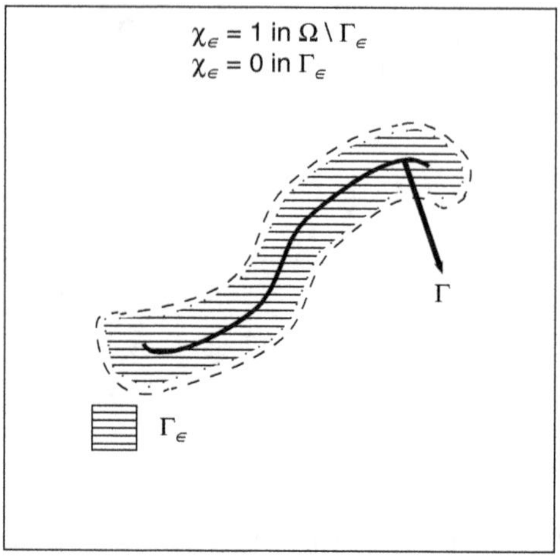

Figure 7.1. Ambrosio-Tortorelli approximation of the discontinuity set Γ.

Theorem 7.1.5 *Let $g \in L^{\infty}(\Omega)$. Let $V = \{(u,\chi) \in H^1(\Omega)^2; 0 \le \chi \le 1\}$. Then the problem*

$$\inf_V \left(\tilde{\mathcal{J}}_\epsilon(u,\chi_\epsilon) + \frac{1}{s}\|u-g\|^s_{L^s(\Omega\setminus D)} \right)$$

admits a solution $(u_\epsilon, \chi_\epsilon)$ with $\|u_\epsilon\|_{L^\infty} \le \|g\|_{L^\infty}$.

Proof The proof is similar to before – an application of the direct method of calculus of variation. ■

We also have the following approximation result:

Theorem 7.1.6 Approximation Properties in Terms of Γ-Convergence *Let $\tilde{\mathcal{J}}_\epsilon \colon L^1(\Omega) \times L^1(\Omega) \to [0,+\infty]$ be defined by*

$$\tilde{\mathcal{J}}_\epsilon(u,\chi) = \begin{cases} \frac{1}{2}\int_{\Omega\setminus D}(u-g)^2\,dx & \text{if } (u,\chi) \in H^1(\Omega)^2, \\ \quad +\frac{\alpha}{2}\int_\Omega(\chi^2+h(\epsilon))|\nabla u|^2\,dx & \quad 0 \le \chi \le 1, \\ \quad +\beta\int_\Omega\left(\epsilon|\nabla\chi|^2+\frac{1}{4\epsilon}(\chi-1)^2\right)\,dx, & \\ +\infty, & \text{otherwise,} \end{cases}$$

and let $\mathcal{J} \colon L^1(\Omega) \times L^1(\Omega) \to [0,+\infty]$ be defined by

$$\mathcal{J}(u,\chi) = \begin{cases} \frac{1}{2}\int_{\Omega\setminus D}(u-g)^2\,dx + \frac{\alpha}{2}\int_\Omega|\nabla u|^2\,dx & \text{if } u \in GSBV(\Omega) \text{ and} \\ \quad +\beta\mathcal{H}^1(S_u), & \quad v = 1 \text{ a.e.,} \\ +\infty, & \text{otherwise,} \end{cases} \tag{7.9}$$

where S_u is the jump set of u. If $h(\epsilon) = o(\epsilon)$, then $\tilde{\mathcal{J}}_\epsilon$ Γ-converges to $\mathcal{J}$ in the $L^1(\Omega)^2$-strong topology. Moreover, $\tilde{\mathcal{J}}_\epsilon$ admits a minimiser $(u_\epsilon, \chi_\epsilon)$ such that u_ϵ converges in $L^1(\Omega)$ to a minimiser of $\mathcal{J}$, $u \in SBV(\Omega)$ and $\inf \tilde{\mathcal{J}}_\epsilon \to \inf \mathcal{J}$ *$(\epsilon \to 0)$*

A general disadvantage of the Ambrosio-Tortorelli approximation is that in practice one cannot choose ϵ as small as would be required for obtaining a highly resolved approximation to the edge set Γ. The ϵ-tube that contains Γ has to be at least as large as the grid size Δx. In fact, as discussed in Esedoglu and Shen [ES02], for numerical stability of the transition layer of χ_ϵ, the parameter ϵ even has to be a multiple of Δx. This is an important issue for segmentation, where an accurate detection of the object boundaries is desired. However, for image inpainting, we aim primarily to recover the image u not the edge set χ_ϵ. In fact, the low resolution of the latter is acceptable because it does not influence the resolution of the inpainting result very much. The big advantage of the Ambrosio-Tortorelli approximation, though, is equally important in segmentation and inpainting, namely, that it is a quadratic approximation that lends itself to fast linear solvers on the computer. In the following, we shall discuss its numerical solution in detail.

We start by deriving the corresponding Euler-Lagrange equations for the minimisation problem of (7.7) and (7.8) for χ_ϵ and u, respectively. They read

$$
\begin{aligned}
&\alpha|\nabla u|^2\chi_\epsilon + \beta\left(-2\epsilon\Delta\chi_\epsilon + \frac{\chi_\epsilon - 1}{2\epsilon}\right) = 0, && \text{in } \Omega,\\
&\mathbb{1}_{\Omega\setminus D}(u-g) - \alpha\operatorname{div}((\chi_\epsilon^2 + h(\epsilon))\nabla u) = 0, && \text{in } \Omega,\\
&\frac{\partial u}{\partial \vec{n}} = \frac{\partial \chi_\epsilon}{\partial \vec{n}} = 0, && \text{on } \partial\Omega,
\end{aligned}
$$

where $\mathbb{1}_{\Omega\setminus D}$ is the indicator function for the intact part of the image domain $\Omega \setminus D$, and $\vec{n}$ is the outward-pointing normal to $\partial\Omega$, as before. One way of solving this non-linear system of equations is to approximate it iteratively by a sequence of linear equations as follows: let $u^{(0)} = 0$ and $z^{(0)} = 0$, and solve for $k = 1, 2, \ldots$

$$
A_{u^{(k-1)}}\chi_\epsilon^{(k)} = 1, \quad B_{\chi_\epsilon^{(k)}} u^{(k)} = \frac{\mathbb{1}_{\Omega\setminus D}}{\alpha} g, \tag{7.10}
$$

with elliptic operators

$$
\begin{aligned}
A_u &= 1 + \frac{2\epsilon\alpha}{\beta}|\nabla u|^2 - 4\epsilon^2\Delta,\\
B_\chi &= -\operatorname{div}((\chi_\epsilon^2 + h(\epsilon))\nabla) + \mathbb{1}_{\Omega\setminus D}/\alpha.
\end{aligned}
$$

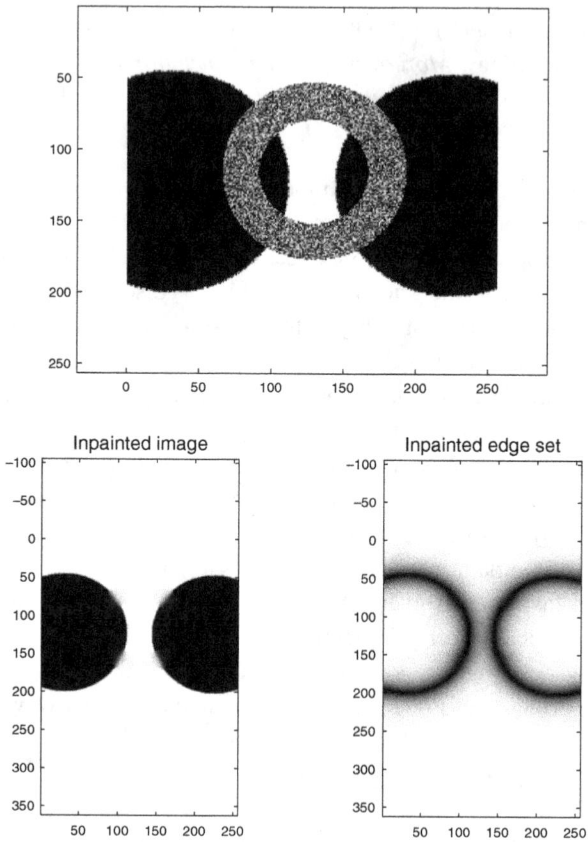

Figure 7.2. Mumford-Shah inpainted image and its inpainted edge set computed by the alternating scheme (7.10) for the Ambrosio-Tortorelli approximation.

The alternating scheme (7.10) also can be parallelised by solving the two linear equations for u and χ simultaneously. In this case, the previous iterate $\chi^{(k-1)}$ rather than the current iterate $\chi^{(k)}$ is used in the computation of B_χ.

Numerical Discussion We present some examples for Mumford-Shah inpainting using the Ambrosio-Tortorelli approximation (7.10). Figures 7.2 through 7.5 show the inpainted image u and its edge set χ in the Ambrosio-Tortorelli approximation. The example with the two rings in Figure 7.2 is reconstructed rather well. In addition, erasing the superimposed writing in Figure 7.3 gives a convincing result. However, as soon as larger parts of image structures are missing, such as in Figures 7.4 and 7.5, the method behaves similarily

Figure 7.3. Erasing superimposed text in an image. The Mumford-Shah inpainted image and its inpainted edge set computed by the alternating scheme (7.10) for the Ambrosio-Tortorelli approximation. Photo courtesy of Kostas Papafitsoros.

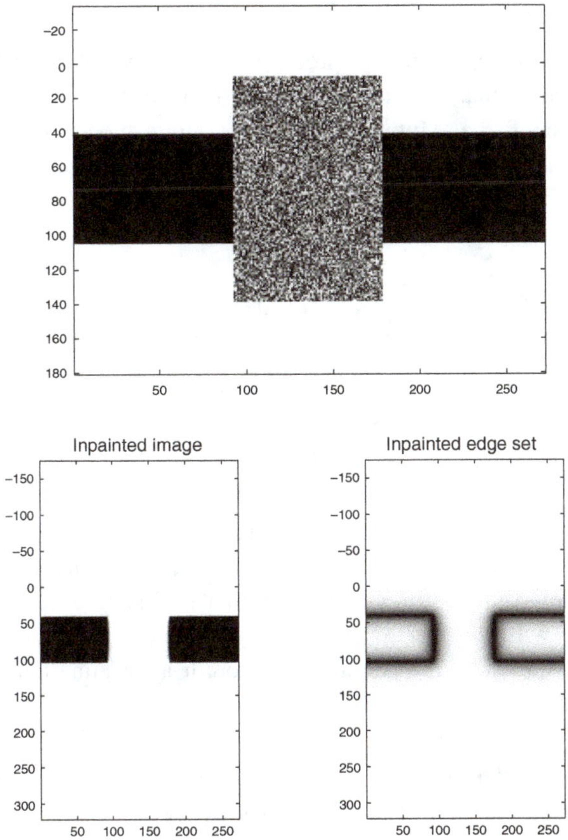

Figure 7.4. The Mumford-Shah inpainted image and its inpainted edge set computed by the alternating scheme (7.10) for the Ambrosio-Tortorelli approximation.

to TV inpainting (compare also Figure 4.8 in Section 4.3 and Figure 5.12 in Section 5.3). In Figure 7.5, the edges of the fault in the couch are not reconstructed correctly.

Figure 7.5. Example of the sleeping dog. The Mumford-Shah inpainted image and its inpainted edge set computed by the alternating scheme (7.10) for the Ambrosio-Tortorelli approximation.

7.2. Mumford-Shah-Euler Inpainting

With the insights gained in the preceding section we can now revisit the inpainting scheme that uses the Mumford-Shah image model. In particular, we will consider the Mumford-Shah inpainting model but replace the insufficient-length model on the edge set with Euler's elastica curve model as proposed in [MD97 and ES02] for image inpainting. That is, we consider the modified energy

$$\mathcal{J}(u,\Gamma) = \frac{\gamma}{2}\int_{\Omega\setminus\Gamma} |\nabla u|^2 \, dx + \int_\Gamma (\alpha + \beta\kappa^2) \, ds$$

and the corresponding Mumford-Shah-Euler inpainting model

$$\mathcal{E}(u,\Gamma) = \frac{1}{2}\int_{\Omega\setminus D} (u-g)^2 \, dx + \mathcal{J}(u,\Gamma). \tag{7.11}$$

As before, finding the right balance between the length and curvature terms in Euler's elastica energy should give us an inpainting method that fulfils the continuation principle of Kanisza. In order to find a numerical minimiser of (7.11), we will consider an elliptic approximation of the elastica curve model that is similar to the Ambrosio-Tortorelli approximation discussed in Section 7.1.1.

Elliptic Approximation à la De Giorgi For numerical purposes, we consider an elliptic approximation of Euler's elastica energy

$$\alpha \int_\Gamma ds + \beta \int_\Gamma \kappa^2 \, ds,$$

that reads

$$\mathcal{J}_\epsilon(\chi) = \alpha \int_\Omega \left(\epsilon|\nabla\chi|^2 + \frac{W(\chi)}{4\epsilon}\right) dx + \frac{\beta}{\epsilon}\int_\Omega \left(2\epsilon\Delta\chi - \frac{W'(\chi)}{4\epsilon}\right) dx. \tag{7.12}$$

Here χ is an ϵ-approximation of the signature function of Γ; that is, Γ is close to the zero level set of χ, and W is the symmetric double-well potential $W(\chi) = (1-\chi^2)^2 = (1-\chi)^2(1+\chi)^2$.

The elliptic approximation of the elastica energy is known as *De Giorgi's conjecture* [MD97] and has been studied extensively both in the differential geometry community and by people working in non-linear partial differential equations (PDEs) and variational calculus. There

$$E[u](t) = \int_\Omega \left(\frac{\epsilon}{2}|\nabla u(x,t)|^2 + \frac{1}{\epsilon}W(u(x,t))\right)dx \tag{7.13}$$

is called the *Ginzburg-Landau free energy*, and

$$G[u](t) = \frac{1}{4\epsilon}\int_\Omega (\epsilon\Delta u(x,t) - \frac{1}{\epsilon}W'(u(x,t)))^2 dx \tag{7.14}$$

is a phase-field approximation of the Willmore functional. The former appears as the free energy for non-linear PDEs that model phase separation and coarsening, as well as a diffuse interface energy for approximating the lengths of curves or surfaces. The Willmore functional is considered to describe the geometrical boundary of two different stable states and the movement of curves under anisotropic flows. It has its origin in differential geometry, where it appears as a phase-field approximation for solutions of the so-called Willmore problem (cf. [Wil93]). The Willmore problem is to find a surface Γ in an admissible class embedded in $\mathbb{R}^3$ which minimises the mean curvature energy $\int_\Gamma H^2\, dS$ under certain constraints on the surface, where $H = (\kappa_1 + \kappa_2)/2$ is the mean curvature and κ_1, κ_2 are the principal curvatures of Γ. For the analytical and computational modelling of a minimising surface of the Willmore problem, the phase-field method is considered, among other approaches. In [DLRW05], the authors consider solutions of a constrained minimisation problem for (7.14) of the form $u_\epsilon(x) = \tanh(d(x)/\sqrt{2}\epsilon) + \epsilon h$ with fixed mass and fixed energy (7.13), where d is the signed distance function to the zero level set of u_ϵ, and h is an arbitrary function in $C^2(\Omega)$ independent of ϵ. The authors show that the level sets $\{u_\epsilon = 0\}$ converge uniformly to a critical point of the Willmore problem as $\epsilon \to 0$. Also in this range of considerations falls a modified De Giorgi conjecture. The authors of [RS06] considered functionals $F_\epsilon : L^1(\Omega) \to \mathbb{R}$ for domains $\Omega \in \mathbb{R}^2$ and $\mathbb{R}^3$ with

$$F_\epsilon[u](t) = E[u](t) + 4G[u](t)$$

if $u \in L^1(\Omega) \cap W^{2,2}(\Omega)$ and $F_\epsilon[u](t) = \infty$ if $u \in L^1(\Omega) \setminus W^{2,2}(\Omega)$. They showed that this sequence of functionals evaluated in characteristic functions $\chi = 2\chi_E - 1$ with $E \subset \Omega$ Γ-converges in $L^1(\Omega)$ as $\epsilon \to 0$ to a functional $F[\chi]$

given by

$$F[\chi](t) = \sigma \mathcal{H}^{n-1}(\partial E \cap \Omega) + \sigma \int_{\partial E \cap \Omega} |\mathbf{H}_{\partial E}|^2 \, d\mathcal{H}^{n-1}.$$

Here $\sigma = \int_{-1}^{1} \sqrt{2F}$ (where F is the double-well potential), $\mathbf{H}_{\partial E}$ denotes the mean curvature vector of ∂E and $\mathcal{H}^{n-1}$ is the $(n-1)$-dimensional Hausdorff measure. For additional considerations of Γ-limits of this type, see [RT08] and especially [C$^+$96].

Replacing the Euler elastica energy with its elliptic approximation (7.12) in (7.11), we obtain the following new inpainting approach:

$$\mathcal{E}_\epsilon(u,\chi) = \frac{1}{s}\|u-g\|^s_{L^s(\Omega\setminus D)} + \underbrace{\frac{\gamma}{2}\int_\Omega \chi^2|\nabla u|^2\,dx + \mathcal{J}_\epsilon(\chi)}_{=:\mathcal{J}_\epsilon(u,\chi)}.$$

As in Section 7.1.1, we are attempting to compute a minimiser (u,χ) of $\mathcal{E}_\epsilon$ via solving the system of its corresponding Euler-Lagrange equations that reads

$$\mathbb{1}_{\Omega\setminus D}(u-g) - \gamma \operatorname{div}((\chi^2 + h(\epsilon))\nabla u) = 0, \tag{7.15}$$

$$\gamma\,|\nabla u|^2\chi - \alpha g - \frac{\beta W''(\chi)}{2\epsilon^2} g + 4\beta\Delta g = 0, \tag{7.16}$$

$$2\epsilon\Delta\chi - \frac{W'(\chi)}{4\epsilon} = g. \tag{7.17}$$

The first equation is the Euler-Lagrange equation for u and is the same as the one in the standard Mumford-Shah model in Section 7.1.1. For a fixed χ, this is a linear elliptic equation in u and can be solved with an efficient direct linear solver. The second equation is the Euler-Lagrange equation for χ and for a fixed u is now a fourth-order non-linear equation in χ. Care has to be taken when attempting its numerical solution. In the following, we shall approach the stationary equation (7.16) with a time-marching procedure. In particular, we solve the steepest-descent equation for $\mathcal{J}_\epsilon$ in χ, that is,

$$\chi_t = -\gamma\,|\nabla u|^2\chi + \alpha g + \frac{\beta W''(\chi)}{2\epsilon^2} g - 4\beta\Delta g$$

$$g = 2\epsilon\Delta\chi - \frac{W'(\chi)}{4\epsilon},$$

which we discretise in time – as proposed in [ES02] – with a semi-implicit scheme as

$$(1+\Delta t A_u)\chi^{n+1} = \chi^n + \Delta t f(\chi^n), \tag{7.18}$$

where

$$A_u = \gamma\,|\nabla u|^2 - 2\alpha\epsilon\Delta + 8\beta\epsilon\Delta^2,$$

and $f(\chi)$ collects all the non-linear terms in χ, that is,

$$f(\chi) = -\frac{\alpha}{4\epsilon}W'(\chi) + \frac{\beta}{2\epsilon^2}W''(\chi)g + \frac{\beta}{\epsilon}\Delta W'(\chi).$$

Another possibility for solving (7.16) would be to compute a stationary point via Newton iteration, that is,

$$D_\chi^2 \mathcal{J}_\epsilon(\chi^n) \cdot \chi^{n+1} = -D_\chi \mathcal{J}_\epsilon(\chi^n),$$

where $D_\chi \mathcal{J}_\epsilon$ is the Jacobian of $\mathcal{J}_\epsilon$ in χ and $D_\chi^2 \mathcal{J}_\epsilon$ its Hessian.

8

Inpainting Mechanisms of Transport and Diffusion

In [CKS02], the authors give an interpretation of Euler's elastica inpainting in terms of the mechanisms of transport and diffusion. More precisely, they derive the optimality condition for elastica minimising curves and compare it with the transport equation of Bertalmio et al. [BSCB00] and the curvature-driven diffusion (CDD) inpainting approach of Chan and Shen [CS01b]. Thereby the optimality condition for elastica minimising curves or, preferably, the corresponding flux field shows a natural decomposition into its normal and tangent field (cf. Theorem 5.2.5). In this chapter we want to present this decomposition for Euler's elastica inpainting and make a similar analysis for TV-H^{-1} inpainting and inpainting with low curvature image simplifiers (LCISs) presented in Sections 5.3.2 and 5.4, respectively. To see the connection to the inpainting models of Bertalmio et al. [BSCB00] and Chan and Shen [CS01b], let us recall them briefly. These two inpainting models were presented in Sections 6.1 and 4.5, respectively.

Bertalmio's Transport Inpainting Bertalmio's approach is based on solving the following transport equation inside the inpainting domain D:

$$\frac{\partial u}{\partial t} = \nabla^{\perp} u \cdot \nabla L(u),$$

where $\nabla^{\perp} = (-u_y, u_x) = |\nabla u|\vec{t}$, and $L(u)$ can be any smoothness measure of the image u. For their numerical experiments, the authors in [BSCB00] chose $L(u) = \Delta u$, the Laplacian of u. For the equilibrium state, the equation reads

$$\vec{t} \cdot \nabla L(u) = 0, \qquad \text{that is,} \qquad \frac{\partial L(u)}{\partial \vec{t}} = 0,$$

which means that the smoothness measure remains constant along any completed level line. In other words, assuming available boundary data, boundary smoothness gets transported along the level lines into the missing domain.

Second-Order Diffusion Inpainting Orthogonal to this idea of smoothness transport along level lines, Chan and Shen proposed the CDD inpainting model in [CS01b], that is,

$$\frac{\partial u}{\partial t} = \nabla \cdot \left(\frac{g(\kappa)}{|\nabla u|} \nabla u \right),$$

where g: $B \to [0, +\infty)$ is a continuous function with $g(0) = 0$ and $g(\pm\infty) = +\infty$, and B equals, for example, $C^2(\Omega)$. The function g penalises large curvatures, and encourages diffusion when the curvature is small. This model diffuses the image contents across the level lines (since $\nabla u / |\nabla u|$ is the normal vector to the level lines!), which is completely orthogonal to the behaviour of the Bertalmio et al. approach.

Euler's Elastica Inpainting In [CKS02], the authors showed that Euler's elastica inpainting model unifies these two mechanisms. Theorem 5.2.5 says that the flux $\vec{V}$ for the inpainting energy consists of two components, the normal part

$$\vec{V}_n^{\text{Euler}} = \phi(\kappa) \cdot \vec{n}$$

and the tangential part

$$\vec{V}_t^{\text{Euler}} = -\frac{1}{|\nabla u|} \frac{\partial(\phi'(\kappa)|\nabla u|)}{\partial \vec{t}} \cdot \vec{t}.$$

We immediately see that the normal flux corresponds to the flux of the CDD inpainting equation with $g(\kappa) = \phi(\kappa)$. By rewriting the tangential component, we see that this corresponds to the Bertalmio flux with a special smoothness measure

$$\vec{V}_t = -\left(\frac{1}{|\nabla u|^2} \frac{\partial(\phi'(\kappa)|\nabla u|)}{\partial \vec{t}} \right) \cdot \nabla^{\perp} u.$$

Then its divergence reads

$$\nabla \cdot \vec{V}_t = \nabla^{\perp} u \cdot \nabla \left(\frac{-1}{|\nabla u|^2} \frac{\partial(\phi'(\kappa)|\nabla u|)}{\partial \vec{t}} \right)$$

because the $\nabla^{\perp} u$ is divergence free. Then the tangential component of Euler's elastica flux corresponds to the scheme of Bertalmio et al. with smoothness measure

$$L_{\phi}^{\text{Euler}} = \frac{-1}{|\nabla u|^2} \frac{\partial(\phi'(\kappa)|\nabla u|)}{\partial \vec{t}}.$$

This measure can be further rewritten in a way which makes its connection to the Laplacian visible. In the case $\phi(s) = |s|$, we get

$$L_{\phi}^{\text{Euler}} = \frac{\pm 1}{|\nabla u|^2}[\nabla \times \nabla u](\vec{n},\vec{t}),$$

which closely resembles Bertalmio's choice of the Laplacian, which can written as

$$L^{\text{Bertalmio}}(u) = \Delta u = \text{tr}(\nabla \times \nabla u) = [\nabla \times \nabla u](\vec{n},\vec{n}) + [\nabla \times \nabla u](\vec{t},\vec{t}).$$

TV-H^{-1} Inpainting Now we want to do a similar analysis for TV-H^{-1} inpainting. The gradient descent of the TV-H^{-1} inpainting regulariser is given by

$$\frac{\partial u(x,t)}{\partial t} = -\Delta\left(\nabla \cdot \left(\frac{\nabla u}{|\nabla u|}\right)\right) = \nabla \cdot \left[-\nabla\left(\nabla \cdot \left(\frac{\nabla u}{|\nabla u|}\right)\right)\right],$$

with flux field

$$\vec{V}^{\text{TV}-\text{H}^{-1}} = -\nabla\left(\nabla \cdot \left(\frac{\nabla u}{|\nabla u|}\right)\right).$$

Now, to write this flux as the sum of its normal and tangential component, we project it onto both the normal and the tangential directions and use the fact that the sum of these projections gives the identity, that is,

$$\vec{n} \otimes \vec{n} + \vec{t} \otimes \vec{t} = Id,$$

where $\otimes$ denotes the tensor product. With this, the steepest descent for TV-H^{-1} inpainting reads

$$\begin{aligned}\frac{\partial u(x,t)}{\partial t} &= \nabla \cdot (-\nabla \kappa),\\ &= \nabla \cdot \left(-\nabla \kappa (\vec{t} \otimes \vec{t} + \vec{n} \otimes \vec{n})\right),\\ &= \nabla \cdot \left(-\left(\frac{\partial \kappa}{\partial \vec{t}}\vec{t} + \frac{\partial \kappa}{\partial \vec{n}}\vec{n}\right)\right).\end{aligned}$$

Now we want to compare the normal and tangential components with the components of Euler's elastica inpainting approach and with the approaches of CDD and Bertalmio et al., respectively. Let us start with the normal part of the TV-H^{-1} flow, which reads

$$\nabla \cdot \vec{V}_n^{\text{TV}-\text{H}^{-1}} = \nabla \cdot \left(-\frac{\partial \kappa}{\partial \vec{n}}\vec{n}\right).$$

At the same time, the normal part in Euler's elastica is

$$\nabla \cdot \vec{V}_n^{\text{Euler}} = \nabla \cdot (\phi(\kappa)\vec{n})$$

or, more specifically,

$$\nabla \cdot \vec{V}_n^{\text{Euler}} = \nabla \cdot ((a + b\kappa^2)\vec{n}),$$

and the CDD flux reads

$$\nabla \cdot \vec{V}^{\text{CDD}} = \nabla \cdot (g(\kappa)\vec{n}).$$

Note that the strength of the diffusion in all three approaches depends on the size of the diffusivity constant. In the case of TV-H^{-1} inpainting $D = (-\partial\kappa/\partial\vec{n})/|\nabla u|$, the diffusion depends on the change in the curvature across the level lines. The larger the curvature changes across the level lines, the more diffusive is this approach. Thus, the sign of the diffusion depends on the sign of the covariant derivative (forward or backward diffusion!). In particular, this means that there is no diffusion in areas of the image with homogeneous grey values and strong diffusion of edge information into the missing domain.

Next, we do the same for the tangential component in $\vec{V}$. For TV-H^{-1} inpainting, this reads

$$\begin{aligned} \nabla \cdot \vec{V}_t^{\text{TV}-\text{H}^{-1}} &= \nabla \cdot \left(-\frac{\partial\kappa}{\partial\vec{t}}\vec{t}\right) \\ &= \nabla \cdot \left(\frac{\partial\kappa}{\partial\vec{t}} \frac{-1}{|\nabla u|} \cdot \nabla^{\perp} u\right) \\ &= \nabla^{\perp} u \cdot \nabla \left(\frac{-1}{|\nabla u|} \frac{\partial\kappa}{\partial\vec{t}}\right), \end{aligned}$$

because $\nabla^{\perp} u$ is divergence free. This corresponds to the scheme of Bertalmio et al. with smoothness measure

$$L^{\text{TV}-\text{H}^{-1}}(u) = \frac{-1}{|\nabla u|} \frac{\partial\kappa}{\partial\vec{t}},$$

whereas Euler's elastica tangent component has the smoothness measure

$$L^{\text{Euler}}(u) = \frac{-2b}{|\nabla u|^2} \frac{\partial(\kappa|\nabla u|)}{\partial\vec{t}}.$$

LCIS Inpainting The steepest descent for the LCIS regulariser reads

$$\frac{\partial u(x,t)}{\partial t} = \nabla \cdot \vec{V}^{\text{LCIS}} = \nabla \cdot (-g(\Delta u)\nabla\Delta u).$$

Taking $g(s) = 1/(1+s^2)$, the divergence of the flux changes to

$$\nabla \cdot \vec{V}^{\text{LCIS}} = \nabla \cdot (\nabla(\arctan(\Delta u))).$$

Similarly as for Euler's elastica inpainting and the TV-H^{-1} approach, we split the flux of LCIS into its normal and tangential components and compare it with the CDD and Bertalmio et al. approaches respectively. For the normal component, we get

$$\nabla \cdot \vec{V}_n^{\mathrm{LCIS}} = \nabla \cdot \left(-\frac{\partial \arctan(\Delta \mathrm{u})}{\partial \vec{n}} \vec{n} \right).$$

That is, the normal flux diffuses the grey values with diffusivity constant

$$D = -\frac{\partial \arctan(\Delta \mathrm{u})}{\partial \vec{n}} \frac{1}{|\nabla u|}.$$

Next, we compare the tangential component of LCIS with the approach of Bertalmio et al. The divergence of the tangential flux of LCIS reads

$$\begin{aligned}
\nabla \cdot \vec{V}_t^{\mathrm{LCIS}} &= \nabla \cdot \left(-\frac{\partial \arctan(\Delta \mathrm{u})}{\partial \vec{t}} \vec{t} \right) \\
&= \nabla \cdot \left(-\frac{\partial \arctan(\Delta \mathrm{u})}{\partial \vec{t}} \frac{1}{|\nabla u|} \nabla^{\perp} u \right) \\
&= \nabla^{\perp} u \cdot \nabla \left(-\frac{\partial \arctan(\Delta \mathrm{u})}{\partial \vec{t}} \frac{1}{|\nabla u|} \right).
\end{aligned}$$

This reassembles Bertalmio's smoothness transport along the level lines with smoothness measure

$$\begin{aligned}
L^{\mathrm{LCIS}}(u) &= -\frac{\partial \arctan(\Delta \mathrm{u})}{\partial \vec{t}} \frac{1}{|\nabla u|} \\
&= -\frac{1}{|\nabla u|} \frac{1}{1 + (\Delta u)^2} \frac{\partial (\Delta u)}{\partial \vec{t}}.
\end{aligned}$$

Summary and Numerical Results Table 8.1 summarises a comparison of the flux field of the various inpainting approaches. For a better understanding of the derived geometrical properties of TV-H^{-1} inpainting and inpainting with LCIS, we present a comparison of those with Euler's elastica inpainting for the inpainting of a straight line in Figure 8.1. For all three inpainting approaches, several intermediate steps of the iteration (evolution) are shown. The processes seem quite similar. Firstly, the inpainting domain is filled in where homogeneous boundary condition are given; that is, the intensity values are transported into the inpainting domain very quickly. At the edges (black-white boundary), the evolution takes longer until the edge is continued.

Table 8.1. *A comparison of the normal flux field $\vec{V}_n$ (which represents the diffusive part of the flux), and the smoothness measure $L(u)$, which is transported along the tangential flux, for all discussed inpainting approaches*

	$\nabla \cdot \vec{V}_n$	$L(u)$
Bertalmio et al.'s transport	0	Δu
Cahn and Shen's CDD	$\nabla \cdot (g(\kappa)\vec{n})$	0
Euler's elastica	$\nabla \cdot ((a+b\kappa^2)\vec{n})$	$\frac{-2b}{\lvert\nabla u\rvert^2}\frac{\partial(\kappa\lvert\nabla u\rvert)}{\partial \vec{t}}$
TV-H^{-1}	$\nabla \cdot (-\frac{\partial \kappa}{\partial \vec{n}}\vec{n})$	$\frac{-1}{\lvert\nabla u\rvert}\frac{\partial \kappa}{\partial \vec{t}}$
LCIS	$\nabla \cdot \left(-\frac{\partial \arctan(\Delta u)}{\partial \vec{n}}\vec{n}\right)$	$-\frac{1}{\lvert\nabla u\rvert}\frac{1}{1+(\Delta u)^2}\frac{\partial(\Delta u)}{\partial \vec{t}}$

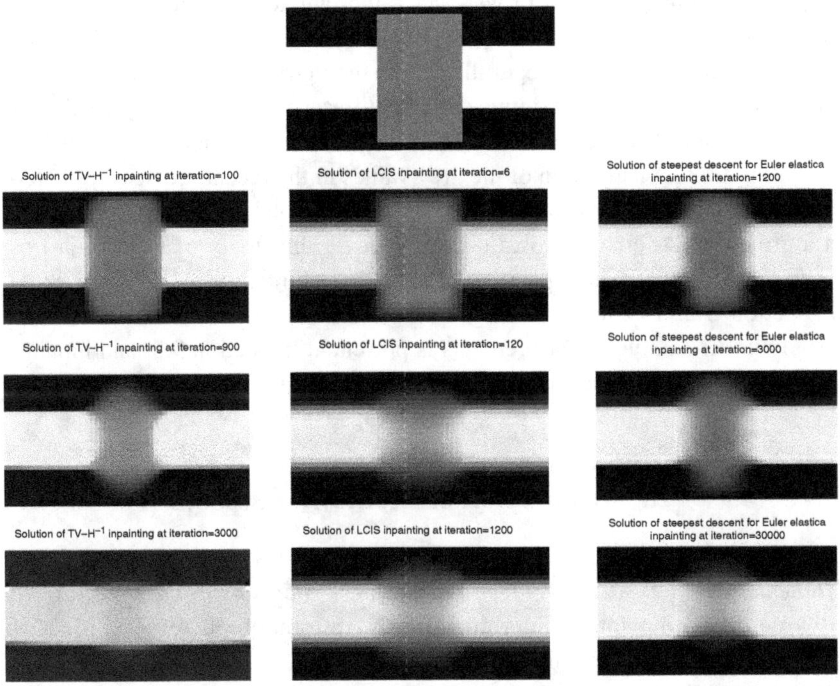

Figure 8.1. A comparison of the evolution of TV-H^{-1} inpainting, inpainting with LCIS and Euler's elastica inpainting at three different time steps. Left: The evolution of TV-H^{-1} inpainting at time steps $t = 100$, 900 and 3,000. Center: The evolution of LCIS inpainting at time steps $t = 6$, 120 and 1,200. Right: The evolution of Euler's elastica inpainting with $a = 0.1$ and $b = 2$ at time steps $t = 1,200$, 3,000 and 30,000.

Note that at the edge the change of the curvature in the normal direction is equal to 0. It becomes non-zero when approaching the boundary of the inpainting domain in between the two edges.

9
Applications

Digital inpainting methods provide an important tool in the restoration of images in a wide range of applications. In this book we present four special applications which hopefully reflect the myriad of potential inpainting methods. The first is the inpainting of ancient frescoes. In particular, we discuss the Cahn-Hilliard equation for the inpainting of binary structure and TV-H^{-1} inpainting for reconstruction of the grey values in the recently discovered Neidhart frescoes in Vienna (cf. Section 9.1). The second application originated in a project of Andrea Bertozzi taking place at University of California Los Angeles and is about the inpainting of roads in satellite images. It is presented in Section 9.2. Inpainting aiding tomographic image reconstruction from heavily under-sampled measurements is presented in Section 9.3, and inpainting for three-dimensional conversion of movies is outlined in Section 9.4.

9.1. Restoration of Medieval Frescoes

In the course of an interdisciplinary project (WWTF Five Senses-Call 2006, *Mathematical Methods for Image Analysis and Processing in the Visual Arts*), we used digital inpainting algorithms for the restoration of frescoes. Most of the results presented in this section can be found in [BFM$^+$08]. However, please also refer to a more light-weight article, 'Restoring Profanity', which was published online in *Cambridge Plus Magazine* [Sch09].

Particular consideration has been extended to the newly found Neidhart frescoes (Tuchlauben 19, 1010 Vienna). These medieval frescoes from the fourteenth century depict a cycle of songs of the thirteenth-century minnesinger Neidhart von Reuental. Hidden behind a wall over years, the frescoes have been damaged during exposure. Advanced mathematical tools were developed specifically for so-called mathematical inpainting/retouching

Figure 9.1. Part of the Neidhart frescoes. Photo courtesy of Andrea Baczynski.

of digital images. To this end, variational methods and third- and fourth-order partial differential equations (PDEs) have been investigated. Efficient numerical methods for solution of the devised PDEs have been designed. In the following, we discuss our mathematical inpainting methods and present numerical results from their application to the Neidhart frescoes.

Neidhart Frescoes Fragments of fourteenth-century wall frescoes found beneath the crumbling plaster of an old apartment in the heart of Vienna depict a popular medieval cycle of songs of the thirteenth-century minnesinger Neidhart von Reuental. In the very late fourteenth century, Michel Menschein, a wealthy Viennese council member and cloth merchant, commissioned local artists to paint the stories in Neidhart's songs on the walls of his *Festsaal* (banquet hall). The Neidhart frescoes provide a unique peek into medieval humor and, at the same time, the taste of a medieval man.

Figure 9.1 shows a part of the Neidhart frescoes.[1] The white holes in this fresco are due to the wall which covered the fresco until a few years ago. The holes occurred when the wall was removed. In the following, we want to apply digital restoration methods to these frescoes. Therefore, the main challenge

[1] I thank Andrea Baczynski for providing the fresco data.

is to capture the structures in the preserved parts of the fresco and transport them into the damaged parts continuously. Given their great age and almost 600 years of living by owners and tenants in the apartment, saturation, hue and contrast quality of the colours in the frescoes suffered. Digital grey value (i.e., colour) interpolation in the damaged parts of the frescos therefore demands sophisticated algorithms that take these detrimental factors into account.

9.1.1. Methods

In the following, we present the mathematical methods we used to reconstruct the damaged parts of the frescos. In particular, two inpainting methods based on higher-order PDEs, that is, Cahn-Hilliard (5.18) and TV-H^{-1} inpainting (5.20), are to be presented. We finalise this section by proposing a possible strategy to adapt these two inpainting approaches to the requirements of the Neidhart frescoes.

Let us start by briefly recalling the Cahn-Hilliard inpainting approach (5.18). The inpainted version u of $g \in L^2(\Omega)$ is constructed by following the evolution equation:

$$\begin{cases} u_t = \Delta(-\epsilon\Delta u + \frac{1}{\epsilon}F'(u)) + \mathbb{1}_{\Omega\setminus D}\lambda(g-u), & \text{in } \Omega, \\ \frac{\partial u}{\partial \nu} = \frac{\partial \Delta u}{\partial \nu} = 0, & \text{on } \partial\Omega, \end{cases}$$

with $F(u)$ a so-called double-well potential, for example, $F(u) = u^2(u-1)^2$, whereas

$$\mathbb{1}_{\Omega\setminus D}(x) = \begin{cases} \lambda, & x \in \Omega\setminus D \\ 0, & x \in D \end{cases}$$

is the indicator function of $\Omega \setminus D$, and $\lambda \gg 1$ is a large constant.

A generalisation of the Cahn-Hilliard inpainting approach to an approach for grey value images was presented in Section 5.3.2 (also cf. [BHS09]); that is, we consider TV-H^{-1} inpainting (5.20): let $g \in L^2(\Omega)$, $|g| \leq 1$ be the given grey value image. The inpainted version $u(x)$ of $g(x)$ evolves in time like

$$u_t = -\Delta\left(\nabla\cdot\left(\frac{\nabla u}{|\nabla u|}\right)\right) + \mathbb{1}_{\Omega\setminus D}\lambda(g-u), \tag{9.1}$$

where $p \in \partial TV(u)$ is replaced by a relaxed version

$$p \approx \nabla\cdot\left(\frac{\nabla u}{|\nabla u|}\right) = \nabla\cdot\left(\frac{\nabla u}{\sqrt{|\nabla u|^2+\epsilon}}\right),$$

for an $0 < \epsilon \ll 1$.

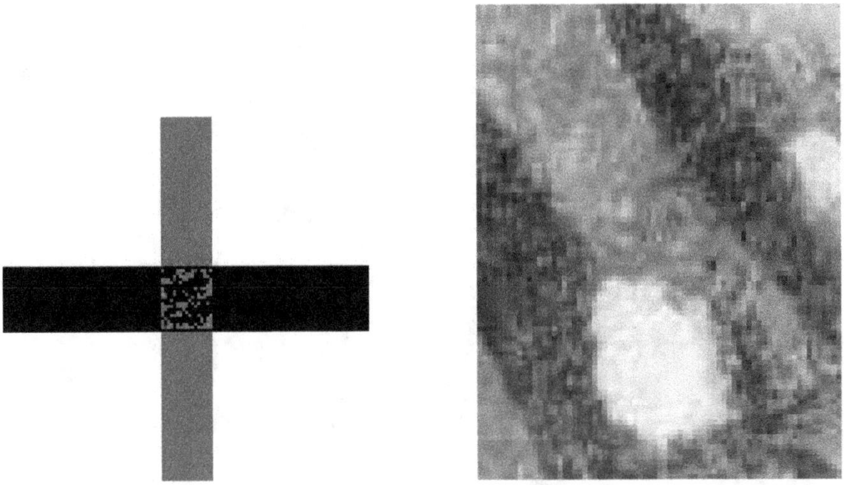

Figure 9.2. Left: What is the right solution? Right: Part of the Neidhart fresco. How should the inpainting algorithm be decided in this case? Photo courtesy of Andrea Baczynski.

The Neidhart frescoes pose a special challenge concerning their digital restoration. The main issues can be summerised as follows:

1. Lack of grey value contrast
2. Low colour saturation and hue
3. Damaged parts that are rather large (i.e., the diameter of the damaged domain can be larger than the width of lines which are to be continued into the damaged part)

Hence, we need an inpainting approach which takes into account these possible difficulties and solves (or circumvents) them. As pointed out earlier in this section, the third issue can be solved by using a higher-order inpainting method such as (5.18) and (5.20). Unfortunately, difficulties 1 and 3 prevent the effective application of these methods. As the contrast between grey values is low, the edges (which identify the main structure of an image) are not clearly defined. Because inpainting lives or dies with the uniqueness of edge continuation (cf. Figure 9.2), we may run into trouble if we do not pre-process the digital images of the fresco in an adequate manner.

Specifically we follow two strategies:

- *Strategy 1: Structure inpainting on binary images with the Cahn-Hilliard equation.* Based on the recovered binary structure, the fresco is colourised. We discuss the re-colourisation, that is, the filling in of grey values based on

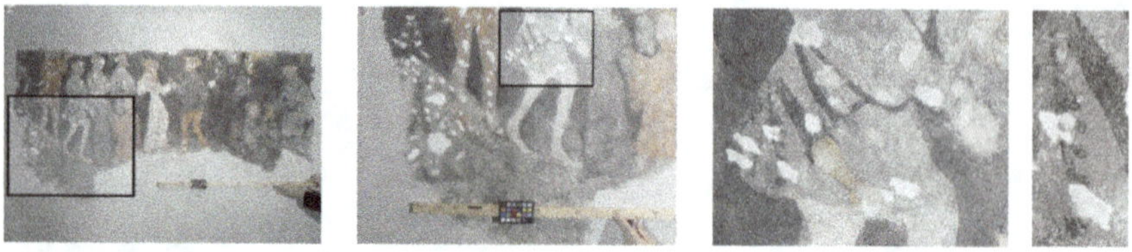

Figure 9.3. Part of the Neidhart frescoes. Photo courtesy of Andrea Baczynski.

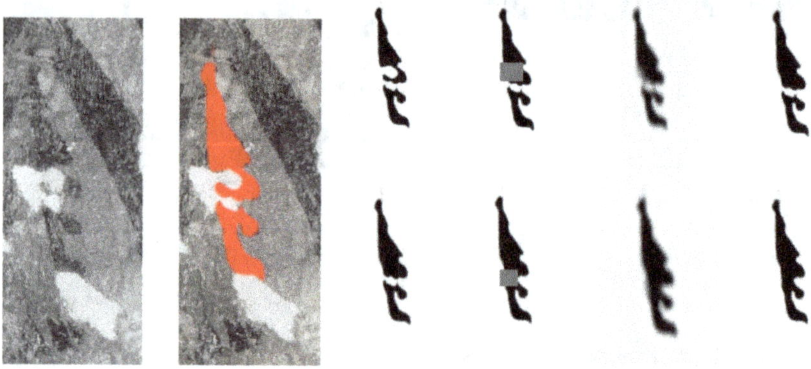

Figure 9.4. Cahn-Hilliard inpainting with $\lambda = 10^7$. From left to right: Part of the fresco, binary selection in red, binary selection in black and white, initial condition for the inpainting algorithm where the inpainting region is marked with a grey rectangle, inpainting result after 200 time steps with $\epsilon = 3$ and inpainting result after an additional 800 time steps with $\epsilon = 0.01$.

the given binary structure, in more detail later. Also see [For06] for a similar approach.

- *Strategy 2: Apply TV-H^{-1} inpainting in two steps.* Firstly, with a small λ, for example, $\lambda = 1$, merge together fine artefacts in the fresco by diffusion. In the second step, we choose a large $\lambda \gg 1$, for example, $\lambda = 10^3$, to reconstruct the fresco inside the damaged parts.

In the following, we present the numerical results following these two strategies. We focus on the details of the Neidhart frescoes in Figures 9.2 and 9.3. For both inpainting methods (5.18) and (5.20), we use the convexity splitting algorithms presented in Section 5.3 for their discretisation in time.

Strategy 1: Binary-Based Fresco Inpainting We begin with the inpainting of the binary structure of the frescoes by means of (5.18) (cf. Figures 9.4 and 9.5). In our numerical examples we applied (5.18) in two steps (cf. [BEG07]).

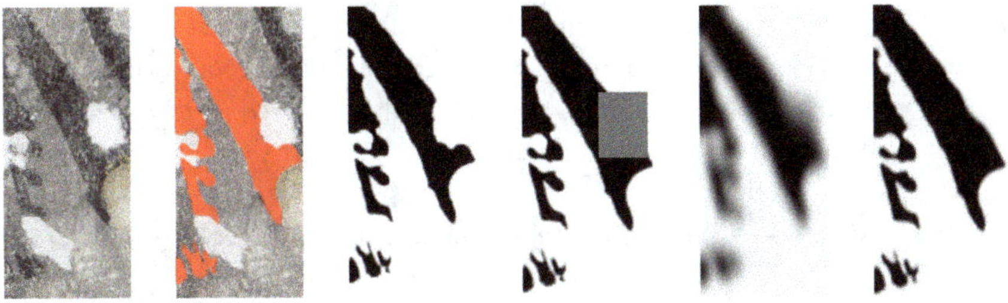

Figure 9.5. Cahn-Hilliard inpainting with $\lambda = 10^6$. From left to right: Part of the fresco, binary selection in red, binary selection in black and white, initial condition for the inpainting algorithm where the inpainting region is marked with a grey rectangle, inpainting result after 200 time steps with $\epsilon = 4$ and inpainting result after an additional 800 time steps with $\epsilon = 0.01$.

In the first few time steps, we solve (5.18) with a rather big ϵ, for example, $\epsilon = 3$. We stop when we are sufficiently close to a steady state. Then we switch the ϵ value to a smaller one, for example, $\epsilon = 0.01$. Using the steady state from the first few time steps of (5.18) with a large ϵ as an initial condition, we now apply the iteration for the switched ϵ. Again, we stop when we are sufficiently close to the steady state.

The next step is to re-colourise the damaged parts by using the recovered binary structure as underlying information. This can be done in the following way: motivated by previous work by Fornasier [For06] and Fornasier and March [FM07], we propose an inpainting model for grey value images which uses a given (or previously obtained) binary structure inside the missing domain D. Therefore, the binary structure of the image is usually obtained by a pre-processing step with Cahn-Hilliard inpainting [BEG07, BEglG07]. Let us describe this method in more detail.

Let $g \in L^2(\Omega)$ be a given image with grey values in $\Omega \setminus D$ and binary information in D. We wish to recover the grey value information in D based on the binary structure given by g by means of the following minimisation problem:

$$u^* = \text{argmin}\left\{\frac{\mu}{2}\int_{\Omega\setminus D} |u(x)-g(x)|^2\, dx + \frac{\lambda}{2}\int_D |\mathcal{L}_{\text{bin}}(u(x)) - g(x)|^2\, dx \right.$$
$$\left. + |Du|(\Omega),\ u \in L^2(\Omega)(\Omega,\mathbb{R}_+)\right\}. \tag{9.2}$$

In our case, $\mathcal{L}_{\text{bin}}$ is a transformation which projects the grey value range of u, for example, $[0,1]$, on the binary range $\{0,1\}$.

The corresponding Euler-Lagrange equation (9.2) then reads

$$0 = p + \mu \cdot \mathbb{1}_{\Omega\setminus D}(u-g) + \lambda \cdot \mathbb{1}_D(\mathcal{L}_{\text{bin}}(u) - g)\frac{\partial \mathcal{L}_{\text{bin}}}{\partial u}(u), \quad p \in \partial\, |Du|\,(\Omega), \tag{9.3}$$

where $\mathbb{1}_{\Omega\setminus D}$ and $\mathbb{1}_D$ are the characteristic functions of $\Omega \setminus D$ and D, respectively. In (9.3) we approximate the subgradient of the total variation by a relaxed version $p \approx -\nabla\cdot(\nabla u/|\nabla u|_\epsilon)$, where $|\nabla u|_\epsilon := \sqrt{|\nabla u|^2+\epsilon}$. The relaxed version of (9.3) then reads

$$0 = -\nabla\cdot\left(\frac{\nabla u}{|\nabla u|_\epsilon}\right) + \mu\cdot\mathbb{1}_{\Omega\setminus D}(u-g) + \lambda\cdot\mathbb{1}_D(\mathcal{L}_{\text{bin}}(u)-g)\frac{\partial\mathcal{L}_{\text{bin}}}{\partial u}(u).$$

For our purpose, $\mathcal{L}_{\text{bin}}$ is modelled by a relaxed version of the Heaviside function depending on a (presumably) given threshold τ. Recall that the Heaviside function is given by

$$H(x) = \begin{cases} 1, & x \geq 0, \\ 0, & x < 0. \end{cases}$$

In order to have a differentiable transformation, we approximate H by H^δ with

$$H^\delta(x) = \frac{1}{2}\left(1 + \frac{2}{\pi}\arctan\left(\frac{x}{\delta}\right)\right), \quad 0 < \delta \ll 1.$$

Let us assume now that the binary part of g in D was obtained such that it is 1 where the lost grey value information $g_g \geq \tau$ and 0 otherwise; that is, for $x \in D$, we assume that

$$g(x) = \begin{cases} 1, & g_g(x) \geq \tau, \\ 0, & g_g(x) < \tau. \end{cases}$$

Then we want the binarization $\mathcal{L}_{\text{bin}}$ of u to be modelled subject to the same assumption, and we finally define

$$\mathcal{L}_{\text{bin}}(u) := H^\delta(u-\tau) \approx \begin{cases} 1, & u \geq \tau, \\ 0, & u < \tau. \end{cases} \tag{9.4}$$

The derivative of $\mathcal{L}_{\text{bin}}$ is then a smooth approximation of the Dirac δ-function in $u = \tau$, that is,

$$\frac{dH^\delta}{dx}(x) = \delta^\delta(x) = \frac{1}{\pi\delta}\frac{1}{1+(x/\delta)^2}, \quad 0 < \delta \ll 1,$$

and

$$\frac{\partial\mathcal{L}_{\text{bin}}}{\partial x}(u) = \delta^\delta(u-\tau). \tag{9.5}$$

Figure 9.6 shows two numerical examples of binary-based grey value inpainting.

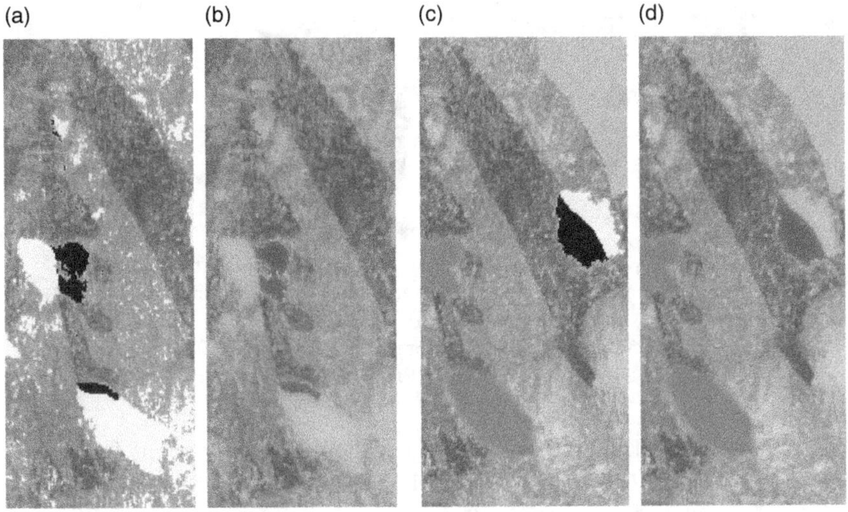

Figure 9.6. Two examples of binary-based grey value inpainting with $\mu = \lambda = 10^2$ and $5,000$ time steps. The grey values inside the inpainting domain are initialised with the results from Cahn-Hilliard inpainting in Figures 9.4 and 9.5. (a) Initial condition from the example in Figure 9.4. (b) Inpainting result for part a. (c) Initial condition from the example in Figure 9.5. (d) Inpainting result for part c. Binary-based inpainting solving (9.2) via a steepest-descent approach constitutes parts b and c.

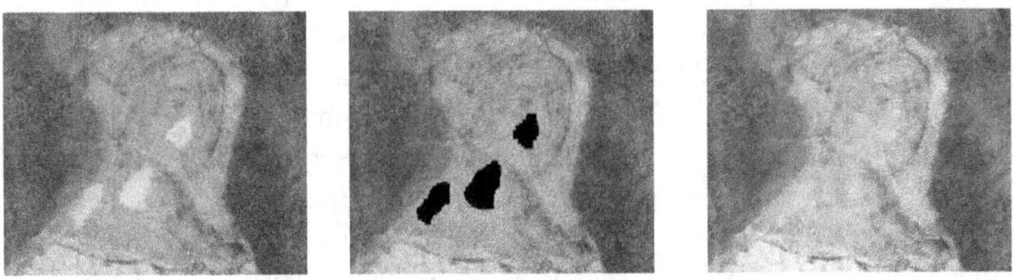

Figure 9.7. TV-H^{-1} inpainting applied to a part of the Neidhart frescos.

Strategy 2: Grey Value Fresco Inpainting We consider (9.1) and apply it for the grey value inpainting of the Neidhart frescoes. In Figure 9.7, the algorithm (9.1) has been applied to a small part of the Neidhart frescoes. In this particular case we did not even have to pre-process the image because only plain grey value information was to be imported into the inpainting domain, whereas in Figure 9.8 we acted on strategy 2. That is, we primarily de-noised the image by (9.1) with data fidelity $\lambda(g - u)$ on the whole image domain and applied

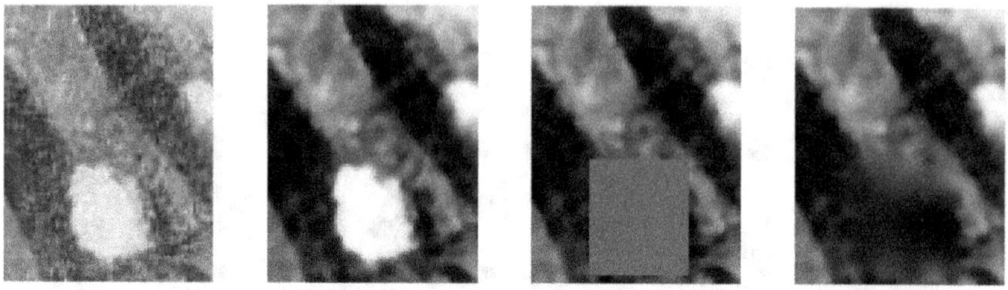

Figure 9.8. TV-H^{-1} inpainting following strategy 2. From left to right: Part of the Neidhart fresco, pre-processed image, initial condition for the inpainting algorithm, where the inpainting domain is marked as a grey rectangle, and preliminary inpainting result (algorithm carried out until $err \approx 10^{-4}$). Inpainting difficulties due to the reasons indicated in Figure 9.2 are clearly visible.

the inpainting algorithm [equation (9.1) without a fidelity term] inside the inpainting domain D on the 'cleaned' image in a second step. Clearly, our binary approach in Figure 9.6 produces superior visual results in the presence of edges than the result in Figure 9.8, produced by the following strategy.

Conclusion We succeeded in developing methods for the restoration of digital images using mathematical concepts of PDEs. We showed reconstruction of the grey value information in the Neidhart frescoes using basically two strategies. In the first strategy, Cahn-Hilliard inpainting was used to reconstruct the binary structure in the frescoes. Then the grey values were filled into the missing parts based on the binary information available from the Cahn-Hilliard reconstruction. The main idea of this strategy is to exploit the good reconstruction qualities of Cahn-Hilliard inpainting (smooth connection of edges even across large distances) as much as possible. This approach turned out to be the method of choice in the presence of edges in the neighborhood of the missing parts and when the gaps in the frescoes were large. The second strategy uses TV-H^{-1} inpainting for reconstruction of the grey value information in the frescoes. This approach produced fairly good results for the inpainting of homogeneous areas in the frescoes. As one immediately observes, parts of the surface of the Neidhart frescoes are more texture-like. Since the inpainting methods we used cannot reproduce texture, they can only deliver visually good results to a limited extent.

Figure 9.9 shows a part of the fully restorated frescoes.[2] A direct comparison between the restorated frescoes and our digital results defines the next step within this project.

[2] I thank Wolfgang Baatz for providing these data.

Figure 9.9. The restored fresco. Photo courtesy of Wolfgang Baatz.

Figure 9.10. Road data from Los Angeles.

9.2. Road Reconstruction

This project is about the continuation of roads in aerial images and takes place in Andrea Bertozzi's group at UCLA.[3] Some of the ideas in this section also appear in [DB08]. The roads are partially occluded by trees (cf. Figure 9.10), and now the challenge is to reconstruct the roads such that one is able to follow them in aerial images.

Our first approach is to binarise the road data and apply Cahn-Hilliard inpainting [cf. (5.18) in Section 5.3] to the binary roads.[4] Figure 9.11 presents two examples of Cahn-Hilliard inpainting of binary roads. Note that the Cahn-Hilliard inpainting approach is applied to the corrupted road images in two steps. Firstly the large gap(s) in the road are filled by choosing a rather large parameter ϵ in (5.18) and letting the inpainting algorithm run until it reaches a steady state. Using the result from this first step, Cahn-Hilliard inpainting is applied again with a small ϵ to sharpen the edges in the image.

[3] I thank the UCLA Mathematics Department and Alan Van Nevel and Gary Hewer from the Naval Air Weapons Station in China Lake, CA, for providing the data.

[4] I thank Shao-Ching Huang (UCLA) for preparation of the data.

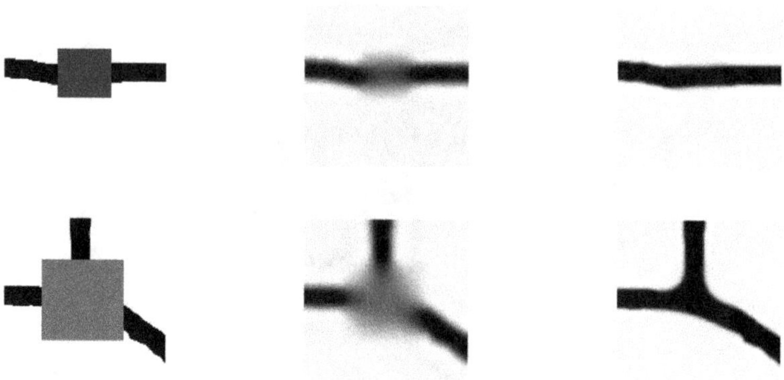

Figure 9.11. Cahn-Hilliard inpainting in two steps, namely, with $\epsilon = 0.1$ and $\epsilon = 0.01$ in the top row and $\epsilon = 1.6$ and $\epsilon = 0.01$ in the bottom row.

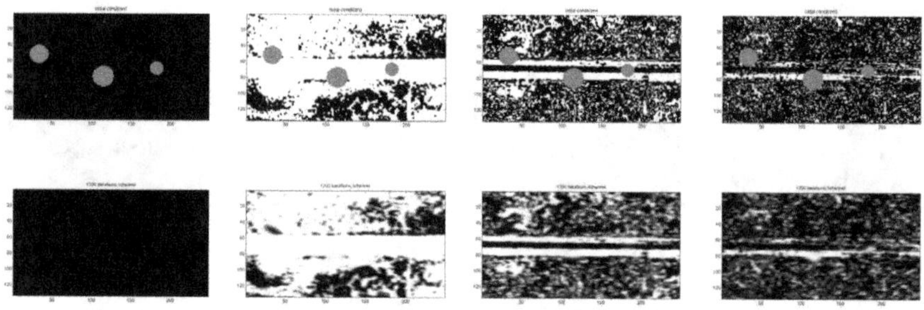

Figure 9.12. What is going on in the channels? The given image (top row) and the Cahn-Hilliard inpainting result (bottom row) for the first, third, fourth and fifth channels.

9.2.1. Bitwise Cahn-Hilliard Inpainting

In a second approach, we shall reconstruct the roads with bitwise Cahn-Hilliard inpainting. Specifically, one possible generalisation of Cahn-Hilliard inpainting for grey scale images is to split the grey scale image bitwise into channels, that is,

$$u(x) \rightsquigarrow \sum_{k=1}^{K} u_k(x) 2^{-(k-1)},$$

where $K > 0$. The Cahn-Hilliard inpainting approach is then applied to each binary channel u_k separately (cf. Figure 9.12). At the end of the inpainting process, the channels are assembled again, and the result is the inpainted grey value image in lower grey value resolution (cf. Figure 9.13). Figures 9.12 and 9.13 demonstrate the application of bitwise Cahn-Hilliard inpainting to the restoration of satellite images of roads. One can imagine that the black dots in

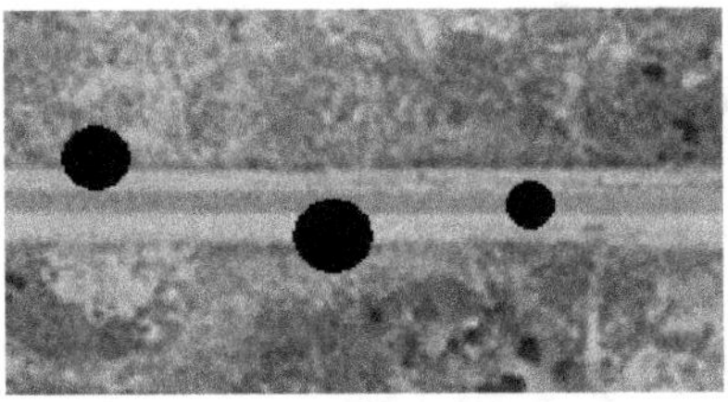

Figure 9.13. Bitwise Cahn-Hilliard inpainting with $K = 8$ binary channels applied to road restoration. Left: Given distorted aerial image of a road. Right: Result of the inpainting process with the assembled binary channels of Figure 9.12.

the first picture in Figure 9.13 represent trees that cover parts of the road. The idea of bitwise Cahn-Hilliard inpainting was proposed in [DB08] for inpainting with wavelets based on Allen-Cahn energy.

9.3. Sinogram Inpainting for Limited Angle Tomography

X-ray computed tomographic (CT) scanning plays a very important role in daily biomedical practice for diagnosis and therapy and in materials science for accurate data analysis and quality control. Besides accurate and robust image reconstruction, there is a growing interest in methods for limited angle tomography. The idea is to reduce the X-ray dose delivered to patients or to cope with physical angle limitations in materials science via a significant subsampling of projection data.

Many of such transmission imaging system are based on the Radon transform as the underlying imaging operator. Reconstruction methods can suffer from severe artefacts if the underlying noise, the experimental design (e.g., sparse or limited angle sampling) or prior information on measurements or reconstructions is not taken into account sufficiently.

In this section we present a novel variational method proposed in [BBLS14] for joint image reconstruction and directional total generalised variation (TGV)–based sinogram inpainting to address the problem of limited angle tomography.

9.3.1. Problem Formulation

In many applications in biomedicine and materials science, accurate image reconstruction from X-ray CT studies is of fundamental importance for reliable data analysis. This reconstruction is complicated by the presence of imperfections in the data. The overall goal of this work is to study a novel variational method based on simultaneous sinogram inpainting and image reconstruction to address the problem of limited angle tomography.

The standard model that interrelates two-dimensional images and projections onto one-dimensional lines of arbitrary orientation is the Radon transform (9.6). This forward operator maps desired images onto sinogram data (representing measured intensities along line integrals) and acts as a basis component in modelling various inverse reconstruction methods in CT scanning. Recently, driven by practical applications from the applied sciences (e.g., [KSJ$^+$03]), there is a growing interest in robust and accurate reconstruction methods which not only deal with uncertainty such as noise but also a significant subsampling of sinograms. At the moment, in literature on CT scanning, there are two main classes of important subsampling strategies related to limited angle tomography. On the one hand, sparsely sampled projection data, that is, coarse sampling of angles, enable rapid scanning (compressive sensing) and usually result in a reduced X-ray dose delivered to patients. On the other hand, often physical experimental hardware limitations necessitate limited angle tomography, where a certain connected range of angles in the sinogram data is missing. The latter is particularly common in tomography for the material sciences [SHL$^+$11] but also in hybrid optical tomography systems (e.g., in transmission electron microscopy) [Ökt15, FÖ08]. The difficulty or, on the contrary, the informativeness of different experimental designs for compressive sampling is related to the coherence [CW08] (compressive sampling in CT scanning [HLXZ14, Fri13]) of sparse angle measurements. Hence, dealing with large, connected gaps of missing angles (i.e., likely coherent information) is very challenging in tomographic reconstruction methods and one of their main challenges. For an overview of state-of-the-art approaches to limited angle tomography, we refer readers to [BBLS14]. Let us continue here with a discussion of the reconstruction method proposed in our work [BBLS14].

The main goal of the work of [BBLS14] presented here is the development and numerical investigation of a reconstruction model for limited angle tomography with joint higher-order non-linear (TGV) sinogram inpainting.

9.3.2. Preliminaries and Notation

The Radon Transform The model that inter-relates two-dimensional images and projections onto one-dimensional lines of arbitrary orientation is the Radon transform. Let $\Omega \subset \mathbb{R}^2$ be an open and bounded Lipschitz domain in the following. The Radon transform $\mathcal{R}\colon L^1(\Omega;\mathbb{R}) \to L^1(S^1 \times \mathbb{R};\mathbb{R})$ of a function u

is defined as

$$(\mathcal{R}u)(s,\theta) := \int_{x\cdot\theta=s} u(x)\,dx, \tag{9.6}$$

for $\theta \in S^1$ being a vector on the two-dimensional unit sphere [usually parameterised in terms of an angle $\varphi \in [0,\pi[$ via $\theta(\varphi) = (\cos(\varphi), \sin(\varphi))^{\mathrm{T}}$] and $s \in \mathbb{R}$ being the offset from the centre of the coordinate system. The function $y = \mathcal{R}u$ is called the *sinogram of u*.

Back in 1917 [Rad17], Johann Radon proved that for $u \in C_0^\infty(\mathbb{R}^2)$, the function u can be recovered from $y = \mathcal{R}u$ via the two-dimensional inverse Radon transform

$$(\mathcal{R}^{-1}y)(x) = \frac{1}{4\pi^2}\int_{S^1}\int_{\mathbb{R}} \frac{\frac{\partial}{\partial s}y(s,\theta)}{x\cdot\theta - s}\,ds\,d\sigma(\theta). \tag{9.7}$$

Due to the ill-posedness of $\mathcal{R}$ and measurements taken only for discrete and finite values of φ and s, (9.7), however, is not applicable in practise.

Instead, filtered back-projection (see, e.g., [NW01]) is a standard approach to obtain an image u from a given sinogram y. For a filter $v \in L^1(S^1 \times \mathbb{R}, \mathbb{R})$, *filtered back-projection* (FBP) is defined as

$$u = \mathcal{R}^*(v * y), \tag{9.8}$$

with $*$ denoting the (Fourier) convolution operator. The filter v is usually chosen such that $(\mathcal{R}^*v) * u \approx u$, which is motivated by the result $(\mathcal{R}^*v) * u = \mathcal{R}^*(v * \mathcal{R}u)$ for $u \in C_0^\infty(\mathbb{R}^2)$ and $v \in C_0^\infty(S^1 \times \mathbb{R})$ [NW01]. Classical examples of v are the Ram-Lak filter, the Shepp-Logan filter and the cosine filter.

In the setup of limited angle tomography, however, we potentially face the absence of a large, connected range of angles, a coarse sampling of angles and/or noisy projections, as described earlier. These issues have a severe impact on the quality of reconstructions computed via (9.8). For this reason, non-linear variational methods, which allow the incorporation of a priori information based on sparsity assumptions, have become popular tools to handle the issue of missing or noisy information. In the following, we briefly recall basic non-linear variational methods.

Total Generalised Variation A recently proposed second-order total variation (TV) model is called total generalised variation (TGV) [BKP10]. The TGV^2 regulariser is of the form

$$TGV_\beta^2(u) = \sup\Big\{\int_\Omega u\,\mathrm{div}^2\xi\,dx : \xi \in C_c^2(\Omega, \mathrm{Sym}^2(\mathbb{R}^2)), \\ \|\xi\|_\infty \le \beta,\ \|\mathrm{div}\xi\|_\infty \le 1\Big\}, \tag{9.9}$$

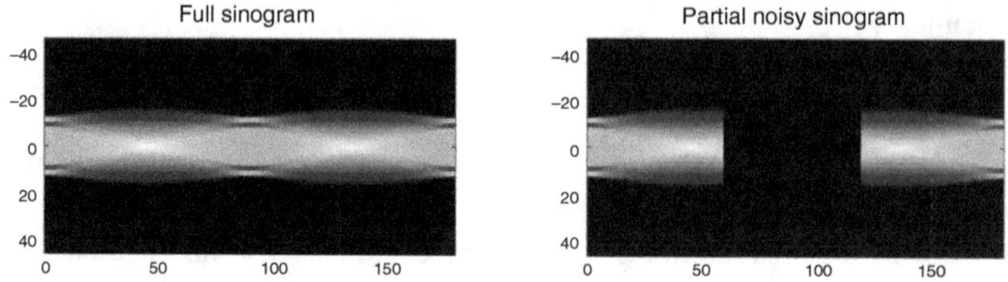

Figure 9.14. Limited angle tomography. (a) The fully sampled sinogram y^*. (b) The partial sinogram y with missing angles $[59, 117]$.

where $\mathrm{Sym}^2(\mathbb{R}^2)$ denotes the space of symmetric tensors of order 2 with arguments in $\mathbb{R}^2$, and β is a fixed positive parameter. For $u \in L^1(\Omega)$, we have the following characterisation for TGV^2 [BV11]:

$$TGV_\beta^2(u) = \min_{w \in BD(\Omega)} \|Du - w\|_{\mathcal{M}} + \beta \|\mathcal{E}w\|_{\mathcal{M}}, \tag{9.10}$$

where $BD(\Omega)$ denotes the space of vector fields of bounded deformation, that is, $w \in L^1(\Omega, \mathbb{R}^2)$ such that the distributional symmetrised derivative $\mathcal{E}w = \frac{1}{2}(\nabla w + \nabla w^{\mathrm{T}})$ is a $\mathrm{Sym}^2(\mathbb{R}^2)$-valued Radon measure.

9.3.3. TGV Inpainting and Sinograms

One way of phrasing the limited angle problem in tomography is in the form of an inpainting problem for the sinogram, defined in (9.6) as the Radon transform of an image in (9.6). Let $y = y(s, \theta)$ be a sinogram with $s \in \mathbb{R}^d$ and $\theta \in [0, 2\pi] \setminus [a, b]$, where $0 \le a < b \le 2\pi$ and $[a, b]$ constitute the wedge of missing angles. This setup is illustrated in Figure 9.14. Based on the limited angle sinogram y, we seek a sinogram y^* with

$$y^* = \begin{cases} y, & \theta \in [0, 2\pi] \setminus [a, b], \\ I(y), & \theta \in [a, b], \end{cases}$$

where $I\colon \Omega \times ([0, 2\pi] \setminus [a, b]) \to \Omega \times [0, 2\pi]$ is a generic inpainting operator. If y is additionally corrupted by noise, then the condition $y^* = y$ in the given angles has to be relaxed.

In this book we have discussed various methods for image inpainting, mainly focussing on structure inpainting only. Anisotropic TGV inpainting proposed in [BBLS14] for the limited angle problem also falls within this regime. That is, given $y \in L^2(\Omega \times [0, 2\pi] \setminus [a, b])$, we can compute the inpainted

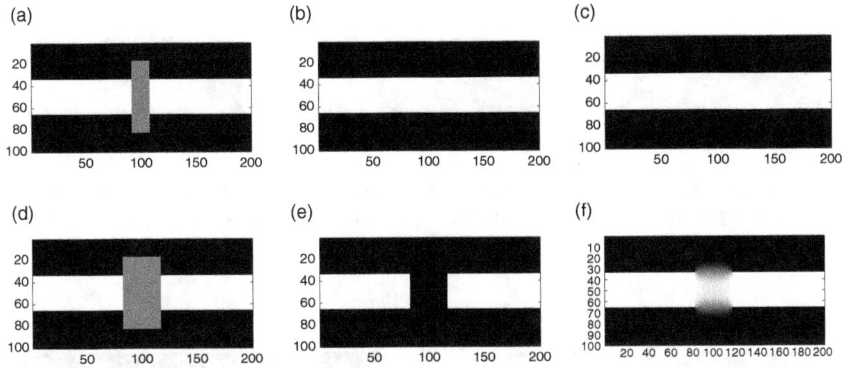

Figure 9.15. TV versus TGV inpainting for a broken constant bar with different widths w of the gap. Here l is the length of the bar. (a) $w = l/12$. (b) TV inpainting. (c) TGV inpainting with $\beta = 5$. (d) $w = l/6$. (e) TV inpainting. (f) TGV inpainting with $\beta = 1.5$.

sinogram y^* as a solution of

$$\min_z \{I(z), \text{ s. t. } Sz = y\}. \tag{9.11}$$

To motivate this approach, Figures 9.15 through 9.17 show the inpainting results for the connection of two broken lines using TV and TGV inpainting. The convex problems related to these two methods have been solved using CVX, a software package for specifying and solving convex programs [GB14, GB08]. The performance of TV inpainting $I(z) = TV(z)$, as discussed in Section 4.3, and TGV inpainting $I(z) = TGV_\beta^2(z)$, as defined in (9.10), are discussed. The constant bar with a small gap in Figure 9.15a can be perfectly inpainted using TV as the inpainter in Figure 9.15b. If the gap, however, becomes too large, 9.15c the TV inpainter fails (compare Figure 9.15e). This is due to the fact that TV inpainting occurs by connecting level lines over their shortest distance (see Section 4.3). In this case, TGV inpainting in Figure 9.15f with a careful choice of parameters α, β gives an inpainting result that, although not perfect, is still closer to the connected bar. This example suggests that TV inpainting performs well for small gaps, whereas more elaborate, in particular, higher-order, inpainting functionals such as TGV need to be consulted if structures must be reconnected across large gaps.

This conclusion changes, however, if the constant bar is replaced by a bar whose intensity increases from left to right in Figure 9.16. Here TV inpainting fails in the presence of a small gap. It cannot preserve the intensity increase within the bar, whereas TGV inpainting does the job perfectly. If the width of

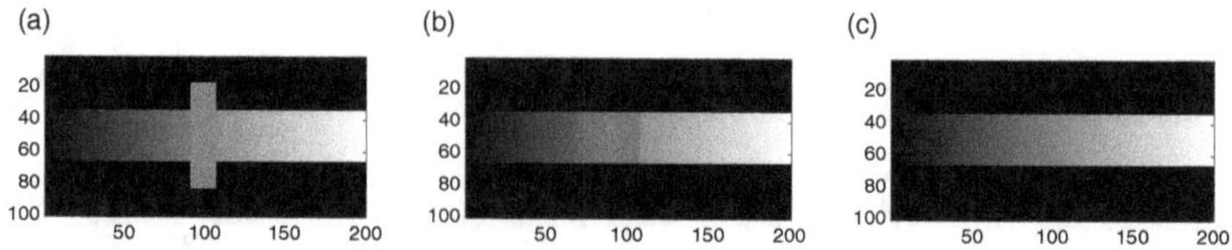

Figure 9.16. TV versus TGV method for a linear bar for a fixed gap of width $w = l/12$. Here l is the length of the bar. (a) $w = l/12$. (b) TV inpainting. (c) TGV inpainting with $\beta = 6.5$.

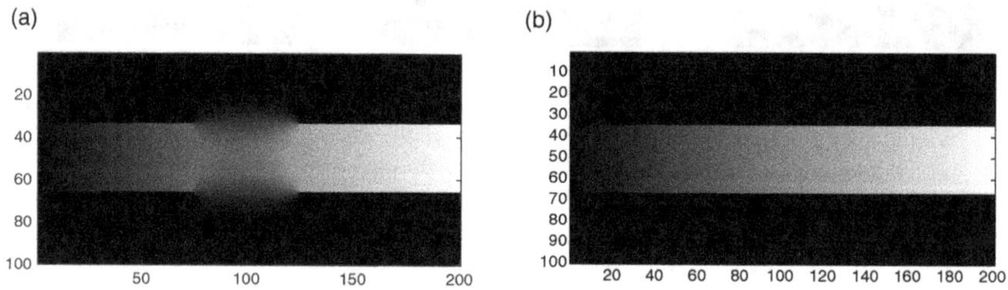

Figure 9.17. TGV versus directional TGV inpainting, where Du in (9.10) is replaced by u_y; that is, the TGV inpainter only regularises in y-direction, for a linear bar and different gap widths w. (a) TGV for $w = l/4$, $\beta = 1.5$. (b) Directional TGV for $w = l/4$, $\beta = 1.5$.

the gap for this example is increased as in Figure 9.17, the TGV inpainter also breaks down if the gap gets too large.

A simple modification of TGV inpainting seems to improve this result. Here Du in (9.10) is replaced by u_y; that is, the TGV inpainter only regularises in the y-direction. This new inpainting functional can perfectly connect the bar and interpolate the linearly, increasing intensity values inside (compare the right side in Figure 9.17). This change, from an isotropic TGV inpainter to an anisotropic directional TGV inpainter, carries the information that the missing structures can be explained by propagating intensity values in y-direction only. This constitutes a first example of a directional inpainting procedure that will be introduced later for sinogram inpainting.

To propose an inpainting procedure (9.11) for the limited angle problem, we need to investigate the qualitative properties of limited angle sinograms that arise as the Radon transform of typical images recorded in tomographic applications. In [BBLS14], the authors make the following observations:

- The limited angle problem in tomography typically results in large gaps in the sinogram in angular direction. This means that we are in the framework of the example in Figure 9.14.

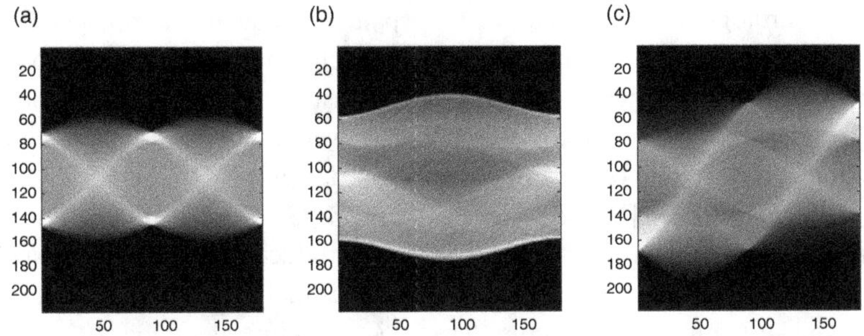

Figure 9.18. Sinograms are typically very structured functions: (a) sinogram 1; (b) sinogram 2; (c) sinogram 3.

- The Radon transform of an image is always more regular than the image itself [NW01, theorem 2.10]. In particular, this means that if we are modelling images to be in BV, then their corresponding sinogram should be more regular than that.
- Sinograms usually constitute very structured functions (cf. Figure 9.18). They possess constant and linear parts, as well as structures of higher order. Moreover, sinograms feature linelike structures related to sines and cosines.

This list motivates an inpainting procedure that

- Is able to represent and preserves structural properties of the sinogram, suggesting a non-smooth inpainting functional such as TV and TGV;
- Encodes higher-order derivatives that can represent structures beyond piecewise constancy, for example, TGV^k, $k \geq 2$; and
- Has a sense of directionality and is able to propagate intensity values in the intact part of the sinogram along dominant structures in the image.

This motivates a directional TGV inpainting for sinogram interpolation that collectively models these desired inpainting properties.

Directional TV and TGV Inpainting Based on the TV functional as used in Section 4.3 (compare also Appendix B.11) and the TGV^2 functional defined in (9.10), in [BBLS14], the authors defined directional TV_d and $\text{TGV}^{d,2}$ as

$$TV_d(u) = |D_{\text{T}}u|(\Omega) \tag{9.12}$$

and

$$TGV_\beta^{d,2}(u) = \min_{w \in BD(\Omega)} \|D_{\text{T}}u - w\|_{\mathcal{M}} + \beta \|\mathcal{E}w\|_{\mathcal{M}}, \tag{9.13}$$

respectively, where $D_{\text{T}} = TD$ is a weighted directional gradient whose orientation and size in each direction are determined by a tensor $T \in \mathbb{R}^{2\times 2}$.

The two orthogonal eigenvectors of T span the rotated coordinate system in which the gradient is computed and, as such, encode the orientations of the two derivatives. The eigenvalues of T measure the degree of anisotropy in TV_d and $\mathrm{TGV}^{d,2}$ and weight the first derivatives in the two orthogonal directions given by the eigenvectors of T. For designing TV_d and $\mathrm{TGV}^{d,2}$ for sinogram inpainting in [BBLS14], the authors proposed using a tensor T whose orientation depends on the structure tensor introduced by Weickert [Wei98] and whose weights depend on the coherence of linelike structures as defined next. Weickert defines the structure tensor $\mathcal{J}_\rho$ of an image u to be

$$\mathcal{J}_\rho(\nabla u_\sigma) := K_\rho * (\nabla u_\sigma \otimes \nabla u_\sigma), \quad \rho > 0. \tag{9.14}$$

Here K_ρ is a Gaussian kernel with variance ρ, and u_σ is the image u convolved with K_σ. The use of $\nabla u_\sigma \otimes \nabla u_\sigma := \nabla u_\sigma \cdot \nabla u_\sigma^\perp$ as a structure descriptor is aimed at making $\mathcal{J}_\rho$ insensitive to noise and sensitive to change in orientation only; that is, the sign of the gradient is not taken into account in the definition of $\mathcal{J}_\rho$. The tensor $\mathcal{J}_\rho$ is positive semi-definite and has two orthonormal eigenvectors $v_1 \parallel \nabla u_\sigma$ (points in the gradient direction) and $v_2 \parallel \nabla u_\sigma^\perp$ (points in the direction of the level lines) and corresponding eigenvalues μ_1, μ_2, which can be computed as

$$\begin{aligned} \mu_1 &= \frac{1}{2}\left(j_{11} + j_{22} + \sqrt{(j_{11} - j_{22})^2 + 4j_{12}^2}\right), \\ \mu_2 &= \frac{1}{2}\left(j_{11} + j_{22} - \sqrt{(j_{11} - j_{22})^2 + 4j_{12}^2}\right), \end{aligned} \tag{9.15}$$

where the j_{ik} values are the components of $\mathcal{J}_\rho$, that is,

$$\begin{aligned} j_{11} &= K_\rho * \left(\frac{\partial}{\partial x} u_\sigma\right)^2, \\ j_{12} = j_{21} &= K_\rho * \left(\frac{\partial}{\partial x} u_\sigma \cdot \frac{\partial}{\partial y} u_\sigma\right), \\ j_{22} &= K_\rho * \left(\frac{\partial}{\partial y} u_\sigma\right)^2. \end{aligned} \tag{9.16}$$

The eigenvalues of $\mathcal{J}_\rho$ describe the ρ-averaged contrast in the eigendirections; for example, if $\mu_1 = \mu_2 = 0$, it means that the image is homogeneous in this area; if $\mu_1 \gg \mu_2 = 0$, we are sitting on a straight line; and finally, if $\mu_1 \geq \mu_2 \gg 0$, we are at a corner of an object. Based on the eigenvalues, we can define the quantity

$$Coh = (\mu_1 - \mu_2)^2 = (j_{11} - j_{22})^2 + 4j_{12}^2 \tag{9.17}$$

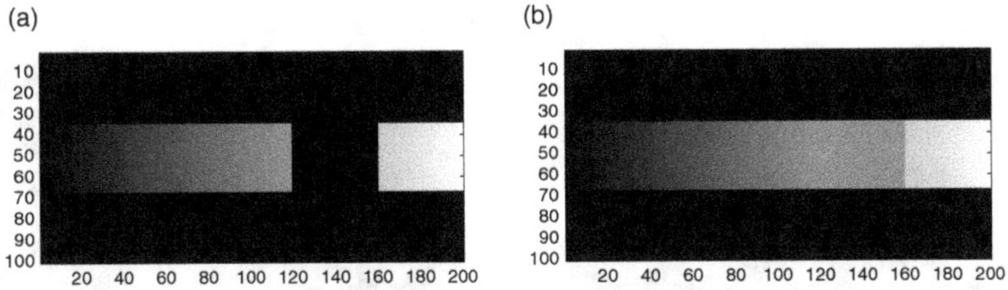

Figure 9.19. Linear bar gap inpainting: *TV* versus directional TV_d in (9.12) inpainting. (a) $I = TV$. (b) $I = TV_d$.

as the local coherence of structures: this quantity is large for linelike structures, whereas it is small for constant areas in the image.

With the derived structure tensor, we consider a tensor T where the eigenvectors of $T\big(\mathcal{J}_\rho(\nabla u_\sigma)\big) : \Omega \mapsto \mathbb{R}^{2\times 2}$ are parallel to the ones of $\mathcal{J}_\rho(\nabla u_\sigma)$, and its eigenvalues λ_1 and λ_2 are chosen depending on the desired enhancement method. For inpainting sinograms, we want to preserve linelike structures, and hence, regularisation along the lines of the sinogram seems appropriate. For this purpose, the authors considered the weights Weickert used for enhancing the coherence of flowlike structures. With μ_1, μ_2 being the eigenvalues of $\mathcal{J}_\rho$ as before, we define (for $C > 0$)

$$\lambda_1 = \gamma, \quad \lambda_2 = \begin{cases} \gamma, & \text{if } \mu_1 = \mu_2 \\ \gamma + (1-\gamma)\cdot e^{-\lambda/Coh}, & \text{else} \end{cases} \tag{9.18}$$

for $x \in \Omega$, where $\gamma \in (0,1)$, $\gamma \ll 1$. The constant $\gamma > 0$ determines the steepness of the exponential function. With this choice for the eigenvalues, the regularisation is stronger in the neighbourhood of coherent structures (where the radius of the neighbourhood is determined by ρ) while reducing regularisation in homogeneous areas, at corners and in general in incoherent (random) areas of the image.

For the application of (9.12) and (9.13) as directional inpainters I in (9.11), we use

$$T = T\big(\mathcal{J}_\rho(\nabla y_\sigma^*)\big), \tag{9.19}$$

where for the numerical discussion in their paper [BBLS14] the authors computed the structure tensor from the original fully sampled sinogram y^*. The latter eventually should be replaced by a function that is computed from the limited angle sinogram directly.

We start with the artificial example of the linear bar with increasing intensities inside in Figure 9.19. While TV inpainting cannot connect the bar via the missing angles, directional TV_d produces a connection, but the connection does not preserve the linearly increasing intensities but rather

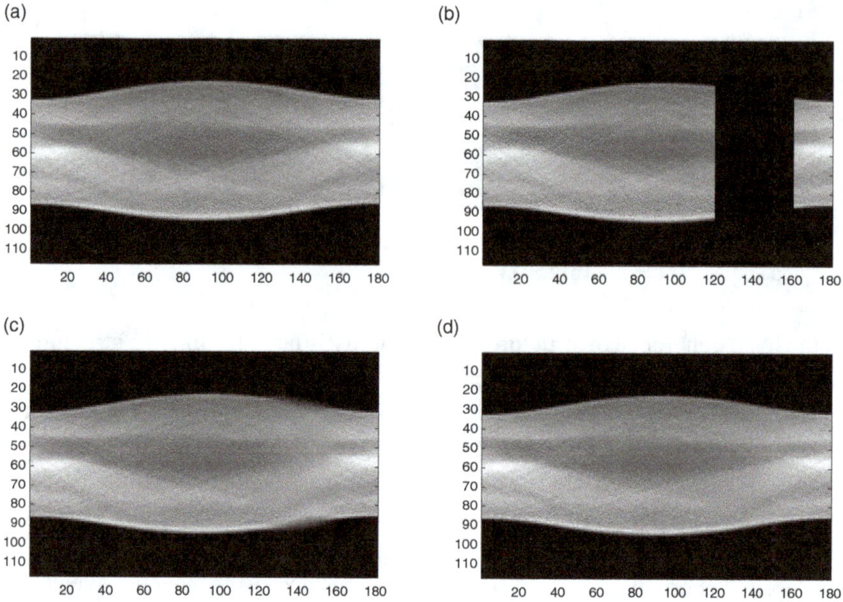

Figure 9.20. Sinogram inpainting. Standard TGV^2 versus directional $\mathrm{TGV}^{d,2}$ inpainting. (a) Ground truth. (b) Angles (120, 160 degrees) missing. (c) $I = TGV^2_{1.5}$. (d) $I = TGV^{d,2}_{1.5}$.

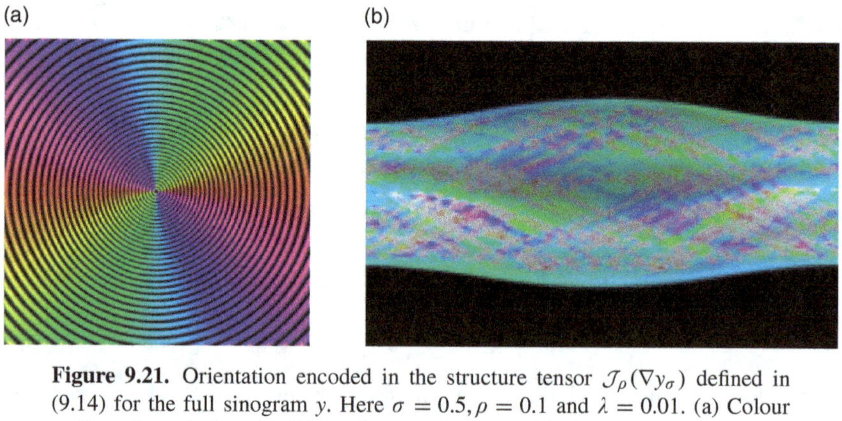

Figure 9.21. Orientation encoded in the structure tensor $\mathcal{J}_\rho(\nabla y_\sigma)$ defined in (9.14) for the full sinogram y. Here $\sigma = 0.5, \rho = 0.1$ and $\lambda = 0.01$. (a) Colour coding of orientation. (b) Orientation encoded in $\mathcal{J}_\rho(\nabla y_\sigma)$.

inpaints with a constant function. As already suggested by the preceding example in Figure 9.16, this issue can be resolved by replacing TV by TGV^2 and combining it with the directional idea. Finally, Figure 9.20 compares TGV^2 and $\mathrm{TGV}^{d,2}$ for inpainting a sinogram. We observe that the directional regulariser helps to propagate linelike structures into the gap. Figure 9.21

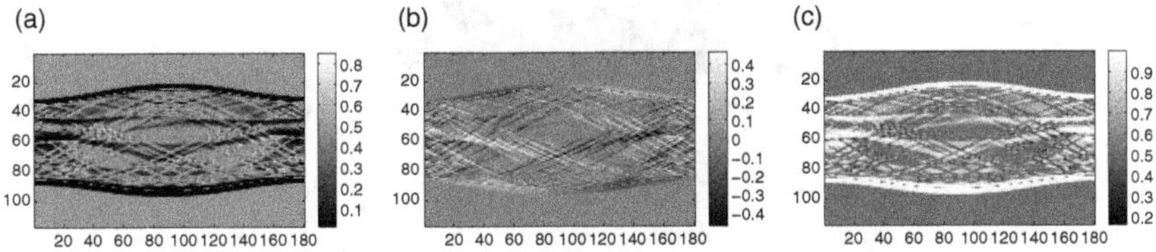

Figure 9.22. Components in the directional tensor T in (9.19) corresponding to the components in the structure tensor (9.16) with eigenvalues defined in (9.18). Here $\sigma = 0.5, \rho = 0.1$ and $\lambda = 0.01$. (a) Directional information related to the x-direction. (b) Directional information related to the mixed direction. (c) Directional information related to the y-direction.

shows the orientation encoded in the first eigenvector of the structure tensor (9.14) for the full sinogram in Figure 9.20. Figure 9.22 shows the components of the directional tensor T for the same sinogram. The component in Figure 9.22 which corresponds to the y-direction in the sinogram is the strongest, suggesting that the horizontal direction is the dominant direction for TGV^{d,2} inpainting.

We observe that the three desired properties that we stated an inpainting functional for a sinogram should have – that is, non-smoothness, higher order and directionality – seem to be nicely combined in the directional TGV^{d,2} inpainting procedure proposed in [BBLS14].

9.3.4. Simultaneous Inpainting and Inversion

Up to now we have dealt with the question of how to inpaint a sinogram in a meaningful way. Another aspect considered in [BBLS14] is the fact that an inpainting solution y^* actually needs to represent a feasible sinogram $\mathcal{R}u$ (cf. Figure 9.23). Indeed, we want a sinogram y to be in the image of $\mathcal{R}$, that is, $y \in \{\mathcal{R}u \mid u \in L^1(\Omega;\mathbb{R})\} \subsetneq L^1(S^1 \times \mathbb{R};\mathbb{R})$. The question is how limited angle tomography can benefit from joint sinogram inpainting and inversion. To answer this, in [BBLS14], TV inpainting on the sinogram was computed in two ways:

$$z_\alpha \in \operatorname{argmin}_z \left\{ \frac{1}{2} \|Sz - y\|_2^2 + \alpha TV(z) \right\} \tag{9.20}$$

and

$$u_\alpha \in \operatorname{argmin}_u \left\{ \frac{1}{2} \|S\mathcal{R}u - y\|_2^2 + \alpha TV(\mathcal{R}u) \right\}, \tag{9.21}$$

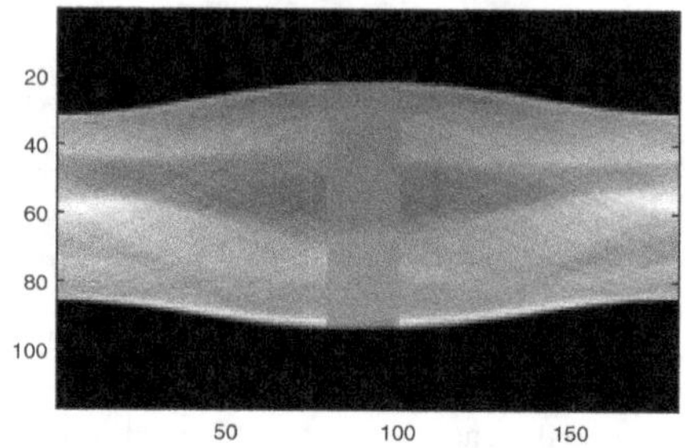

Figure 9.23. TV inpainting of a sinogram as described in (9.20) for $\alpha = 0.01$. There does not seem to exists a function u such that its Radon transform $\mathcal{R}u$ could correspond to this inpainting result (see also [Wu13, section 3]).

for $\alpha = 0.01$ and the same noisy sinogram y. The first approach constitutes pure TV sinogram inpainting, whereas in the second approach, inpainting and Radon inversion are computed simultaneously. This yields the two results shown in Figure 9.24. We discover, similar to Figure 9.23, that the pure TV sinogram inpainting (9.20) yields a piecewise constant approximation for the missing sinogram region. However, the TV joint sinogram inpainting and inversion (9.21) yields an inpainted sinogram $y_\alpha = \mathcal{R}u_\alpha$ with minimal TV for which at least one image (like the computed u_α, for instance) exists such that y_α is a sinogram with $y_\alpha \in \{\mathcal{R}u \mid u \in L^1(\Omega;\mathbb{R})\}$.

In many limited angle applications (e.g., materials sciences applications) one often deals with relatively (piecewise) constant objects that have sharp edges. For this reason, it is sensible to add a regularisation term on the image u such as TV regularisation to enforce piecewise constant solutions that allow for discontinuities. TV regularisation for limited angle tomography has been successfully investigated, for example, in [KSJ+03 and SHL+11] in order to significantly improve reconstructions compared to FBP. In [BBLS14], the sinogram inpainting model (9.21) is combined with TV regularisation on the image. More precisely, the authors considered the model

$$u_\alpha \in \operatorname{argmin}_u \left\{ \frac{1}{2}\|S\mathcal{R}u - y\|_2^2 + \alpha_1 I(\mathcal{R}u) + \alpha_2 TV(u) \right\}. \tag{9.22}$$

The functional I is a placeholder for the inpainter, for example, $I = TV$, as in (9.21). From our previous discussion, we found that either $I(v) = TGV_\beta(v)$ as defined in (9.10) or its structural tensor counterpart $I(v) = TGV_\beta^{d,2}(v)$ as defined in (9.13) is a good inpainting function for sinograms. In the following, we want to demonstrate and discuss numerical results of (9.22).

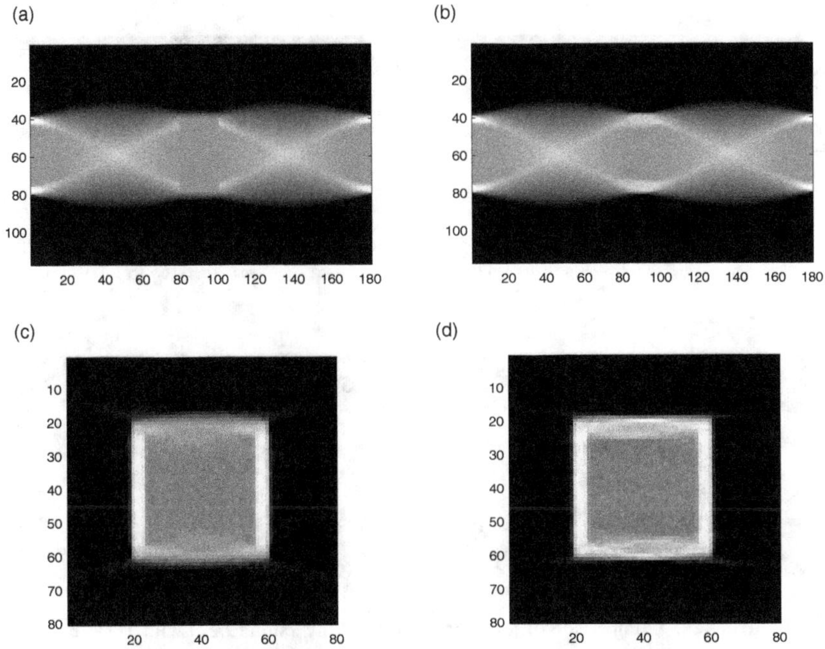

Figure 9.24. Comparison of the two inpainting methods (9.20) and (9.21). (a) Inpainting result z_α from (9.20). (b) Inpainting result $\mathcal{R}u_\alpha$ from (9.21). It is quite evident that forcing the sinogram to be in the image of $\mathcal{R}$ yields a completely different sinogram. (c) Result of applying FBP (9.8) to part a. (d) Result of applying FBP (9.8) to b.

Figure 9.25 shows a comparison of a joint inpainting and inversion approach of a noise-free missing-gap sinogram of the Shepp-Logan phantom via (9.22) for $I = TGV_\beta^2$ and its directional counterpart $I = TGV_\beta^{d,2}$. The missing gap in this example consists of all angles in $[120, 160]$. The regularisation parameters chosen for this example are $\alpha_1 = 0.01$, $\alpha_2 = 1$ and $\beta = 1.5$. By comparing Figures 9.25c and 9.25d, it becomes clear that the directional tensor allows us recover fine structures that go missing in the plain TGV reconstruction.

The directional TGV-TV seems to outperform conventional TGV-TV joint inpainting and inversion. Further, Figure 9.26 compares the directional TGV-TV approach (9.22) with the same model setting $\alpha_1 = 0$ (i.e., no direct inpainting on the sinogram). The solution of (9.22) with the directional TGV functional (9.13) shows significant visual improvement over the solution of (9.22) with $\alpha_1 = 0$. The set of missing angles in this example is $[60, 120]$. The regularisation parameters are $\alpha_1 = 0.1$ (or 0, respectively), $\alpha_2 = 1$ and $\beta = 1.5$.

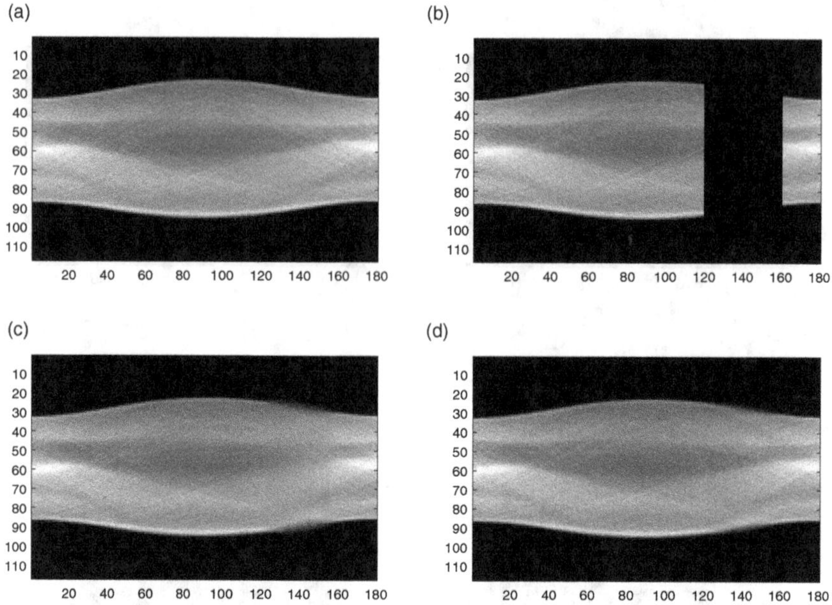

Figure 9.25. Comparison of joint inpainting and inversion (9.22) with TGV and directional TGV inpainting. (a) Ground truth. (b) Limited angle sinogram. (c) TGV^2-TV inpainting. (d) $\mathrm{TGV}^{d,2}$-TV inpainting.

9.3.5. Conclusion

In this section we discussed the application of image inpainting for solution of the limited angle problem in tomography. In particular, we present a novel variational image reconstruction model (9.22) proposed in [BBLS14] that jointly inpaints the limited angles in the partial sinogram with a directional TGV regulariser and reconstructs an image with a TV prior. Cumulatively designing the directional TGV inpainting procedure – taking into account qualitative properties of sinograms – and the TV inversion procedure for the Radon transform – which is based on the assumption that the desired image can be roughly explained by sharp edges and constant areas in between – model (9.22) is created. As discussed, is seems that indeed the resulting simultaneous inpainting and inversion are crucial to obtaining image reconstructions whose quality improves on inpainting the sinogram and regularising the image alone.

9.4. Inpainting for 3D Conversion

The discussion in this section was provided by Rob Hocking, who gained his experience on 3D conversion while working for the company Gener8 (www.gener8.com).

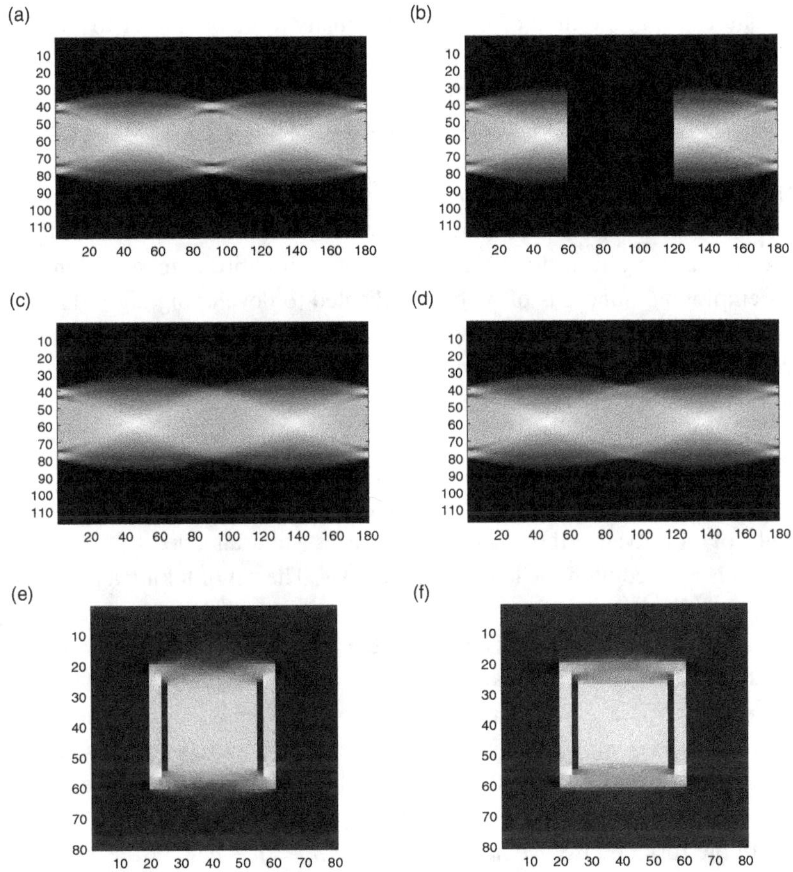

Figure 9.26. Comparison between active and inactive ($\alpha_1 = 0$) sinogram inpainting in model (9.22) for $I = TGV_{\beta}^{d,2}$. (a) Ground truth. (b) Limited angle sinogram. (c) Sinogram of TV inversion ($\alpha_1 = 0$). (d) Sinogram of $\mathrm{TGV}^{d,2}$-TV inpainting. (e) Image of TV inversion ($\alpha_1 = 0$). (f) Image of $\mathrm{TGV}^{d,2}$-TV inpainting.

Suppose that we are given a photograph (or video) u_0 taken by a camera at position $p \in \mathbb{R}^3$ with orientation $O \in SO(3)$. Is it possible to construct the photograph (video) u that would be taken by the same camera at the perturbed position and orientation $p + \delta p$, $O + \delta O$?

This problem lies at the heart of *3D conversion*. Once niche, 3D feature-length films are now mainstream. However, footage shot with a traditional camera is not automatically 3D viewable, so how are 3D films made? One option is to shoot directly using a specialised 3D camera, which is really just a rig of two cameras simulating a set of eyes. However, for various reasons at the time of this writing, many directors prefer to shoot the old-fashioned way and add the 3D in post-production via a process called *3D conversion*. It is used to convert not only new films as they come out but also old films and even cartoons, such as the 2012 3D remake of *Beauty and the Beast* [Pul12]. 3D conversion amounts to solving a special case of the aforementioned problem where $\delta O = 0$ and δp corresponds to the (slight) lateral displacement between a pair of eyes. At the time of this writing, 3D conversion is a multi-million-dollar industry with entire companies, in some cases employing hundreds of people, dedicated to developing algorithms that address conversion issues.

We will see shortly how image and/or video inpainting arises as a subproblem in the 3D conversion pipeline. Firstly, we put things in context by giving a brief account of the steps leading up to inpainting.

The first step in the pipeline is to determine the camera parameters, which include the position p and orientation O. The matrix $O \equiv [\hat{e}_1, \hat{e}_2, \hat{e}_3]$ should be thought of as a local orthonormal coordinate system attached to the camera, where $\hat{e}_3$ is aligned with the latter's optical axis. The act of taking a photo then corresponds mathematically to a type of projection into the plane spanned by $\hat{e}_1, \hat{e}_2$. Camera parameters may be known but more often must be calculated or else set manually (or a combination of the two).

Once the camera parameters are known, the problem is to compute the images or videos u_L, u_R viewed through the left and right eyes at their respective positions p_L, p_R. It is almost always assumed that the camera position either coincides with one of the eyes or else sits halfway between them. In the former case, we have $p_L = p$ and $p_R = p + \delta\hat{e}_1$ (or the equivalent with the roles of p_R, p_L reversed), whereas in the latter case $p_L = p - \frac{\delta}{2}\hat{e}_1$, $p_R = p + \frac{\delta}{2}\hat{e}_1$. Then $\delta > 0$ is a parameter controlling the separation of the eyes, which may be tweaked to get the right artistic feel.

Image (video) inpainting arises as a subproblem of 3D conversion because an object or portion thereof hidden behind something when viewed from p might be visible from one or more of p_L, p_R. For example, in Figure 9.27 (left), part of a blackboard is hidden behind a pair of cartoon characters when viewed from p (The title of the book clutched under the arm of the individual on the left is *Probability with Nightingales*). However, a thin sliver of the occluded area is visible when viewed from p_L, and a similar sliver can be seen from p_R (in this case, the original camera position is assumed to be midway between the eyes). This is illustrated in Figure 9.27 (right), where the silhouette of the blackboard is colour coded so as to convey occlusion properties. Pixels which are occluded when viewed from p but visible from p_L or p_R are called *revealed* and are coloured red and blue, respectively (green pixels are visible from both

Figure 9.27. 3D conversion of a cartoon: In this example, a pair of cartoon characters (left) occludes a portion of a blackboard. It is assumed that the original image is photographed from a point p half way between the desired left and right eye positions p_L, p_R. On the right we see the silhouette of the blackboard colour coded according to visibility. The area of the board consisting of pixels visible from p is coloured light grey. Revealed pixels visible from one or more of p_L, p_R but occluded at p are coloured red (visible to p_L only), blue (p_R only) and green (visible to p_L and p_R). Fully occluded pixels visible to neither p_L nor p_R are coloured dark grey. In this case, plausible camera parameters had to be invented as the image is not a real photograph. Photo courtesy of Rob Hocking.

p_L and p_R). (This terminology is due to Russell MacKenzie, who coined the term while working as a software developer for Gener8, then a 3D conversion startup.) Revealed pixels clearly have to be inpainted in order to construct u_L and u_R. However, light grey pixels are visible from p (and hence known), whereas dark grey pixels (dubbed *fully occluded*) are visible from neither p_L nor p_R and hence do not require inpainting.

Given artist-provided data on the boundary of an object (specified by, for example, Bézier splines), geometrical data (triangle meshes or 3D planes), along with stereoscopic camera data (briefly, p_L, p_R and O, together with a few other parameters such the field of view), the revealed areas and hence the inpainting domain may be calculated automatically.

It is worth pointing out that this inpainting problem is subtly different from the problems we have encountered in this book so far. In particular, if we view the inpainting domain D as the *entire* occluded portion of the blackboard,[5] then our problem is unusual in that D need not be filled in its entirety – only the revealed pixels need to be filled. This has a number of consequences:

[5] We may equivalently regard D to be just the revealed pixels, but with two different types of boundary conditions – also a situation we have not encountered so far.

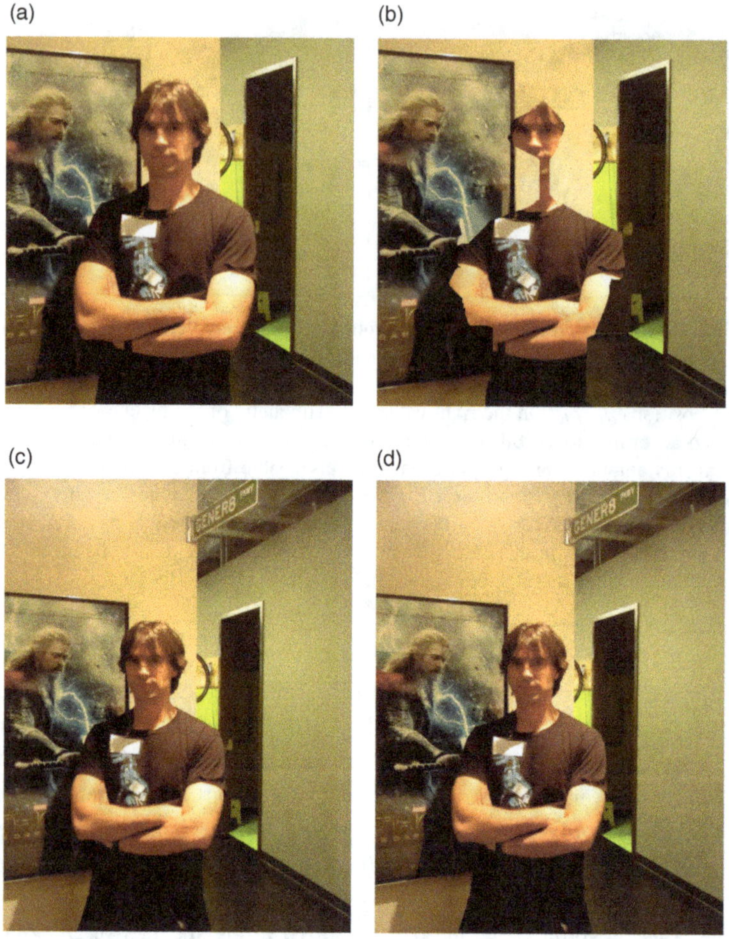

Figure 9.28. A sampling of steps in the 3D conversion pipeline from the original image to resulting left and right eye views. (a) Original image of Rob. (b) Inpainting with GuideFill stopped early once the revealed pixels were filled. (c) The resulting right eye view. (d) The resulting left eye view. Photo courtesy of Rob Hocking.

1. Iterative methods such as TV or Euler's elastica discussed in Chapters 4 and 5, set up to act on the whole of D, may be very wasteful because the revealed area is typically only a very small portion of the occluded region.[6]

[6] That is, unless these iterative methods were adapted to act only on the revealed pixels, taking into account the presence of two types of boundary conditions.

2. Single-pass methods that fill D in from the boundary inwards are natural candidates, for they can simply be stopped early once the revealed pixels have all been filled.
3. Since these methods tend to perform very well close to the boundary of the inpainting domain, the fact that the revealed area is typically quite thin makes them even more attractive.
4. PDE-based methods, which can be much cheaper than exemplar-based methods, are natural candidates as they are best suited to thin inpainting domains.

These observations suggest that methods such as the exemplar-based approaches of Criminisi et al. [CPT03] and Arias et al. [ACS09, AFCS11], as well as transport-based inpainting, such as coherence transport and GuideFill (Chapter 6), might be attractive options. Indeed, the last of these was originally developed with 3D conversion in mind at the 3D conversion company Gener8 (while Rob Hocking was doing contract work there) and is in use by its stereo artists. Figure 9.28 shows a sampling of steps in the 3D conversion pipeline using GuideFill for the inpainting step.

Appendix A

Exercises

1. Show that solutions $u(x,t)$ of the heat equation

$$u_t = \Delta u,$$
$$u(x,0) = f(x),$$

for a given function $f \in L^\infty(\mathbb{R}^2)$ form a family of linearly filtered versions of f which can be explicitly computed as

$$u(x,t) = \frac{1}{4\pi t} \exp\left(-\frac{|x|^2}{4t}\right).$$

2. Prove that the Green's function of the grounded Poisson equation (4.14) fulfils the following comparison principle: let $D_1 \subset D_2$ and $G_1(z_0,z)$ and $G_2(z_0,z)$ be the Green's functions solving (4.14) on D_1 and D_2, respectively. Then $G_1(z_0,z) \leq G_2(z_0,z)$ for all $z_0, z \in D_1$. (*Hint: Consider the difference $G_2 - G_1$, and use the fact that harmonic functions fulfil the extremum principle.*)
3. Let B_1 denote the unit disc with centre in the origin and $G_{B_1}(z_0,z)$ be the Green's function of the grounded Poisson equation (4.14) defined on B_1. Show that then

$$\int_{B_1} G_{B_1}(z_0,z)\, dz = \frac{1-|z_0|^2}{4}, \qquad \text{for all } z_0 \in B_1.$$

4. Let $1 < p,q < \infty$. Prove that functions $u \in W^{p,q}(0,1)$ are Hölder continuous. What does this imply for a function $u \in W^{p,q}((0,1)^2)$?
5. Prove that the problem $\Delta u = 0$ in $B(0,r) \setminus \{0\}$, $u = 0$ on $\partial B(0,r)$ and $u(0) = 1$ does not have a solution. Prove that the same result holds for

any linear combination of the form

$$\alpha D^2u\left(\frac{\nabla u}{|\nabla u|},\frac{\nabla u}{|\nabla u|}+\beta D^2u\left(\frac{\nabla^\perp u}{|\nabla u|},\frac{\nabla^\perp u}{|\nabla u|}\right)\right),\quad \alpha,\beta>0.$$

In contrast to the diffusion equation, the AMLE equation does indeed have a solution to the preceding problem. Prove that

$$\begin{aligned} D^2u\left(\frac{\nabla u}{|\nabla u|},\frac{\nabla u}{|\nabla u|}\right) &= 0, \quad \text{in } B(0,r)\setminus\{0\}\\ u &= 0, \quad \text{on } \partial B(0,r)\\ u(0) &= 1 \end{aligned}$$

obtains the solution $u(x) = 1-|x|$.

6. For every $u\in L^1_{\text{loc}}(\Omega)$, the total variation of u is given by

$$TV(u)=|Du|(\Omega)=\sup\left\{-\int_\Omega u\operatorname{div}\varphi\,dx:\varphi\in C_c^\infty(\Omega;\mathbb{R}^2),|\varphi(x)|\le 1\ \forall x\in\Omega\right\}.$$

For a function u with $TV(u)<\infty$, prove (by using the Riesz representation theorem or otherwise) that the distributional derivative Du is a finite Radon measure. Show further that in this case $TV(u)\ge 0$.
7. Some special forms of total variation:
   (a) Let $u\in W^{1,1}(\Omega)$. Prove that in this case $|Du|(\Omega)=\|Du\|_{L^1(\Omega)}$.
   (b) Let $\Omega\subset\mathbb{R}^2$, and let $B_r=B(0,r)\subset\Omega$ denote the disc centred at the origin with radius $r>0$. Define $\mathbb{1}_{B_r}$, the indicator function of B_r, and compute its total variation.
8. Let C be a convex subset of $L^2(\Omega)$, the space of all square integrable functions on $\Omega\subset\mathbb{R}^2$, open and bounded. We define

$$\chi_C(u)=\begin{cases}0, & u\in C\\ +\infty, & \text{otherwise,}\end{cases}$$

the characteristic function of the set C. Prove that χ_C is lower semi-continuous with respect to weak convergence in L^2.
9. Verify the identity

$$|\nabla u|\cdot\operatorname{div}\left(\frac{\nabla u}{|\nabla u|}\right)=D^2u\left(\frac{\nabla^\perp u}{|\nabla u|},\frac{\nabla^\perp u}{|\nabla u|}\right)$$

in (4.2). This gives the interpretation of the mean curvature flow of a function u as a directional diffusion equation of u, the direction given by the orthogonal to the gradient of the function. Prove by Taylor expansion that the directional diffusion equation

$$D^2u(\nabla u, \nabla u) = 0$$

is the asymptotic limit for $h \to 0$ of

$$u(x) = \tfrac{1}{2}(u(x+h\nabla u) + u(x-h\nabla u)) + ø(h^2).$$

What is the geometrical difference between the two directional diffusions? What do they do to an image function u?

10. This exercise is taken from [CMS98a]. Let $D = B(0,1)$ and $\phi(x) = \alpha_1 x^2 + \alpha_2 y^2$, $\alpha_1 > \alpha_2$. Prove that both functions

$$u_1(x,y) = \phi\left(\sqrt{1-y^2}, y\right) \quad \text{and} \quad u_2(x,y) = \phi\left(x, \sqrt{1-x^2}\right)$$

are viscosity solutions of (4.11), that is,

$$D^2(u)\left(\frac{\nabla^{\perp}u}{|\nabla u|}, \frac{\nabla^{\perp}u}{|\nabla u|}\right) = 0, \quad \text{in } B(0,1).$$

$$u|_{\partial B(0,1)} = \phi.$$

[*Hint: First prove that u_1 and u_2 are classical solutions of the regularised equation $D^2(u)\left(\nabla^{\perp}u/|\nabla u|_\epsilon, \nabla^{\perp}u/|\nabla u|_\epsilon\right) + \epsilon\Delta u = 0$, where $|p|_\epsilon = \sqrt{p_1^2 + p_2^2 + \epsilon^2}$ for $p = (p_1, p_2)$ and $\epsilon > 0$ with boundary data ϕ. Then take the limit as $\epsilon \to 0$ and conclude with the assertion.*]

11. Bertalmio's transport inpainting (Section 6.1). Explicitly compute solutions to the transport equation in Section 6.1. Can there be crossing of characteristics in two space dimensions? Can you formulate Bertalmio's inpainting approach in level-line terminology?
12. Bertalmio's transport inpainting (Section 6.1). Implement the inpainting scheme proposed in Section 6.1. Play around with different parameters, and compare the different outcomes. Non-linear diffusion versus pure transport. What kind of non-linear diffusion should be chosen?
13. Let $\tilde{S} := (0,1)$, and for $i = 1, 2, \ldots,$

$$S_i := \bigcup_{k=0}^{2^{i-1}-1}\left(k\frac{1}{2^{i-1}}, k\frac{1}{2^{i-1}} + \frac{1}{2^i}\right),$$

that is, $S_1 = (0, 0.5)$, $S_2 = (0, 0.25) \cup (0.5, 0.75)$, $S_3 = (0, 0.125) \cup (0.25, 0.375) \cup (0.5, 0.625) \cup (0.75, 0.875)$, ... Verify that the sequence S_i converges in Hausdorff measure to $\tilde{S}$ for $i \to \infty$, whereas $\mathcal{H}^1(S_i) = 0.5$ for all $i \geq 1$ and $\mathcal{H}^1(\tilde{S}) = 1$. What does this fact tell you about the semi-continuity of $\mathcal{H}^1$ with respect to the Hausdorff topology?

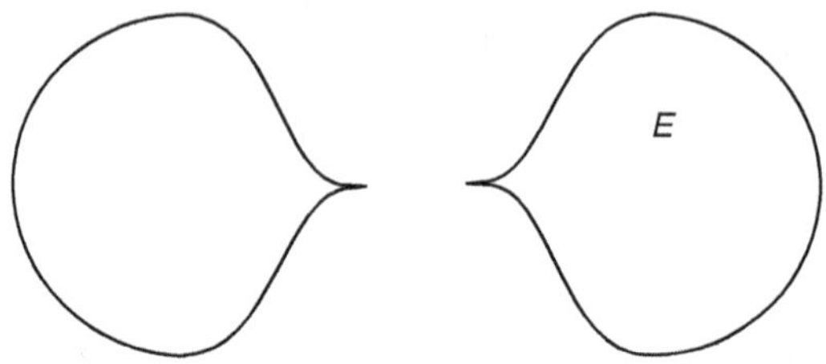

Figure A.1. For this set E, the energy $\mathcal{J}(E) = \infty$ because of the two cusp points.

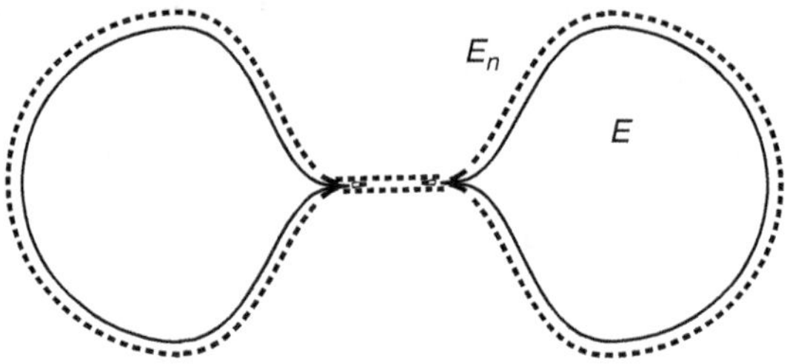

Figure A.2. A sequence of approximating sets E_h with $|E_n \Delta E| \to 0$ as $n \to \infty$ for which $\bar{\mathcal{J}}(E) < \infty$.

14. Let γ be a twice continuously differentiable curve in the plane. Check that the modulus of the curvature of $\gamma = \gamma(s)$ with arc length s is given by $|\kappa(s)| = |\gamma''(s)|$. In particular prove that the latter is independent of the parametrisation, that is, $\gamma(s(t)) = \gamma(s(\tilde{t}))$.
[*Hint: Use the definition of curvature in* (B.3).]
15. This example is supplementary to the discussion in Section 5.2 of the topic of relaxation and the analysis carried out in [BDP93]. Consider

$$\mathcal{J}(E) = \int_{\partial E} (1 + |\kappa(z)|^p)\, d\mathcal{H}^1(z),$$

for $E \in \mathbb{R}^2$ a bounded and open set of class C^2, $p > 1$ and curvature $\kappa(z) = \kappa_{\partial E}(z)$ for a $z \in \partial E$. We can extend the definition of this variational criterion from C^2 to the more general class of functions $\mathcal{M}$ (the space of all measurable functions) via the method of relaxation. For this purpose, we define a new functional $\bar{\mathcal{J}}(E)$: $\mathcal{M} \to [0, +\infty]$ as the lower semi-continuous envelope of $\mathcal{J}$ w.r.t. $L^1(\mathbb{R}^2)$, that is,

$$\bar{\mathcal{J}}(E) := \inf\left\{\liminf_{n\to\infty} \mathcal{J}(E_n) : \; E_n \to E \text{ in } L^1(\mathbb{R}^2)\right\}.$$

Here the L^1 topology on sets E means that a sequence $\{E_n\}_{n\in\mathbb{N}}$ approximates E if and only if the symmetric difference between the sets $|E_n \Delta E| \to 0$ as $n \to \infty$. The new functional $\bar{\mathcal{J}}$ is called the *relaxed functional* of $\mathcal{J}$.

To understand the concept of relaxation, lets consider the set E in Figure A.1. This set is not of class C^2 because of the two cusp points, and hence $\mathcal{J}(E) = +\infty$. However, we can find an approximating sequence E_n in Figure A.2 such that $\bar{\mathcal{J}}(E) < \infty$. Can you explain why? Moreover, can you think of other examples of approximating sets E_n such that $\mathcal{J}(E_n) \to \infty$ for $n \to \infty$?

Appendix B

Mathematical Preliminaries

B.1. Distributional Derivatives

Take $\Omega \subset \mathbb{R}^d$, $d \geq 1$, to be an open set. We define $\mathcal{D}(\Omega) := C_c^\infty(\Omega)$, where $C_c^\infty(\Omega)$ is the set of C^∞ functions with compact support in Ω. Then $\mathcal{D}'(\Omega)$ denotes the set of real continuous linear functionals on $\mathcal{D}(\Omega)$ which we call *the space of distributions* on Ω. We will denote the pairing of $\mathcal{D}'(\Omega)$ and $\mathcal{D}(\Omega)$ by $\langle \cdot, \cdot \rangle_{\mathcal{D}',\mathcal{D}}$. In the following, we define partial derivatives of a distribution.

Definition B.1.1 ($\partial^\alpha u$) For $u \in \mathcal{D}'(\Omega)$ and for any multi-index α, we define $\partial^\alpha u$ by

$$\langle \partial^\alpha u, \phi \rangle_{\mathcal{D}',\mathcal{D}} := (-1)^{|\alpha|} \langle u, \partial^\alpha \phi \rangle_{\mathcal{D}',\mathcal{D}}, \qquad \text{for all } \phi \in \mathcal{D}(\Omega).$$

see [Fol99] for more details on distribution theory.

B.2. Subgradients and Subdifferentials

Let H be a real Hilbert space with norm $\|\cdot\|$ and inner product $(\cdot,\cdot)$. Given $\mathcal{J}\colon H \to \mathbb{R}$ and $u \in H$, we say that $\mathcal{J}$ is *Fréchet-differentiable* at $u \in H$ if

$$\nabla \mathcal{J}(u) = \lim_{t=0} \frac{d}{dt} \mathcal{J}(u + tv) < \infty.$$

Then $\nabla \mathcal{J}(u)$ is called the *Fréchet derivative* (or *first variation*) of $\mathcal{J}$.

In some cases, the preceding limit does not exist; that is, the function $\mathcal{J}$ is not differentiable. Then we introduce the notion of the subdifferential of a function (cf. [Eva98]).

Definition B.2.1 Let X be a locally convex space, X' its dual, $\langle \cdot, \cdot \rangle$ the bilinear pairing over $X \times X'$ and $\mathcal{J}$ a mapping of X into $\mathbb{R}$. The *subdifferential* of $\mathcal{J}$ at

$u \in X$ is defined as

$$\partial \mathcal{J}(u) = \{ p \in X' \,|\, \langle v - u, p \rangle \leq \mathcal{J}(v) - \mathcal{J}(u), \forall v \in X \}.$$

B.3. Special Differential Operators

We define the following special differential operators for a smooth function $u\colon \Omega \to \mathbb{R}$, $\Omega \subset \mathbb{R}^d$:

$$\nabla u = \left(\frac{\partial u}{\partial x_1}, \ldots, \frac{\partial u}{\partial x_d} \right) = \textit{gradient vector}$$

$$\Delta u = \sum_{i=1}^{d} \frac{\partial^2 u}{\partial x_i^2} = \textit{Laplacian of } u.$$

For a vector-valued function $u\colon \Omega \to \mathbb{R}^d$, $u = (u^1, \ldots, u^d)$, we define

$$\nabla \cdot u = \sum_{i=1}^{d} \frac{\partial u^i}{\partial x_i} = \textit{divergence of } u.$$

B.4. Sobolev Spaces

The following definitions can be found in [Eva98].

Definition B.4.1 ($W^{m,p}$) For m a non-negative integer and $p \in [1, \infty)$, we define

$$\|u\|_{W^{m,p}(\Omega)} := \left\{ \sum_{0 \leq |\alpha| \leq m} \|\partial^\alpha u\|^p_{L^p(\Omega)} \right\}^{1/p}$$

whenever the right-hand side makes sense and where the derivatives are taken in the sense of distributions (cf. Definition B.1.1).

Then the *Sobolev spaces* $W^{m,p}(\Omega)$ are defined as

$$W^{m,p}(\Omega) := \{ u \in L^p(\Omega) : \partial^\alpha u \in L^p(\Omega), \quad \text{for all } |\alpha| \leq m \},$$

and are Banach spaces with the preceding defined norm. Further, we denote

$$W_0^{m,p} := \text{ closure of } C_c^\infty(\Omega) \text{ in } W^{m,p}(\Omega).$$

In the special cases $p = 2$, we denote $W^{m,2}(\Omega)$ and $W_0^{m,2}(\Omega)$ by $H^m(\Omega)$ and $H_0^m(\Omega)$, respectively. In these cases, the Sobolev spaces $H^m(\Omega)$ are Hilbert spaces.

B.5. Functional Analysis

Theorem B.5.1 Rellich-Kondrachov Compactness Theorem (see [Alt85], theorem 8.7, p. 243) *Assume that Ω is a bounded and open subset of $\mathbb{R}^d$ with Lipschitz boundary. Suppose that $1 \leq r < d$. Then*

$$W^{1,r}(\Omega) \hookrightarrow\hookrightarrow L^q(\Omega),$$

for each $1 \leq q < dr/(d-r)$.

Theorem B.5.2 Fatou's Lemma *If $f_1, f_2, \ldots$ is a sequence of non-negative measurable functions defined on a measure space (S, Σ, μ), then*

$$\int_S \liminf_{n\to\infty} f_n \, d\mu \leq \liminf_{n\to\infty} \int_S f_n \, d\mu .$$

Let X denote a real Banach space.

Definition B.5.3 A set $K \subset X$ is *convex* if for all $u, v \in K$ and constants $0 \leq \lambda \leq 1$, $\lambda u + (1-\lambda)v \in K$.

Theorem B.5.4 Characterisation of Compact Sets *A closed subset K of a Banach space X is compact if and only if there is a sequence (x_n) in X such that $||x_n|| \to 0$ and K is a subset of the closed convex hull of (x_n).*

Theorem B.5.5 Schauder's Fixed Point Theorem *Suppose that $K \subset X$ is compact and convex, and assume also that A: $K \to K$ is continuous. Then A has a fixed point.*

Proof See [Eva98, section 9.2.2. pp. 502–507]. ■

Theorem B.5.6 Poincaré Inequality in L^1. *Assume that $\Omega \subset \mathbb{R}^d$ is an open, bounded Lipschitz domain. Then there exists a constant C, depending only on Ω, such that for every function u in the Sobolev space $W^{1,1}(\Omega)$,*

$$\|u - u_\Omega\|_{L^1(\Omega)} \leq C\|\nabla u\|_{L^1(\Omega)},$$

where $u_\Omega = (1/|\Omega|)\int_\Omega u(y)\,\mathrm{d}y$ is the average value of u over Ω, with $|\Omega|$ denoting the Lebesgue measure of the domain Ω.

B.6. The Space H^{-1} and the Inverse Laplacian Δ^{-1}

We denote by $H^{-1}(\Omega)$ the dual space of $H_0^1(\Omega)$ with corresponding norm $\|\cdot\|_{-1}$. For a function $f \in H^{-1}(\Omega)$, the norm is defined as

$$\|f\|_{-1}^2 = \left\|\nabla\Delta^{-1}f\right\|_2^2 = \int_\Omega \left(\nabla\Delta^{-1}f\right)^2\, dx.$$

Thereafter, the operator Δ^{-1} denotes the inverse to the negative Dirichlet Laplacian; that is, $u = \Delta^{-1}f$ is the unique solution to

$$\begin{aligned} -\Delta u &= f, \quad \text{in } \Omega, \\ u &= 0, \quad \text{on } \partial\Omega. \end{aligned}$$

B.7. Measure Theory

The general form of the Riesz representation theorem is as follows:

Theorem B.7.1 *Take $\Omega \subset \mathbb{R}^d$, $d \geq 1$, to be an open set. Let L be a bounded linear functional on the space of compactly supported continuous functions on Ω; then there exists a measure μ such that*

$$L\varphi = \int \varphi\, d\mu.$$

We also make use of the Hausdorff measure in this book.

Definition B.7.2 Let $0 \leq k \leq +\infty$ and S be a subset of $\mathbb{R}^d$. The k-dimensional Hausdorff measure of S is given by

$$\mathcal{H}^k(S) = \lim_{\rho\to 0}\left[n(k)\cdot \inf\left\{\sum_{i\in I} |\text{diam}(A_i)|^k :\ \text{diam}(A_i) \leq \rho,\ A \subset \bigcup_{i\in I} A_i\right\}\right],$$

where $n(k)$ is a normalisation factor, and $\text{diam}(A_i)$ denotes the diameter of the set A_i. Then the Hausdorff dimension of a set S is defined by $\mathcal{H}--\text{dim}(S) = \inf\{k \geq 0 :\ \mathcal{H}^k(S) = 0\}$.

Remark B.7.3 Let k be a positive integer less than or equal to the dimension d and $S \subset \mathbb{R}^d$ be a C^1 k–dimensional manifold in $\mathbb{R}^d$. Then the Hausdorff measure $\mathcal{H}^k(S)$ equals the classical k-dimensional area of S. Moreover, $\mathcal{H}^d(S)$ equals the Lebesgue measure dx on S.

B.8. Γ-convergence

Definition B.8.1 Let $X = (X,d)$ be a metric space and (F_n), $n \in \mathbb{N}$, be family of functions F_n: $X \to [0,+\infty]$. We say that (F_n) *Γ-converges* to a function F: $X \to [0,+\infty]$ on X as $n \to \infty$ if $\forall x \in X$, we have

(i) For every sequence x_n with $d(x_n,x) \to 0$, we have

$$F(x) \leq \liminf_n F_n(x_n);$$

(ii) There exists a sequence $\bar{x_n}$ such that $d(\bar{x_n},x) \to 0$ and

$$F(x) = \lim_n F_n(\bar{x_n})$$

[or, equivalently, $F(x) \geq \limsup_n F_n(\bar{x_n})$].

Then F is the Γ-limit of (F_n) in X, and we write $F(x) = \Gamma - \lim_n F_n(x)$, $x \in X$.

The formulation of the Γ-limit for $\epsilon \to 0$ is analogous by defining a sequence ϵ_n with $\epsilon_n \to 0$ as $n \to \infty$. We also have the following property:

Theorem B.8.2 [DM93, proposition 6.21] *Let* G: $X \to \mathbb{R}$ *be a continuous function and* (F_h) *Γ-converges to* F *in* X*; then* $(F_h + G)$ *Γ-converges to* $F + G$ *in* X.

The important property of Γ-convergent sequences of functions F_n is that their minima converge to minima of the Γ-limit F. In fact, we have the following theorem:

Theorem B.8.3 *Let* (F_n) *be as in Definition B.8.1 and additionally equi-coercive; that is, there exists a compact set* $K \subset X$ *(independent of* n*) such that*

$$\inf_{x\in X}\{F_n(x)\} = \inf_{x\in K}\{F_n(x)\}.$$

If F_n *Γ-converges on* X *to a function* F*, we have*

$$\min_{x\in X}\{F(x)\} = \lim_n \inf_{x\in X}\{F_n(x)\}.$$

For more details on Γ-convergence, we refer readers to [DM93 and Bra02].

B.9. Level sets

Let u be an integrable function on $\Omega \subset \mathbb{R}^2$ open and bounded with Lipschitz boundary. We define the sub-level sets Γ_λ of u as

$$\Gamma_\lambda(u) = \{x \in \Omega :\ u(x) \leq \lambda\}. \quad \text{(B.1)}$$

The sup-level sets are defined analogously, replacing the $u(x) \leq \lambda$ in (B.1) with $u(x) \geq \lambda$. Then the level-set representation of u is $\Gamma(u) = \{\Gamma_\lambda(u) : \lambda \in \mathbb{R}\}$. Note that the definition of level sets is not unique. In particular, the preceding definition differs from the classical level-set formulation, where the level sets are defined as curves γ_λ [i.e., $\gamma_\lambda = \{x \in \Omega : u(x) = \lambda\}$]] rather than as sets. In fact, for a continuous image function u, the boundary $\partial\Gamma_\lambda = \gamma_\lambda$. The advantage of the set notion (B.1) is that it makes sense for non-smooth images as well.

B.10. Curves, Length and Curvature

We recall some facts about planar curves, their length and their curvature. Let $\gamma : [0,1] \to \mathbb{R}^2$ be a simple (i.e., without self-intersections) curve parameterised from the interval $[0,1]$ to $\mathbb{R}^2$. Then the length of γ is defined as

$$\text{Length}(\gamma) := \sup\left\{\sum_{i=1}^{n} \|\gamma(t_i) - \gamma(t_{i-1})\| : n \in \mathbb{N} \text{ and } 0 = t_0 < t_1 < \cdots < t_n = 1\right\},$$

where $\|\cdot\|$ is the Euclidean norm on $\mathbb{R}^2$. A rectifiable curve is a curve with finite length. Moreover, if γ is Lipschitz continuous on $[a,b]$, then the metric derivative (the speed) of the curve γ is defined by

$$|\gamma'|(t) := \lim_{s \to 0} \frac{\|\gamma(t+s) - \gamma(t)\|}{|s|}.$$

With that, the length of γ is equivalently defined by

$$\text{Length}(\gamma) = \int_0^1 |\gamma'|(t)\, dt. \tag{B.2}$$

Note that a generalised notion of length of a curve appears in the context of the co-area formula for functions of bounded variation (cf. Theorem 4.3.6). The arc length $s(t)$ of γ is given in the same flavour as (B.2) with

$$s(t) = \int_0^t |\gamma'|(\tilde{t})\, d\tilde{t}, \quad t \in [0,1].$$

Re-parametrising γ in terms of its arc length is called the *natural parametrisation* and yields the unit tangent vector

$$\gamma'(s) = \frac{\gamma'(t)}{|\gamma'|(t)}.$$

If γ is twice continuously differentiable, then the signed curvature of γ at t is given by

$$\kappa(t) = \frac{\det(\gamma'(t), \gamma''(t))}{|\gamma'(t)|^3} \tag{B.3}$$

and

$$|\kappa(s)| = |\gamma''(s)|. \tag{B.4}$$

B.11. Functions of Bounded Variation

The following results can be found in [AFP00] (see also [Giu84, and EG91]. Let $\Omega \subset \mathbb{R}^2$ be an open and bounded Lipschitz domain. As in [AFP00], the space of functions of bounded variation $BV(\Omega)$ in two space dimensions is defined as follows:

Definition B.11.1 ($BV(\Omega)$) Let $u \in L^1(\Omega)$. We say that u is a *function of bounded variation in* Ω if the distributional derivative of u is representable by a finite Radon measure in Ω, that is, if

$$\int_\Omega u \frac{\partial \phi}{\partial x_i}\, dx = -\int_\Omega \phi d D_i u \quad \forall \phi \in C_c^\infty(\Omega), \quad i = 1,2,$$

for some $\mathbb{R}^2$-valued measure $Du = (D_1 u, D_2 u)$ in Ω. The vector space of all functions of bounded variation in Ω is denoted by $BV(\Omega)$.

Further, the space $BV(\Omega)$ can be characterised by the total variation of Du. For this we first define the so-called variation $V(u,\Omega)$ of a function $u \in L^1_{\text{loc}}(\Omega)$.

Definition B.11.2 [Variation] Let $u \in L^1_{\text{loc}}(\Omega)$. The *variation* $V(u,\Omega)$ of u in Ω is defined by

$$V(u,\Omega) := \sup\left\{\int_\Omega u \operatorname{div}\phi\, dx : \phi \in \left(C_c^1(\Omega)\right)^2, \|\phi\|_{L^\infty(\Omega)} \le 1\right\}.$$

A simple integration by parts proves that

$$V(u,\Omega) = \int_\Omega |\nabla u|\, dx,$$

if $u \in C^1(\Omega)$. By a standard density argument, this is also true for functions $u \in W^{1,1}(\Omega)$. Before we proceed with the characterisation of $BV(\Omega)$, let us recall the definition of the total variation of a measure.

Definition B.11.3 Total Variation of a Measure Let $(X,\mathcal{E})$ be a measure space. If μ is a measure, we define its *total variation* $|\mu|$ as follows:

$$|\mu|(E) := \sup\left\{\sum_{h=0}^{\infty}|\mu(E_h)| : E_h \in \mathcal{E} \text{ pairwise disjoint }, E = \bigcup_{h=0}^{\infty} E_h\right\}, \quad \forall E \subset \mathcal{E}.$$

With Definition B.11.2, the space $BV(\Omega)$ can be characterised as follows:

Theorem B.11.4 *Let $u \in L^1(\Omega)$. Then u belongs to $BV(\Omega)$ if and only if $V(u,\Omega) < \infty$. In addition, $V(u,\Omega)$ coincides with $|Du|(\Omega)$, the total variation of Du, for any $u \in BV(\Omega)$, and $u \mapsto |Du|(\Omega)$ is lower semi-continuous in $BV(\Omega)$ with respect to the $L^1_{\text{loc}}(\Omega)$ topology.*

Note that $BV(\Omega)$ is a Banach space with respect to the norm

$$\|u\|_{BV(\Omega)} = \|u\|_{L^1(\Omega)} + |Du|(\Omega).$$

Now we introduce so-called weak* convergence in $BV(\Omega)$, which is useful for its compactness properties. Note that this convergence is much weaker than the norm convergence.

Definition B.11.5 Weak* Convergence Let $u, u_h \in BV(\Omega)$. We say that (u_h) *weakly* converges in $BV(\Omega)$ to u* (denoted by $u_h \overset{*}{\rightharpoonup} u$) if (u_h) converges to u in $L^1(\Omega)$ and (Du_h) weakly* converges to Du in all (Ω), that is,

$$\lim_{h\to\infty}\int_\Omega \phi\, dDu_h = \int_\Omega \phi\, dDu, \quad \forall \phi \in C_0(\Omega).$$

A simple criterion for weak* convergence is the following:

Theorem B.11.6 *Let $(u_h) \subset BV(\Omega)$. Then (u_h) weakly* converges to u in $BV(\Omega)$ if and only if (u_h) is bounded in $BV(\Omega)$ and converges to u in $L^1(\Omega)$.*

Further, we have the following compactness theorem:

Theorem B.11.7 Compactness for $BV(\Omega)$

- *Let Ω be a bounded domain with compact Lipschitz boundary. Every sequence $(u_h) \subset BV_{\text{loc}}(\Omega)$ satisfying*

$$\sup\left\{\int_A |u_h|\, dx + |Du_h|(A) : h \in \mathbb{N}\right\} < \infty, \quad \forall A \subset\subset \Omega \text{ open},$$

admits a subsequence (u_{h_k}) converging in $L^1_{\text{loc}}(\Omega)$ to $u \in BV_{\text{loc}}(\Omega)$. If the sequence is further bounded in $BV(\Omega)$, then $u \in BV(\Omega)$, and a subsequence converges weakly to u.*

- *Let Ω be a bounded domain in $\mathbb{R}^d$ with Lipschitz boundary. Then every uniformly bounded sequence $(u_k)_{k\geq 0}$ in $BV(\Omega)$ is relatively compact in $L^r(\Omega)$ for $1 \leq r < d/(d-1)$, $d \geq 1$. Moreover, there exists a subsequence u_{k_j} and u in $BV(\Omega)$ such that $u_{k_j} \rightharpoonup u$ weakly* in $BV(\Omega)$. In particular, for $d = 2$, this compact embedding holds for $1 \leq r < 2$.*

Let $u \in L^1(\Omega)$. We introduce the mean value u_Ω of u as

$$u_\Omega := \frac{1}{|\Omega|}\int_\Omega u(x)\, dx.$$

A generalisation of the Poincaré inequality gives the so-called Poincaré-Wirtinger inequality for functions in $BV(\Omega)$.

Theorem B.11.8 Poincaré-Wirtinger Inequality for BV Functions *For $u \in BV(\Omega)$, let*

$$u_\emptyset := \frac{1}{|\Omega|}\int_\Omega u(x)\, dx.$$

Then there exists a constant $K_\Omega > 0$ such that

$$\|u - u_\emptyset\|_2 \leq K_\Omega |Du|(\Omega).$$

Finally, since every normed vector space is a locally convex space, the theory of subdifferentials from Section B.2 applies to the framework where $X = BV(\Omega)$. For a characterisation of elements in the subdifferential $\partial\,|Du|\,(\Omega)$, we refer readers to the very detailed analysis of L. Vese [Ves01].

We also have the following theorem on the approximability of BV functions by smooth functions:

Theorem B.11.9 *Let $\Omega \subset R^2$ be open and bounded with Lipschitz boundary and $u \in BV(\Omega)$. Then there exists a sequence $(u_n) \in C^\infty(\Omega) \cap W^{1,1}(\Omega)$ such that*

(i) $u_n \to u$ in $L^1(\Omega)$, and
(ii) $\int_\Omega |Du_n|\, dx \to |Du|(\Omega)$ as $n \to \infty$.

Lebesgue Decomposition of *Du* Let $u \in BV(\Omega)$. Then, from the general theory of the Lebesgue decomposition of measures (see, e.g., [AFP00, p. 14, theorem 1.28]), we have

$$Du = \nabla u\, dx + D_s u,$$

where $\nabla u(x) = (d(Du))/dx \in L^1(\Omega)$, the absolute continuous part of Du, and $D_s u \perp dx$ is the singular part of Du. The latter can be further decomposed into a jump part J_u and a Cantor part C_u (cf. [AFP00, section 3.9]). Before we specify what these parts are exactly, we have to introduce some additional terminology first.

For $\lambda \in \mathbb{R}$, $z \in \Omega$ and a small $\rho > 0$, we define the following subsets of the disc $B_{z,\rho} = \{x : |x - z| < \rho\}$:

$$\{u > \lambda\}_{z,\rho} := \{x \in \Omega \cap B_{z,\rho} : u(x) > \lambda\},$$
$$\{u < \lambda\}_{z,\rho} := \{x \in \Omega \cap B_{z,\rho} : u(x) < \lambda\}.$$

Definition B.11.10 We call a function u essentially not greater than λ in a point $x \in \Omega$ and write $u(x) \preceq \lambda$ if

$$\lim_{\rho \to 0+} \frac{dx(\{u > \lambda\}_{x,\rho})}{dx(B_{x,\rho})} = 0$$

and analogously, u is essentially not smaller than λ in x, and we write $u(x) \succeq \lambda$ if

$$\lim_{\rho \to 0+} \frac{dx(\{u < \lambda\}_{x,\rho})}{dx(B_{x,\rho})} = 0.$$

Then we define the approximate upper and lower limits of a measurable function u in Ω as

$$u^+(x) := \inf\{\lambda \in \mathbb{R} : u(x) \preceq \lambda\},$$
$$u^-(x) := \sup\{\lambda \in \mathbb{R} : u(x) \succeq \lambda\},$$

respectively.

For a function $u \in L^1(\Omega)$, we have

$$\lim_{\rho \to 0} \frac{1}{dx(B_{x,\rho})} \int_{B_{x,\rho}} |u(x) - u(y)| \, dy = 0, \quad a.e.\ x \in \Omega.$$

Points x for which the preceding holds are called *Lebesgue points* of u, which have the properties

$$u(x) = \lim_{\rho \to 0} \frac{1}{dx(B_{x,\rho})} \int_{B_{x,\rho}} u(y) \, dy,$$
$$u(x) = u^+(x) = u^-(x).$$

The complement of the set of Lebesgue points (up to a set of $\mathcal{H}^1$ measure zero) is called the *jump set* S_u, that is,

$$S_u = \{x \in \Omega; u^-(x) < u^+(x)\}.$$

The set S_u is countable rectifiable, and for $\mathcal{H}^1$ a.e. $x \in \Omega$, we can define a normal $\vec{n}_u(x)$.

B.12. Viscosity Solutions

For the convenience of readers, we briefly recall the notion of viscosity solutions introduced by Crandall and Lions [CL83]. For a general presentation on viscosity solutions, we refer readers to [CIL92]. We consider a generic equation of the form

$$G(x,u,Du,D^2u) = 0, \quad \text{in } \Omega, \tag{B.5}$$

where G fulfils a fundamental monotonicity condition

$$G(x,r,p,X) \leq G(x,s,p,Y), \qquad \text{whenever} \quad r \leq s \quad \text{and} \quad Y \leq X,$$

where $r,s \in \mathbb{R}$, $x,p \in \mathbb{R}^d$ and $X,Y \in SM(d)$. In the following, let z^* and z_* denote the upper semi-continuous and lower semi-continuous envelopes of the function z: $C \to \mathbb{R}$, where C is a closed subset of $\mathbb{R}$, defined by

$$z^*(x) = \limsup_{y \to x,\ y \in C} z(y), \quad z_*(x) = \liminf_{y \to x,\ y \in C} z(y).$$

Definition B.12.1 A locally bounded function u: $\Omega \to \mathbb{R}$ is a viscosity subsolution (respectively, supersolution) of (B.5) if and only if for all $\varphi \in C^2(\Omega)$ and for all maximum (respectively, minimum) points x of $u^* - \varphi$ (respectively, $u_* - \varphi$), one has

$$G_*(x,u^*,D\varphi,D^2\varphi) \leq 0 \quad [\text{respectively } G^*(x,u^*,D\varphi,D^2\varphi) \geq 0].$$

A locally bounded function is a viscosity solution of (B.5) if it is a viscosity subsolution and a viscosity supersolution.

B.13. Miscellaneous

Definition B.13.1 A continuous and simple Jordan curve Γ is a continuous bijection Γ: $(a,b) \to \mathbb{R}^2$ with $\Gamma(a) = \Gamma(b)$. By the Jordan curve theorem, Γ is the boundary of a bounded and simply connected domain $D = D(\Gamma)$.

Appendix C
MATLAB Implementation

This book is supplemented with a number of MATLAB implementations of some of the inpainting methods. Supplemented MATLAB code includes

- The code `harmonic_inpainting.m`, which implements harmonic inpainting from Section 4.2. This code works for colour and grey value images.
- The code `amle_inpainting.m` is an implementation of AMLE inpainting for surface interpolation, as discussed in Section 4.4.
- The codes `bvl2_inpainting_convs.m` and `bvnegh_inpainting_convs.m` are implementations of TV inpainting from Section 4.3 and TV-H^{-1} inpainting from Section 5.3 using the convexity splitting method discussed in [SB11]. These two codes work for grey value images only.
- The code `cahnhilliard_inpaint_conv.m` is an implementation of Cahn-Hilliard inpainting from Section 5.3. This implementation accepts grey value images only.
- Euler's elastica inpainting from Section 5.2 using the explicit gradient-descent scheme from the original Chan, Kang and Shen paper [CKS02] is implemented in `euler_elastica_inpainting`. For more sophisticated solvers, please see the numerical discussion in Section 5.2 for further references.
- The code `transport_inpainting.m` is an implementation of Bertalmio's transport inpainting method presented in Section 6.1. It works on grey value and colour images.
- Mumford-Shah and Mumford-Shah-Euler inpainting discussed in Chapter 7 are implemented in `mumford_shah_inpaint.m` and `mumford_shah_euler_inpaint.m`, respectively.

Moreover, we refer the reader to the following online sources for additional implementations for discussed inpainting methods:

- A general source of image processing methods supplemented with demos and C-implementations is the *Image Processing On Line* (IPOL) journal, which can be found at http://www.ipol.im. For image inpainting, we currently have

- TV inpainting from Section 4.3, implemented using the split Bregman algorithm [GO09]; see http://www.ipol.im/pub/art/2012/g-tvi/.
- TV-TV^2 inpainting from Section 5.5, also implemented with split Bregman, can be found at http://www.ipol.im/pub/art/2013/40/.

We are currently preparing demos for harmonic, AMLE, Mumford-Shah and transport inpainting for IPOL which will be online soon.

- The implementations of harmonic, TV and TV-H^{-1} inpainting as provided in this book also can be found on MATLAB Central at http://www.mathworks.com/matlabcentral/fileexchange/34356-higher-order-total-variation-inpainting.
- Tom März's implementation of coherence-transport inpainting discussed in Section 6.2, can be found at https://github.com/maerztom/inpaintBCT.

Appendix D
Image Credits

The inpainting examples in figures 5.5, 6.3, and 6.4 have been taken from Marcelo Bertalmio's webpage http://www.dtic.upf.edu/~mbertalmio/restoration.html and have been used in this book with his kind permission. © Marcelo Bertalmio

The turtle photographs in figures 4.3, 5.8 (bottom left), 5.17, 5.19, and 7.3 are courtesy of Kostas Papafitsoros

http://www.flickr.com/photos/bugatsas/6074839062/in/set-72157627361097822/
http://www.flickr.com/photos/bugatsas/8387559202/in/set-72157632539910014
http://www.flickr.com/photos/bugatsas/8404646233/in/photostream
http://www.flickr.com/photos/bugatsas/6144755271/in/set-72157623431601539
and have been used with his kind permission. © Konstantinos Papafitsoros

The inpainting examples in figures 4.7 and 4.8 are courtesy of Kostas Papafitsoros and are used in this book with his kind permission. © Konstantinos Papafitsoros

The inpainting examples in figures 5.2, 5.5, 5.6, 5.7, and 5.8 (bottom right) are courtesy of Jooyoung Hahn and have been used with his kind permission. © Jooyoung Hahn

The inpainting examples in figures 2.4, 2.5, 6.13, 9.27, and 9.28 are courtesy of Rob Hocking and have been used with his kind permission. © Rob Hocking

The image in figure 9.1 is courtesy of Andrea Baczynski and has been used with her kind permission. © Andrea Baczynski

The image in figure 9.9 is courtesy of Wolfgang Baatz and has been used with his kind permission. © Wolfgang Baatz

The images in figures 9.10 are courtesy of Andrea Bertozzi and have been used with her kind permission. © Andrea Bertozzi

Glossaries

Glossary: Function Spaces and Norms

Let Ω be an open and bounded subset of $\mathbb{R}^d$ and X a Banach space with norm $\|\cdot\|_X$. Moreover, let v be a function with v: $(0,T) \to X$ and $\mathcal{J}: X \to (-\infty, +\infty]$ a functional on X.

$\operatorname{argmin}\{\mathcal{J}\}$	$:= \{u \in X : \mathcal{J}(u) = \inf_X \mathcal{J}\}$.
$BV(\Omega)$	Space of functions of bounded variation with semi-norm $\|Df\|(\Omega)$, the total variation of f in Ω.
$BV - w^*$	The weak* topology of $BV(\Omega)$.
$C^m(\Omega)$	The space of functions on Ω, which are m times continuously differentiable.
$C_0^m(\Omega)$	$\{f \in C^m(\Omega) : f\|_{\partial\Omega} = 0\}$.
$C^m(0,T;X)$	With $m \geq 0$, $0 < T < \infty$: space of functions from $[0,T]$ to X, which are m times continuously differentiable. It is a Banach space with the norm $\|v\|_{C^m(0,T;X)} = \max_{0 \leq l \leq m} \left(\sup_{0 \leq t \leq T} \left\| \frac{\partial^l v}{\partial t^l}(t) \right\|_X \right)$.
$\mathbb{R}^d$	The Euclidean space of dimension d with the Euclidean norm $\|\cdot\|$.
$\langle \cdot, \cdot \rangle$	The inner product in $\mathbb{R}^d$, $d \geq 1$, defined for $x, y \in \mathbb{R}^d$ as $\langle x, y \rangle = \sum_{i=1}^d x_i y_i$.
$H^{-1}(\Omega)$	$\left(H_0^1(\Omega)\right)^*$, that is, the dual space of $H_0^1(\Omega)$ with corresponding norm $\|\cdot\|_{H^{-1}(\Omega)} = \left\|\nabla \Delta^{-1} \cdot\right\|_2$ and inner product $\langle \cdot, \cdot \rangle_{H^{-1}(\Omega)} = \left\langle \nabla \Delta^{-1} \cdot, \nabla \Delta^{-1} \cdot \right\rangle_2$. Thus, Δ^{-1} is the inverse of the negative Laplacian $-\Delta$ with zero Dirichlet boundary conditions.
$\|\cdot\|_{-1}$	$:= \|\cdot\|_{H^{-1}(\Omega)}$.
$\langle \cdot, \cdot \rangle_{-1}$	$:= \langle \cdot, \cdot \rangle_{H^{-1}(\Omega)}$.

$H^p(\Omega)$	$W^{p,2}(\Omega)$. This is a Hilbert space with corresponding inner product $\langle f,g\rangle_{H^p(\Omega)} = \sum_{k=1}^p \int_\Omega D^k f \cdot D^k g\, dx$. For this special Sobolev space, we write $\|\cdot\|_{H^p(\Omega)} := \|\cdot\|_{W^{p,2}(\Omega)}$ for its corresponding norm.
$H_0^p(\Omega)$	$W_0^{p,2}(\Omega)$.
$\|\cdot\|_2$	$:= \|\cdot\|_{L^2(\Omega)}$ for the norm in $L^2(\Omega)$.
$\langle\cdot,\cdot\rangle_2$	The inner product in $L^2(\Omega)$ defined for $u,v \in L^2(\Omega)$ as $\langle u,v\rangle_{L^2(\Omega)} = \int_\Omega uv\, dx$.
$L^\infty(\Omega)$	Space of Lebesque measurable functions f such that there exists a constant C with $\|f(x)\| \le C$, a.e. $x \in \Omega$. The space $L^\infty(\Omega)$ is a Banach space with corresponding norm $\|f\|_{L^\infty(\Omega)} = \sup_{x\in\Omega}\{\|f(x)\|\}$.
$L^p(\Omega)$	With $1 \le p < \infty$: space of Lebesque measurable functions f such that $\int_\Omega \|f\|^p\, dx < \infty$. The space $L^p(\Omega)$ is a Banach space with corresponding norm $\|f\|_{L^p(\Omega)} = \left(\int_\Omega \|f\|^p\, dx\right)^{1/p}$. In the case $p=2$, it is a Hilbert space with corresponding inner product $\langle f,g\rangle_{L^2(\Omega)} = \int_\Omega f\cdot g\, dx$.
$L^p_{\mathrm{loc}}(\Omega)$	$L^p_{\mathrm{loc}}(\Omega) = \{f : \Omega \to \mathbb{R} : f \in L^p(D) \text{ for each } D \Subset \Omega\}$.
l.s.c. (sequentially)	lower semi-continuous: $\mathcal{J}$ is called l.s.c. if for every sequence (u^n) converging to u we have $\liminf_{n\to\infty} \mathcal{J}(u^n) \ge \mathcal{J}(u)$.
$\mathbb{R}_+$	The non-negative real numbers.
$W^{p,q}(\Omega)$	With $1 \le p,q \le \infty$: Sobolev space of functions $f \in L^q(\Omega)$ such that all derivatives up to order p belong to $L^q(\Omega)$. The space $W^{p,q}(\Omega)$ is a Banach space with norm $\|f\|_{W^{p,q}(\Omega)} = \left(\sum_{k=1}^p \int_\Omega \|D^k f\|^q\, dx\right)^{1/q}$, where $D^k f$ denotes the kth distributional derivative of f.
$W_0^{p,q}(\Omega)$	$\{f \in W^{p,q}(\Omega) : f\|_{\partial\Omega} = 0\}$.

Glossary: About Functions

Let f be a function with $f\colon \Omega \subset \mathbb{R}^d \to \mathbb{R}$ and $(f^n)_{n\in\mathbb{N}}$ a sequence of functions belonging to a Banach space X. Moreover, let $\mathcal{H}$ be a real separable Hilbert space and $\psi\colon \mathcal{H} \to \mathbb{R}\cup\{+\infty\}$.

Df	Distributional derivative of f.
$\mathrm{div} f = \nabla\cdot f$	Divergence of f, that is, $\nabla\cdot f = \sum_{i=1}^d \frac{\partial f}{\partial x_i}$.
∇f	Gradient of f, that is, absolutely continuous part of Df.
Δf	Laplacian operator, that is, $\Delta f = \sum_{i=1}^d \frac{\partial^2 f}{\partial x_i^2}$.

$\psi^*(\cdot)$	The Legendre-Fenchel transform; that is, convex conjugate $\psi^*\colon \mathcal{H}^* \to \mathbb{R} \cup \{+\infty\}$ is defined by $\psi^*(u) = \sup_{v \in \mathcal{H}}\{\langle v, u\rangle - \psi(v)\}$.
$f^n \to f$ in X	The sequence (f^n) converges strongly to f in X.
$\operatorname{supp}\{f\}$	For a measurable function $f\colon \Omega \subset \mathbb{R}^d \to \mathbb{R}$, let $(w_i)_{i\in I}$ be the family of all open subsets such that $w_i \subseteq \Omega$ and for each $i \in I$, $f = 0$ a.e. on w_i. Then supp (the support of f) is defined by $\operatorname{supp}\{f\} = \Omega \setminus \bigcup_i w_i$.
v_t	$:= \frac{dv(t)}{dt}$. Time derivative of a function $v\colon (0,T) \to X$, for $t > 0$.
$f^n \rightharpoonup f$ in X	The sequence (f^n) converges weakly to f in X.
$f^n \overset{*}{\rightharpoonup} f$ in X	The sequence (f^n) converges to f in the weak* topology of X.
$\\|f\\|_X$	The norm of f in X.

Glossary: Miscellaneous Notation

Let A, B and R be bounded and open sets in $\mathbb{R}^d$.

T^* — The adjoint operator of T in $\mathcal{H}$, that is, $\langle T^*u, v\rangle = \langle u, Tv\rangle$, where $\langle\cdot,\cdot\rangle$ denotes the inner product in $\mathcal{H}$.

$B(x,r)$ — For $x \in \mathbb{R}^d$, $d \geq 1$, the d–dimensional ball of radius $r > 0$ and centre x is $B(x,r) = \{y \in \mathbb{R}^d : \|y - x\| \leq r\}$.

$\mathcal{O}$ — Let f and g be two functions defined on the real numbers. Then $f(x) = \mathcal{O}(g(x))$ as $x \to \infty$ if and only if there exists a constant $C \in \mathbb{R}^+$ and an $x_0 \in \mathbb{R}$ such that $|f(x)| \leq C|g(x)|$ for all $x \geq x_0$.

χ_R — Characteristic function of a bounded and open set R, that is,

$$\chi_R(x) = \begin{cases} 0, & x \in R, \\ +\infty, & \text{otherwise.} \end{cases}$$

$A \hookrightarrow\hookrightarrow B$ — A is compactly embedded into B.

$A \hookrightarrow B$ — A is continuously embedded into B.

V^* — The topological dual for a topological vector space V.

$|\cdot|$ — Euclidean norm in $\mathbb{R}^d$.

$\mathcal{H}_d$ — The d-dimensional Hausdorff measure.

$\mathbb{1}_R$ — Indicator function of a bounded and open set R, that is,

$$\mathbb{1}_R(x) = \begin{cases} 1, & x \in R, \\ 0, & \text{otherwise.} \end{cases}$$

$T \in \mathcal{L}(\mathcal{H})$	T is a bounded linear operator in a Hilbert space $\mathcal{H}$.
$\|T\|$	Operator norm of T.
$a^{\perp}$	Orthogonal vector to $a \in \mathbb{R}^d$.
$\mathrm{sign}(s)$	Sign function, that is, $\mathrm{sign}(s) = \begin{cases} 1, & s > 0, \\ 0, & s = 0, \\ -1, & s < 0. \end{cases}$
$SM(d)$	The space of all d-dimensional symmetric matrices.
A^t	Transposed of an array A.

References

[ABC94] Nicholas D. Alikakos, Peter W. Bates and Xinfu Chen. Convergence of the Cahn-Hilliard equation to the Hele-Shaw model. *Archive for Rational Mechanics and Analysis*, 128(2):165–205, 1994.

[ACF12] Pablo Arias, Vicent Caselles and Gabriele Facciolo. Analysis of a variational framework for exemplar-based image inpainting. *Multiscale Modeling & Simulation*, 10(2):473–514, 2012.

[ACGR02] Andrés Almansa, Frédéric Cao, Yann Gousseau and Bernard Rougé. Interpolation of digital elevation models using AMLE and related methods. *IEEE Transactions on Geoscience and Remote Sensing*, 40(2):314–25, 2002.

[ACS09] Pablo Arias, Vicent Caselles and Guillermo Sapiro. A variational framework for non-local image inpainting. In *Energy Minimization Methods in Computer Vision and Pattern Recognition*, pp. 345–58. Springer, Berlin, 2009.

[AFCS11] Pablo Arias, Gabriele Facciolo, Vicent Caselles and Guillermo Sapiro. A variational framework for exemplar-based image inpainting. *International Journal of Computer Vision*, 93(3):319–47, 2011.

[AFP00] Luigi Ambrosio, Nicola Fusco and Diego Pallara. *Functions of Bounded Variation and Free Discontinuity Problems*, Vol. 254. Clarendon Press, Oxford, UK, 2000.

[AGLM93] Luis Alvarez, Frédéric Guichard, Pierre-Louis Lions and Jean-Michel Morel. Axioms and fundamental equations of image processing. *Archive for Rational Mechanics and Analysis*, 123(3):199–257, 1993.

[AK06] Gilles Aubert and Pierre Kornprobst. *Mathematical Problems in Image Processing: Partial Differential Equations and the Calculus of Variations*, Vol. 147. Springer, Berlin, 2006.

[AKR97] S. Armstrong, A. Kokaram and P.J.W. Rayner. Nonlinear interpolation of missing data using min-max functions. In *IEEE International Conference on Nonlinear Signal and Image Processing*. IEEE, New York, 1997.

[Alm02] A. Almansa. Échantillonnage, interpolation et détection. applications en imagerie satellitaire. Technical report, ENS Cachan, France, 2002.

[ALM10] Jean-François Aujol, Saïd Ladjal and Simon Masnou. Exemplar-based inpainting from a variational point of view. *SIAM Journal on Mathematical Analysis*, 42(3):1246–85, 2010.

[Alt85] Hans Wilhelm Alt. *Lineare Funktionalanalysis*, Vol. 2. Springer, Berlin, 1985.

[AM03] Luigi Ambrosio and Simon Masnou. A direct variational approach to a problem arising in image reconstruction. *Interfaces and Free Boundaries*, 5(1):63–82, 2003.

[Amb89a] Luigi Ambrosio. A compactness theorem for a new class of functions of bounded variation. *Bollettino della Unione Matematica Italiana*, 3(4):857–81, 1989.

[Amb89b] Luigi Ambrosio. Variational problems in SBV and image segmentation. *Acta Applicandae Mathematicae*, 17(1):1–40, 1989.

[Anz83] Gabriele Anzellotti. Pairings between measures and bounded functions and compensated compactness. *Annali di Matematica Pura ed Applicata*, 135(1):293–318, 1983.

[Aro67] Gunnar Aronsson. Extension of functions satisfying Lipschitz conditions. *Arkiv för Matematik*, 6(6):551–61, 1967.

[AT90] Luigi Ambrosio and Vincenzo Maria Tortorelli. Approximation of functional depending on jumps by elliptic functional via t-convergence. *Communications on Pure and Applied Mathematics*, 43(8):999–1036, 1990.

[AT92] Luigi Ambrosio and Vincenzo Maria Tortorelli. On the approximation of free discontinuity problems. *Bollettino della Unione Matematica Italiana*, 6(B):105–23, 1992.

[AV97] Gilles Aubert and Luminita Vese. A variational method in image recovery. *SIAM Journal on Numerical Analysis*, 34(5):1948–79, 1997.

[BB11] Martin Benning and Martin Burger. Error estimates for general fidelities. *Electronic Transactions on Numerical Analysis*, 38(44–68):77, 2011.

[BBC+01] Coloma Ballester, Marcelo Bertalmio, Vicent Caselles Guillermo Sapiro and Joan Verdera. Filling-in by joint interpolation of vector fields and gray levels. *IEEE Transactions on Image Processing*, 10(8):1200–11, 2001.

[BBCS10] Aurélie Bugeau, Marcelo Bertalmío, Vicent Caselles and Guillermo Sapiro. A comprehensive framework for image inpainting. *IEEE Transactions on Image Processing*, 19(10):2634–45, 2010.

[BBLS14] Martin Benning, Christoph Brune, Rien Lagerwerf and Carola-Bibiane Schönlieb. TGV sinogram inpainting for limited angle tomography. Preprint, 2014.

[BBS01] Marcelo Bertalmio, Andrea L Bertozzi and Guillermo Sapiro. Navier-Stokes, fluid dynamics, and image and video inpainting. In *Proceedings of the 2001 IEEE Computer Society Conference on Computer Vision and Pattern Recognition, 2001 (CVPR 2001)*, Vol. 1, pp. 1-355–1-1362. IEEE, New York, 2001.

[BC98] Peter Blomgren and Tony F Chan. Color TV: Total variation methods for restoration of vector-valued images. *IEEE Transactions on Image Processing*, 7(3):304–9, 1998.

[BCG+06] Andrew Blake, Antonio Criminisi, Michel Gangnet, Patrick Perez and Kentaro Toyama. Image region filling by exemplar-based inpainting. U.S. Patent 6,987,520, January 17 2006.

[BCN02] Giovanni Bellettini, Vicent Caselles and Matteo Novaga. The total variation flow in $\mathbb{R}^N$. *Journal of Differential Equations*, 184(2):475–525, 2002.

[BCV03] Coloma Ballester, Vicent Caselles and Joan Verdera. A variational model for disocclusion. In *Proceedings of the 2003 International Conference on Image Processing, 2003 (ICIP 2003)*, Vol. 2, pp. II-677–II-680. IEEE, New York, 2003.

[BDB65] Garrett Birkhoff and Carl R De Boor. Piecewise polynomial interpolation and approximation. *Approximation of Functions, Proc. Sympos. Warren, Michigan, 1964,* pp. 164–90, 1965.

[BDMP93] G. Bellettini, G. Dal Maso and M. Paolini. Semicontinuity and relaxation properties of a curvature depending functional in 2D. *Ann. Scuola Norm. Sup. Pisa Cl. Sci.*, 4(2):247–97, 1993.

[BDP93] G. Bellettini, G. Dal Maso and M. Paolini. Semicontinuity and relaxation properties of a curvature depending functional in 2D. *Ann. Sc. Norm. Super. Pisa, Cl. Sci., IV. Ser.*, 20(2):247–97, 1993.

[BEG07] Andrea L Bertozzi, Selim Esedoglu and Alan Gillette. Inpainting of binary images using the cahn-hilliard equation. *IEEE Transactions on Image Processing*, 16(1):285–91, 2007.

[BEglG07] Andrea Bertozzi, Selim Esedoglu and Alan Gillette.Analysis of a two-scale Cahn-Hilliard model for binary image inpainting. *Multiscale Modeling & Simulation*, 6(3):913–36, 2007.

[Ber01] Marcelo Bertalmio. Processing of flat and non-flat image information on arbitrary manifolds using partial differential equations. Technical report, University of Minnesota, 2001.

[BFM+08] Wolfgang Baatz, Massimo Fornasier, Peter A Markowich, Carola-Bibiane Schönlieb, et al. Inpainting of ancient austrian frescoes. In *Bridges Leeuwarden: Mathematics, Music, Art, Architecture, Culture*, pp. 163–70. Southwestern College, Chula Vista, CA, 2008.

[BG95] Guy Barles and Christine Georgelin. A simple proof of convergence for an approximation scheme for computing motions by mean curvature. *SIAM Journal on Numerical Analysis*, 32(2):484–500, 1995.

[BG04] Andrea L Bertozzi and John B Greer. Low-curvature image simplifiers: Global regularity of smooth solutions and Laplacian limiting schemes. *Communications on Pure and Applied Mathematics*, 57(6):764–90, 2004.

[BHS09] Martin Burger, Lin He and Carola-Bibiane Schönlieb. Cahn-Hilliard inpainting and a generalization for grayvalue images. *SIAM Journal on Imaging Sciences*, 2(4):1129–67, 2009.

[BKP10] Kristian Bredies, Karl Kunisch and Thomas Pock. Total generalized variation. *SIAM Journal on Imaging Sciences*, 3(3):492–526, 2010.

[BL11] Kristian Bredies and Dirk Lorenz. Mathematische Bildverarbeitung. *Einführung in die Grundlagen und moderne Theorie*, 2011.

[BM07] Folkmar Bornemann and Tom März. Fast image inpainting based on coherence transport. *Journal of Mathematical Imaging and Vision*, 28(3):259–78, 2007.

[BO04] Martin Burger and Stanley Osher. Convergence rates of convex variational regularization. *Inverse Problems*, 20(5):1411, 2004.

[Bon96] Alexis Bonnet. On the regularity of the edge set of Mumford-Shah minimizers. In *Variational Methods for Discontinuous Structures*, pp. 93–103. Springer, Berlin, 1996.

[BP10] Maïtine Bergounioux and Loic Piffet. A second-order model for image denoising. *Set-Valued and Variational Analysis*, 18(3–4):277–306, 2010.

[BPW13] Kristian Bredies, Thomas Pock and Benedikt Wirth. A convex, lower semi-continuous approximation of Euler's elastica energy. SFB Report, University of Graz, 2013.

[Bra02] Andrea Braides. *Gamma-Convergence for Beginners*, Vol. 22. Oxford University Press, 2002.

[Bre99] Pierre Bremaud. *Markov Chains: Gibbs Fields, Monte Carlo Simulation, and Queues*, Vol. 31. Springer, Berlin, 1999.

[BRH07] Martin Burger, Elena Resmerita and Lin He. Error estimation for Bregman iterations and inverse scale space methods in image restoration. *Computing*, 81(2–3):109–35, 2007.

[BSCB00] Marcelo Bertalmio, Guillermo Sapiro, Vincent Caselles and Coloma Ballester. Image inpainting. In *Proceedings of the 27th Annual Conference on Computer Graphics and Interactive Techniques*, pp. 417–24. ACM Press/Addison-Wesley, Reading, MA, 2000.

[BV11] Kristian Bredies and Tuomo Valkonen. Inverse problems with second-order total generalized variation constraints. *Proceedings of SampTA*, 201, 2011.

[BVSO03] Marcelo Bertalmio, Luminita Vese, Guillermo Sapiro and Stanley Osher. Simultaneous structure and texture image inpainting. *IEEE Transactions on Image Processing,* 12(8):882–9, 2003.

[BW89] M.I.G. Bloor and M.J. Wilson. Generating blend surfaces using partial differential equations. *Computer-Aided Design*, 21(3):165–71, 1989.

[BW90] Malcolm I.G. Bloor and Michael J. Wilson. Using partial differential equations to generate free-form surfaces. *Computer-Aided Design*, 22(4):202–12, 1990.

[BW96] Malcolm I.G. Bloor and Michael J. Wilson. Spectral approximations to PDE surfaces. *Computer-Aided Design*, 28(2):145–52, 1996.

[BZ87] Andrew Blake and Andrew Zisserman. *Visual Reconstruction*, Vol. 2. MIT Press, Cambridge, MA, 1987.

[C$^+$96] Xinfu Chen et al. Global asymptotic limit of solutions of the Cahn-Hilliard equation. *Journal of Differential Geometry*, 44(2):262–311, 1996.

[Cao03] Frédéric Cao. *Geometric Curve Evolution and Image Processing*. Springer, Berlin, 2003.

[Car88] Stefan Carlsson. Sketch based coding of grey level images. *Signal Processing*, 15(1):57–83, 1988.

[Cas96] Josep R. Casas. Image compression based on perceptual coding techniques. Technical report, UPC, Barcelona, 1996.

[Cas11] Vicent Caselles. Exemplar-based image inpainting and applications. *SIAM News*, 44(10):1–3, 2011.

[CCBT03] Jean Pierre Cocquerez, Laurent Chanas and Jacques Blanc-Talon. Simultaneous inpainting and motion estimation of highly degraded video-sequences. In *Image Analysis: Lecture Notes in Computer Science*, Vol. 2749, pp. 685–92. Springer, Berlin, 2003.

[CCC+10] Antonin Chambolle, Vicent Caselles, Daniel Cremers, Matteo Novaga and Thomas Pock. An introduction to total variation for image analysis. *Theoretical Foundations and Numerical Methods for Sparse Recovery*, 9:263–340, 2010.

[CCN11] Vicent Caselles, Antonin Chambolle and Matteo Novaga. Total variation in imaging. *Handbook of Mathematical Methods in Imaging*, pp. 1016–57. Springer, Berlin, 2011.

[CGMP11] Frédéric Cao, Yann Gousseau, Simon Masnou and Patrick Pérez. Geometrically guided exemplar-based inpainting. *SIAM Journal of Imaging Science*, 4(4):1143–79, 2011.

[CIL92] Michael G. Crandall, Hitoshi Ishii and Pierre-Louis Lions. User's guide to viscosity solutions of second order partial differential equations. *Bulletin of the American Mathematical Society*, 27(1):1–67, 1992.

[CK06] Tony F. Chan and Sung Ha Kang. Error analysis for image inpainting. *Journal of Mathematical Imaging and Vision*, 26(1–2):85–103, 2006.

[CKS02] Tony F. Chan, Sung Ha Kang and Jianhong Shen. Euler's elastica and curvature-based inpainting. *SIAM Journal on Applied Mathematics*, 63(2): 564–92, 2002.

[CL83] Michael G. Crandall and Pierre-Louis Lions. Viscosity solutions of Hamilton-Jacobi equations. *Transaction of the American Mathematical Society*, 277(1):1–42, 1983.

[CL97] Antonin Chambolle and Pierre-Louis Lions. Image recovery via total variation minimization and related problems. *Numerische Mathematik*, 76(2):167–88, 1997.

[CLMC92] Francine Catté, Pierre-Louis Lions, Jean-Michel Morel and Tomeu Coll. Image selective smoothing and edge detection by nonlinear diffusion. *SIAM Journal on Numerical Analysis*, 29(1):182–93, 1992.

[CMS98a] Vicent Caselles, J.-M. Morel and Catalina Sbert. An axiomatic approach to image interpolation. *IEEE Transactions on Image Processing*, 7(3): 376–86, 1998.

[CMS98b] Vicent Caselles, J.-M. Morel and Catalina Sbert. An axiomatic approach to image interpolation. Technical report, Universitat Pompeu Fabra, 1998.

[CPT03] Antonio Criminisi, Patrick Perez and Kentaro Toyama. Object removal by exemplar-based inpainting. In *Proceedings of the 2003 IEEE Computer Society Conference on Computer Vision and Pattern Recognition*, Vol. 2, pp. II-721–II-728. IEEE, New York, 2003.

[CS01a] Tony F. Chan and Jianhong Shen. Morphologically invariant PDE inpaintings. Technical report, UCLA, 2001.

[CS01b] Tony F. Chan and Jianhong Shen. Nontexture inpainting by curvature-driven diffusions. *Journal of Visual Communication and Image Representation*, 12(4):436–49, 2001.

[CS05a] Tony F. Chan and Jianhong Jackie Shen. *Image Processing and Analysis: Variational, PDE, Wavelet, and Stochastic Methods*. SIAM, Philadelphia, 2005.

[CS05b] Tony F. Chan and Jianhong Jackie Shen. Variational image inpainting. *Communications on Pure and Applied Mathematics*, 58(5):579–619, 2005.

[CSZ06] Tony F. Chan, Jianhong Shen and Hao-Min Zhou. Total variation wavelet inpainting. *Journal of Mathematical Imaging and Vision*, 25(1):107–25, 2006.

[CT94] Josep R. Casas and Luis Torres. Coding of details in very low bit-rate video systems. *IEEE Transactions on Circuits and Systems for Video Technology*, 4(3):317–27, 1994.

[CT96] Josep R. Casas and Luis Torres. Strong edge features for image coding. In *Mathematical Morphology and Its Applications to Image and Signal Processing*, pp. 443–50. Springer, Berlin, 1996.

[CW08] E. J. Candes and M. B. Wakin. An introduction to compressive sampling. *Signal Processing Magazine*, 25(2):21–30, March 2008.

[Dac89] Bernard Dacorogna. *Direct Methods in the Calculus of Variations*, Vol. 78. Springer, Berlin, 1989.

[DAG09] Vincent Duval, Jean-François Aujol and Yann Gousseau. The TVL1 model: A geometric point of view. *Multiscale Modeling & Simulation*, 8(1):154–89, 2009.

[DB08] Julia A. Dobrosotskaya and Andrea L. Bertozzi. A wavelet-Laplace variational technique for image deconvolution and inpainting. *IEEE Transactions on Image Processing*, 17(5):657–63, 2008.

[Dem84] Françoise Demengel. Fonctions à hessien borné. In *Annales de l'institut Fourier*, Vol. 34(2), pp. 155–90. Institut Fourier, St. Martin d'Hères, France, 1984.

[MD97] E. De Giorgi, M. Carriero and A. Leaci. Existence theorem for a minimum problem with free discontinuity set. *Archive for Rational Mechanics and Analysis*, 108(4):195–18, 1989.

[DJL+12] Bin Dong, Hui Ji, Jia Li, Zuowei Shen and Yuhong Xu. Wavelet frame based blind image inpainting. *Applied and Computational Harmonic Analysis*, 32(2):268–79, 2012.

[DLRW05] Qiang Du, Chun Liu, Rolf Ryham and Xiaoqiang Wang. A phase field formulation of the willmore problem. *Nonlinearity*, 18(3):1249, 2005.

[DM93] Gianni Dal Maso. *An Introduction to G-Convergence*. Springer, Berlin, 1993.

[DMM07] Agnès Desolneux, Lionel Moisan and Jean-Michel Morel. *From Gestalt Theory to Image Analysis: A Probabilistic Approach*, Vol. 34. Springer, Berlin, 2007.

[DMMS92] Gianni Dal Maso, Jean-Michel Morel and Sergio Solimini. A variational method in image segmentation: existence and approximation results. *Acta Mathematica*, 168(1):89–151, 1992.

[DSC03] Laurent Demanet, Bing Song and Tony Chan. Image inpainting by correspondence maps: a deterministic approach. *Applied and Computational Mathematics*, 1100:217–50, 2003.

[DWB09] S. Didas, J. Weickert and B. Burgeth. Properties of higher order nonlinear diffusion filtering. *Journal of Mathematical Imaging and Vision*, 35:208–26, 2009.

[EG91] Lawrence Craig Evans and Ronald F Gariepy. *Measure Theory and Fine Properties of Functions*, Vo. 5. CRC Press, Bora Raton, FL, 1991.

[EHN96] Heinz Werner Engl, Martin Hanke and Andreas Neubauer. *Regularization of Inverse Problems*, Vol. 375. Springer, Berlin, 1996.

[EL99] Alexei A. Efros and Thomas K. Leung. Texture synthesis by non-parametric sampling. In *Proceedings of the Seventh IEEE International Conference on Computer Vision*, Vol. 2, pp. 1033–38. IEEE, New York, 1999.

[EM76] Gilberte Emile-Mâle. *The Restorer's Handbook of Easel Painting*. Van Nostrand Reinhold, New York, 1976.

[ES02] Selim Esedoglu and Jianhong Shen. Digital inpainting based on the Mumford-Shah-Euler image model. *European Journal of Applied Mathematics*, 13(4):353–70, 2002.

[ESQD05] Michael Elad, J.-L. Starck, Philippe Querre and David L. Donoho. Simultaneous cartoon and texture image inpainting using morphological component analysis (MCA). *Applied and Computational Harmonic Analysis*, 19(3):340–58, 2005.

[ET76] Ivar Ekeland and Roger Temam. *Convex Analysis and Variational Problems*. SIAM, Philadelphia, 1976.

[Eul44] L. Euler. Methodus inveniendi lineas curvas maximi minimive proprietate gaudentes in the additamentum of de curvis elasticis. *Lausanne, Genf: Bousqet & Socios*, 1744.

[Eva93] Lawrence C. Evans. Convergence of an algorithm for mean curvature motion. *Indiana University Mathematics Journal*, 42(2):533–57, 1993.

[Eva98] Lawrence C. Evans. *Partial Differential Equations: Graduate Studies in Mathematics*. Vol. 2. American Mathematical Society, Providence, RI, 1998.

[FACS09] Gabriele Facciolo, Pablo Arias, Vicent Caselles and Guillermo Sapiro. Exemplar-based interpolation of sparsely sampled images. In *Energy Minimization Methods in Computer Vision and Pattern Recognition*, pp. 331–44. Springer, Berlin, 2009.

[FF69] Herbert Federer and Herbert Federer. *Geometric Measure Theory*, Vol. 1996. Springer, New York, 1969.

[FL07] Irene Fonseca and Giovanni Leoni. *Modern Methods in the Calculus of Variations: L^p Spaces*. Springer, Berlin, 2007.

[FM07] Massimo Fornasier and Riccardo March. Restoration of color images by vector valued BV functions and variational calculus. *SIAM Journal on Applied Mathematics*, 68(2):437–60, 2007.

[FÖ08] D. Fanelli and O. Öktem. Electron tomography: A short overview with an emphasis on the absorption potential model for the forward problem. *Inverse Problems*, 24(1):013001, 2008.

[Fol99] Gerald B. Folland. *Real Analysis. Pure and Applied Mathematics*, 2nd ed. Wiley, New York, 1999.

[For06] Massimo Fornasier. Nonlinear projection recovery in digital inpainting for color image restoration. *Journal of Mathematical Imaging and Vision*, 24(3):359–73, 2006.

[FR08] Massimo Fornasier and Holger Rauhut. Recovery algorithms for vector-valued data with joint sparsity constraints. *SIAM Journal on Numerical Analysis*, 46(2):577–613, 2008.

[Fri13] Jürgen Frikel. Sparse regularization in limited angle tomography. *Applied and Computational Harmonic Analysis*, 34(1):117–41, January 2013.

[FSM09] Mohamed-Jalal Fadili, J.-L. Starck and Fionn Murtagh. Inpainting and zooming using sparse representations. *The Computer Journal*, 52(1):64–79, 2009.

[Fun62] P. Funk. Variationsrechnung und ihre Anwendung in Physik und Technik. Springer, Berlin, 1970.

[GB04a] John B. Greer and Andrea L. Bertozzi. H^1 solutions of a class of fourth order nonlinear equations for image processing. *Discrete and Continuous Dynamical Systems*, 10(1–2):349–66, 2004.

[GB04b] John B. Greer and Andrea L. Bertozzi. Traveling wave solutions of fourth order pdes for image processing. *SIAM Journal on Mathematical Analysis*, 36(1):38–68, 2004.

[GB08] Michael C. Grant and Stephen P. Boyd. *Graph Implementations for Nonsmooth Convex Programs*, Vol. 371, pp. 95–110. Springer, London, 2008.

[GB14] Michael Grant and Stephen Boyd. CVX: MATLAB software for disciplined convex programming, Version 2.1. Available at: http://cvxr.com/cvx, March 2014.

[GG84] Stuart Geman and Donald Geman. Stochastic relaxation, Gibbs distributions, and the Bayesian restoration of images. *IEEE Transactions on Pattern Analysis and Machine Intelligence*, 6:721–41, 1984.

[GH$^+$86] Michael Gage, Richard S Hamilton, et al. The heat equation: Shrinking convex plane curves. *Journal of Differential Geometary*, 23(1):69–96, 1986.

[Giu84] E. Giusti. *Minimal Surfaces and Functions of Bounded Variation*, Vol. 80. Springer, Berlin, 1984.

[GK14] Martin Genzel and Gitta Kutyniok. Asymptotic analysis of inpainting via universal shearlet systems. arXiv preprint, arXiv:1405.3747, 2014.

[GM93] Gideon Guy and Gérard Medioni. Inferring global perceptual contours from local features. In *Proceedings of the 1993 IEEE Computer Society Conference on Computer Vision and Pattern Recognition*, pp. 786–7. IEEE, New York, 1993.

[GO07] Guy Gilboa and Stanley Osher. Nonlocal linear image regularization and supervised segmentation. *Multiscale Modeling & Simulation*, 6(2):595–630, 2007.

[GO09] Tom Goldstein and Stanley Osher. The split Bregman method for ℓ_1-regularized problems. *SIAM Journal on Imaging Sciences*, 2(2):323–43, 2009.

[Gra87] Matthew A. Grayson. The heat equation shrinks embedded plane curves to round points. *Journal of Differential Geometry*, 26(2):285–314, 1987.

[Gro04] Harald Grossauer. A combined PDE and texture synthesis approach to inpainting. In *Computer Vision, ECCV 2004*, pp. 214–24. Springer, Berlin, 2004.

[GS03] Harald Grossauer and Otmar Scherzer. Using the complex Ginzburg-Landau equation for digital inpainting in 2D and 3D. In *Scale Space Methods in Computer Vision*, pp. 225–36. Springer, Berlin, 2003.

[GS12] C. Gottschlich and C.-B. Schönlieb. Oriented diffusion filtering for enhancing low-quality fingerprint images. *IET Biometrics*, 1(2):105–13, 2012.

[GTK+12] Miguel Granados, James Tompkin, K. Kim, Oliver Grau, Jan Kautz and Christian Theobalt. How not to be seen: Object removal from videos of crowded scenes. In *Computer Graphics Forum*, vol. 31, pp. 219–28. Wiley Online Library, New York, 2012.

[GZY+06] Jianwei Gu, Li Zhang, Guoqiang Yu, Yuxiang Xing and Zhiqiang Chen. X-ray CT metal artifacts reduction through curvature based sinogram inpainting. *Journal of X-ray Science and Technology*, 14(2):73–82, 2006.

[HLXZ14] Zhanli Hu, Dong Liang, Dan Xia and Hairong Zheng. Compressive sampling in computed tomography: Method and application. *Nuclear Instruments and Methods in Physics Research Section A: Accelerators, Spectrometers, Detectors and Associated Equipment*, 748:26–32, June 2014.

[HS06] Walter Hinterberger and Otmar Scherzer. Variational methods on the space of functions of bounded Hessian for convexification and denoising. *Computing*, 76(1):109–33, 2006.

[HSMM14] L. Robert Hocking, Carola-Bibiane Schönlieb, Thomas März and Russell MacKenzie. GuideFill: Fast artist guided transport inpainting. Preprint, 2014.

[HTBB11] Jooyoung Hahn, Xue-Cheng Tai, Sofia Borok and Alfred Marcel Bruckstein. Orientation-matching minimization for image denoising and inpainting. *International Journal of Computer Vision*, 92(3):308–24, 2011.

[IP97] Homan Igehy and Lucas Pereira. Image replacement through texture synthesis. In *Proceedings of the International Conference on Image Processing*, Vol. 3, pp. 186–9. IEEE, New York, 1997.

[Ise09] Arieh Iserles. *A First Course in the Numerical Analysis of Differential Equations*, Vol. 44. Cambridge University Press, 2009.

[JCL94] Kyeong-Hoon Jung, Joon-Ho Chang and ChoongWoong Lee. Error concealment technique using projection data for block-based image coding. In *Visual Communications and Image Processing '94*, pp. 1466–76. International Society for Optics and Photonics, Bellingham, WA, 1994.

[Jen93] Robert Jensen. Uniqueness of Lipschitz extensions: minimizing the sup norm of the gradient. *Archive for Rational Mechanics and Analysis*, 123(1):51–74, 1993.

[Kan79] Gaetano Kanizsa. *Organization in Vision: Essays on Gestalt Perception.* Praeger, New York, 1979.

[Kan85] G. Kanizsa. Seeing and thinking. *Acta Psychologica*, 59(1):23–33, 1985.

[KCS02] Sung Ha Kang, Tony F. Chan and Stefano Soatto. Inpainting from multiple views. In *Proceedings of the First International Symposium on 3D Data Processing Visualization and Transmission*, pp. 622–5. IEEE, New York, 2002.

[KKL13] Emily J. King, Gitta Kutyniok and Wang-Q Lim. Image inpainting: Theoretical analysis and comparison of algorithms. In *SPIE Optical Engineering & Applications*, pages 885802–885802. International Society for Optics and Photonics, 2013.

[KKZ14] Emily J. King, Gitta Kutyniok and Xiaosheng Zhuang. Analysis of inpainting via clustered sparsity and microlocal analysis. *Journal of Mathematical Imaging and Vision*, 48(2):205–34, 2014.

[KMFR95a] Anil C. Kokaram, Robin D. Morris, William J. Fitzgerald and Peter J. W. Rayner. Detection of missing data in image sequences. *IEEE Transactions on Image Processing*, 4(11):1496–1508, 1995.

[KMFR95b] Anil C. Kokaram, Robin D. Morris, William J. Fitzgerald and Peter J. W. Rayner. Interpolation of missing data in image sequences. *IEEE Transactions on Image Processing*, 4(11):1509–19, 1995.

[KS93] Wilson Kwok and Huifang Sun. Multi-directional interpolation for spatial error concealment. *IEEE Transactions on Consumer Electronics*, 39(3):455–60, 1993.

[KSJ+03] Ville Kolehmainen, Samuli Siltanen, Seppo Järvenpää, J. P. Kaipio, P. Koistinen, M. Lassas, J. Pirttilä and E. Somersalo. Statistical inversion for medical X-ray tomography with few radiographs: II. Application to dental radiology. *Physics in Medicine and Biology*, 48(10):1465, 2003.

[LBU10] S. Lefkimmiatis, A. Bourquard and M. Unser. Hessian-based norm regularization for image restoration with biomedical applications. *IEEE Transactions on Image Processing*, (99):1, 2010.

[LFL96] Hervé Le Floch and Claude Labit. Irregular image sub-sampling and reconstruction by adaptive sampling. In *Proceedings of the International Conference on Image Processing*, Vol. 3, pages 379–82. IEEE, New York, 1996.

[LLT03] Marius Lysaker, Arvid Lundervold and Xue-Cheng Tai. Noise removal using fourth-order partial differential equation with applications to medical magnetic resonance images in space and time. *IEEE Transactions on Image Processing*, 12(12):1579–90, 2003.

[LMS13] Jan Lellmann, Jean-Michel Morel and Carola-Bibiane Schönlieb. Anisotropic third-order regularization for sparse digital elevation models. In *Scale Space and Variational Methods in Computer Vision*, pp. 161–73. Springer, Berlin, 2013.

[Lov13] Augustus Edward Hough Love. *A Treatise on the Mathematical Theory of Elasticity*. Cambridge University Press, 2013.

[Low85] David G. Lowe. *Perceptual Organization and Visual Recognition, 1985*. Kluwer Academic, Boston, 1985.

[LS11] Frank Lenzen and Otmar Scherzer. Partial differential equations for zooming, deinterlacing and dejittering. *International Journal of Computer Vision*, 92(2):162–76, 2011.

[LT06] Marius Lysaker and Xue-Cheng Tai. Iterative image restoration combining total variation minimization and a second-order functional. *International Journal of Computer Vision*, 66(1):5–18, 2006.

[LV08] Linh H. Lieu and Luminita A. Vese. Image restoration and decomposition via bounded total variation and negative Hilbert-Sobolev spaces. *Applied Mathematics and Optimization*, 58(2):167–93, 2008.

[Mal99] Stéphane Mallat. *A Wavelet Tour of Signal Processing*. Academic Press, New York, 1999.

[Mal00] Francois Malgouyres. Increase in the resolution of digital images: Variational theory and applications. Ph.D. thesis, Ecole Normale Superieure de Cachan, France, 2000.

[Mar82] D Marr. *Vision*. Freeman, New York, 1982.

[Mär10] Thomas März. *First Order Quasi-Linear PDEs with BV Boundary Data and Applications to Image Inpainting*. Logos, Berlin, 2010.

[Mär11] Thomas März. Image inpainting based on coherence transport with adapted distance functions. *SIAM Journal on Imaging Sciences*, 4(4): 981–1000, 2011.

[Mär13] Thomas März. A well-posedness framework for inpainting based on coherence transport. *Foundations of Computational Mathematics*, 2013.

[Mas98] Simon Masnou. Filtrage et désocclusion d'images par méthodes d'ensembles de niveau. Ph.D. thesis, L'Université Paris-Dauphine, 1998.

[Mas02] Simon Masnou. Disocclusion: A variational approach using level lines. *IEEE Transactions on Image Processing*, 11(2):68–76, 2002.

[MB02] Andrew J. Majda and Andrea L. Bertozzi. *Vorticity and Incompressible Flow*, Vol. 27. Cambridge University Press, 2002.

[MBBS13] Michael Moeller, Eva-Maria Brinkmann, Martin Burger and Tamara Seybold. Color Bregman TV. arXiv preprint, arXiv:1310.3146, 2013.

[MD97] Riccardo March and Marziano Dozio. A variational method for the recovery of smooth boundaries. *Image and Vision Computing*, 15(9):705–12, 1997.

[Met36] Wolfgang Metzger. *Gesetze des Sehens*. W. Kramer, Frankfurt am Main, 1936.

[Mey01] Yves Meyer. *Oscillating Patterns in Image Processing and Nonlinear Evolution Equations: The Fifteenth Dean Jacqueline B. Lewis memorial Lectures*, Vol. 22. American Mathematical Society, Providence, RI, 2001.

[MG01a] François Malgouyres and Frederic Guichard. Edge direction preserving image zooming: A mathematical and numerical analysis. *SIAM Journal on Numerical Analysis*, 39(1):1–37, 2001.

[MG01b] David Mumford and Basilis Gidas. Stochastic models for generic images. *Quarterly of Applied Mathematics*, 59(1):85–112, 2001.

[MM77a] Luciano Modica and Stefano Mortola. Il limite nella γ-convergenza di una famiglia di funzionali ellittici. *Bollettino dell'Unione Matematica Italiana A (5)*, 14(3):526–9, 1977.

[MM77b] Luciano Modica and Stefano Mortola. Un esempio di γ-convergenza. *Bollettino dell'Unione Matematica Italiana B (5)*, 14(1):285–99, 1977.

[MM98] Simon Masnou and J.-M. Morel. Level lines based disocclusion. In *Proceedings of the 1998 International Conference on Image Processing*, pp. 259–63. IEEE, New York, 1998.

[MO00] Antonio Marquina and Stanley Osher. Explicit algorithms for a new time dependent model based on level set motion for nonlinear deblurring and noise removal. *SIAM Journal on Scientific Computing*, 22(2):387–405, 2000.

[MS63] William W. Mullins and Robert F. Sekerka. Morphological stability of a particle growing by diffusion or heat flow. *Journal of Applied Physics*, 34(2):323–9, 1963.

[MS89] David Mumford and Jayant Shah. Optimal approximations by piecewise smooth functions and associated variational problems. *Communications on Pure and Applied Mathematics*, 42(5):577–685, 1989.

[MS95] Jean Michel Morel and Sergio Solimini. *Variational Methods in Image Segmentation*. Birkhauser, Boston, 1995.

[Mum94] David Mumford. Elastica and computer vision. In *Algebraic Geometry and Its Applications*, pp. 491–506. Springer, Berlin, 1994.

[NAF+14] Alasdair Newson, Andrés Almansa, Matthieu Fradet, Yann Gousseau, and Patrick Pérez. Video inpainting of complex scenes. *SIAM Journal on Imaging Sciences*, 7(4):1993–4954, 2014.

[Nik02] Mila Nikolova. Minimizers of cost-functions involving nonsmooth data-fidelity terms: Application to the processing of outliers. *SIAM Journal on Numerical Analysis*, 40(3):965–94, 2002.

[Nik04] Mila Nikolova. A variational approach to remove outliers and impulse noise. *Journal of Mathematical Imaging and Vision*, 20(1–2):99–120, 2004.

[NM90] Mark Nitzberg and David Mumford. The 2.1-D sketch. In *Proceedings of the Third International Conference on Computer Vision*, pp. 138–44. IEEE, New York, 1990.

[NMS93] Mark Nitzberg, David Mumford and Takahiro Shiota. *Filtering, Segmentation, and Depth*. Springer-Verlag, New York, 1993.

[NW01] Frank Natterer and Frank Wübbeling. *Mathematical Methods in Image Reconstruction* (SIAM Monographs on Mathematical Modeling and Computation) SIAM, Philadelphia, 2001.

[Ökt15] Ozan Öktem. Mathematics of electron tomography. In O. Scherzer, ed., *Handbook of Mathematical Methods in Imaging*, 2nd ed., Chap. 22. Springer-Verlag, New York, 2015.

[OS88] Stanley Osher and James A Sethian. Fronts propagating with curvature-dependent speed: Algorithms based on Hamilton-Jacobi formulations. *Journal of Computational Physics*, 79(1):12–49, 1988.

[OSV03] Stanley Osher, Andrés Solé and Luminita Vese. Image decomposition and restoration using total variation minimization and the H^{-1}. *Multiscale Modeling & Simulation*, 1(3):349–70, 2003.

[Pap14] Konstantinos Papafitsoros. Novel higher order regularisation methods for image reconstruction. Ph.D. thesis, University of Cambridge, 2014.

[Peg89] Robert L. Pego. Front migration in the nonlinear Cahn-Hilliard equation. *Proceedings of the Royal Society of London. A. Mathematical and Physical Sciences*, 422(1863):261–78, 1989.

[PM90] Pietro Perona and Jitendra Malik. Scale-space and edge detection using anisotropic diffusion. *IEEE Transactions on Pattern Analysis and Machine Intelligence*, 12(7):629–39, 1990.

[Pös08] Christiane Pöschl. Tikhonov regularization with general residual term. Ph.D. thesis, Leopold-Franzens-Universität Innsbruck, Austria, 2008.

[PS14] Konstantinos Papafitsoros and Carola-Bibiane Schönlieb. A combined first and second order variational approach for image reconstruction. *Journal of Mathematical Imaging and Vision*, 48(2):308–38, 2014.

[PSB05] Kedar A. Patwardhan, Guillermo Sapiro and Marcelo Bertalmio. Video inpainting of occluding and occluded objects. In *IEEE International Conference on Image Processing*, Vol. 2, pp. 11–69. IEEE, New York, 2005.

[PSB07] Kedar A Patwardhan, Guillermo Sapiro and Marcelo Bertalmío. Video inpainting under constrained camera motion. *IEEE Transactions on Image Processing*, 16(2):545–53, 2007.

[PSS13] Konstantinos Papafitsoros, Carola Bibiane Schoenlieb and Bati Sengul. Combined first and second order total variation inpainting using split Bregman. *Image Processing On Line*, 2013:112–36, 2013.

[Pul12] Andrew Pulver. Beauty and the beast 3D: Review, August 2012. Posted online, 3-May-2012.

[Rad17] Johann Radon. Über die Bestimmung von Funktionen durch ihre Integralwerte längs gewisser Mannigfaltigkeiten. In *Berichte über die Verhandlungen der Königlich-Sächsischen Gesellschaft der Wissenschaften zu Leipzig*, Vol. 69, pp. 262–77. Mathematisch-Physische, Klasse, 1917.

[RF95] Xiaonong Ran and Nariman Farvardin. A perceptually motivated three-component image model: I. Description of the model. *IEEE Transactions on Image Processing*, 4(4):401–15, 1995.

[RO94] Leonid I. Rudin and Stanley Osher. Total variation based image restoration with free local constraints. In *Proceedings of the IEEE International Conference on Image Processing*, Vol. 1, pp. 31–5. IEEE, New York, 1994.

[ROF92] Leonid I. Rudin, Stanley Osher and Emad Fatemi. Nonlinear total variation based noise removal algorithms. *Physica D: Nonlinear Phenomena*, 60(1):259–68, 1992.

[RS06] Matthias Röger and Reiner Schätzle. On a modified conjecture of De Giorgi. *Mathematische Zeitschrift*, 254(4):675–714, 2006.

[RT08] Matthias Röger and Yoshihiro Tonegawa. Convergence of phase-field approximations to the Gibbs–Thomson law. *Calculus of Variations and Partial Differential Equations*, 32(1):111–36, 2008.

[Sap05] Guillermo Sapiro. Inpainting the colors. In *IEEE International Conference on Image Processing,*, Vol. 2, pp. II–698–II-701. IEEE, New York, 2005.

[Sap06] Guillermo Sapiro. *Geometric Partial Differential Equations and Image Analysis*. Cambridge University Press, 2006.

[SB11] Carola-Bibiane Schönlieb and Andrea Bertozzi. Unconditionally stable schemes for higher order inpainting. *Communications in Mathematical Sciences*, 9(2):413–57, 2011.

[SC01] Jianhong Shen and Tony Chan. Variational restoration of nonflat image features: Models and algorithms. *SIAM Journal on Applied Mathematics*, 61(4):1338–61, 2001.

[SC02] Jianhong Shen and Tony F Chan. Mathematical models for local nontexture inpaintings. *SIAM Journal on Applied Mathematics*, 62(3):1019–43, 2002.

[Sch09] Carola-Bibiane Schönlieb. Restoring profanity, March 2009. Posted Line; Plus Magazine, 1 March 2009.

[Sch14] Carola-Bibiane Schönlieb. Mathematically formalising image restoration. *Eureka*, 63, 2014.

[SDC14] Michael Strobel, Julia Diebold and Daniel Cremers. Flow and color inpainting for video completion. In *Pattern Recognition*, pp. 293–304. Springer, Berlin, 2014.

[SGG+08] Otmar Scherzer, Markus Grasmair, Harald Grossauer, Markus Haltmeier and Frank Lenzen. *Variational Methods in Imaging*, Vol. 167. Springer, Berlin, 2008.

[SHL+11] Zineb Saghi, Daniel J. Holland, Rowan Leary, Andrea Falqui, Giovanni Bertoni, Andrew J. Sederman, Lynn F. Gladden and Paul A. Midgley. Three-dimensional morphology of iron oxide nanoparticles with reactive concave surfaces: A compressed sensing-electron tomography (CS-ET) approach. *Nano Letters*, 11(11):4666–73, 2011.

[SS08] S. Setzer and G. Steidl. Variational methods with higher order derivatives in image processing. *Approximation XII*, pp. 360–86. Nashboro Press, Brentwood, TN, 2008.

[Sto96] Barbara E. E. Stoth. Convergence of the Cahn–Hilliard equation to the Mullins–Sekerka problem in spherical symmetry. *Journal of Differential Equations*, 125(1):154–83, 1996.

[Tel04] Alexandru Telea. An image inpainting technique based on the fast marching method. *Journal of Graphics Tools*, 9(1):23–34, 2004.

[THC11] Xue-Cheng Tai, Jooyoung Hahn and Ginmo Jason Chung. A fast algorithm for Euler's elastica model using augmented lagrangian method. *SIAM Journal on Imaging Sciences*, 4(1):313–44, 2011.

[TOH07] Xue-Cheng Tai, Stanley Osher and Randi Holm. Image inpainting using a TV-Stokes equation. In *Image Processing Based on Partial Differential Equations*, pp. 3–22. Springer, Berlin, 2007.

[Tsc06] David Tschumperlé. Fast anisotropic smoothing of multi-valued images using curvature-preserving PDEs. *International Journal of Computer Vision*, 68(1):65–82, 2006.

[TT99] Jack Tumblin and Greg Turk. Lcis: A boundary hierarchy for detail-preserving contrast reduction. In *Proceedings of the 26th Annual Conference on Computer Graphics and Interactive Techniques*, pp. 83–90. ACM Press/Addison-Wesley, Reading, MA, 1999.

[TYJW01] Andy Tsai, Anthony Yezzi, Jr. and Alan S. Willsky. Curve evolution implementation of the mumford-shah functional for image segmentation, denoising, interpolation, and magnification. *IEEE Transactions on Image Processing*, 10(8):1169–86, 2001.

[US88] Shimon Ullman and Amnon Sha'ashua. Structural saliency: The detection of globally salient structures using a locally connected network. In *Proceedings of the International Conference on Computer Vision*, pp. 321–27, 1988.

[Ves01] Luminita Vese. A study in the bv space of a denoising-deblurring variational problem. *Applied Mathematics and Optimization*, 44(2):131–61, 2001.

[VO03] Luminita A. Vese and Stanley J. Osher. Modeling textures with total variation minimization and oscillating patterns in image processing. *Journal of Scientific Computing*, 19(1–3):553–72, 2003.

[Wal85] Sarah Walden. *The Ravished Image: Or How to Ruin Masterpieces by Restoration*. Weidenfeld and Nicholson, St. Martins, 1985.

[Wei98] Joachim Weickert. *Anisotropic Diffusion in Image Processing*, Vol. 1. Teubner, Stuttgart, 1998.

[Wer23] Max Wertheimer. Untersuchungen zur lehre von der gestalt, Part II, *Psychological Research*, 4(1):301–50, 1923.

[WG04] Sarah Walden and Ernst Gombrich. *The Ravished Image: An Introduction to the Art of Picture Restoration and Its Risks*. Gibson Square Books, London, 2004.

[Wil93] Thomas James Willmore. *Riemannian Geometry*, Vol. 33. Clarendon Press, Oxford, UK, 1993.

[WL00] Li-Yi Wei and Marc Levoy. Fast texture synthesis using tree-structured vector quantization. In *Proceedings of the 27th Annual Conference on Computer Graphics and Interactive Techniques*, pp. 479–88. ACM Press/Addison-Wesley, Reading, MA, 2000.

[WPUB11] Manuel Werlberger, Thomas Pock, Markus Unger and Horst Bischof. Optical flow guided TV-ℓ^1 video interpolation and restoration. In *Energy Minimization Methods in Computer Vision and Pattern Recognition*, pp. 273–86. Springer, Berlin, 2011.

[WSI04] Yonatan Wexler, Eli Shechtman and Michal Irani. Space-time video completion. In *Proceedings of the 2004 IEEE Computer Society Conference on Computer Vision and Pattern Recognition*, Vol. 1, pp. 1–120. IEEE, New York, 2004.

[WSI07] Yonatan Wexler, Eli Shechtman and Michal Irani. Space-time completion of video. *IEEE Transactions on Pattern Analysis and Machine Intelligence*, 29(3):463–76, 2007.

[Wu13] Chunlin Wu. On the continuity of images by transmission imaging. *Communications in Mathematical Sciences*, 11(2), 2013.

[WZ98] Yao Wang and Qin-Fan Zhu. Error control and concealment for video communication: A review. *Proceedings of the IEEE*, 86(5):974–97, 1998.

[ZC10] Xiaoqun Zhang and Tony F. Chan. Wavelet inpainting by nonlocal total variation. *Inverse Problems and Imaging*, 4(1):191–210, 2010.

[Zie89] William P. Ziemer. *Weakly Differentiable Functions*. Springer-Verlag, New York, 1989.

[ZM97] Song Chun Zhu and David Mumford. Prior learning and Gibbs reaction-diffusion. *IEEE Transactions on Pattern Analysis and Machine Intelligence*, 19(11):1236–50, 1997.

Index

For EU product safety concerns, contact us at Calle de José Abascal, 56–1°,
28003 Madrid, Spain or eugpsr@cambridge.org.

www.ingramcontent.com/pod-product-compliance
Lightning Source LLC
LaVergne TN
LVHW011515200926
848484LV00001B/46

* 9 7 8 1 1 0 7 0 0 1 0 0 8 *